MW00654642

Environmental Magnetism

This is Volume 86 in the
INTERNATIONAL GEOPHYSICS SERIES
A series of monographs and textbooks
Edited by RENATA DMOWSKA, JAMES R. HOLTON,
and H. THOMAS ROSSBY

A complete list of books in this series appears at the end of this volume.

Environmental Magnetism

Principles and Applications of Enviromagnetics

Michael E. Evans
University of Alberta
Edmonton, T6G 2J1 Canada

Friedrich Heller
Swiss Federal Institute of Technology Zürich
8093 Zürich, Switzerland

ACADEMIC PRESS

An imprint of Elsevier Science

Amsterdam Boston London New York Oxford Paris
San Diego San Francisco Singapore Sydney Tokyo

Acquisitions Editor: Frank Cynar
Senior Project Manager: Angela G. Dooley
Editorial Coordinator: Jennifer Helé
Marketing Manager: Linda Beattie
Cover Design: Gary Ragaglia
Project Management: Graphic World Publishing Services
Composition: Kolam Information Services Pvt. Ltd.
Printer: Maple-Vail

This book is printed on acid-free paper. ∞

Cover photos: Gary Ragaglia/Metro Design

Academic Press
An imprint of Elsevier Science
525 B Street, Suite 1900, San Diego, California 92101-4495, USA
http://www.academicpress.com

Academic Press
An imprint of Elsevier Science
84 Theobald's Road, London WC1X 8RR, UK
http://www.academicpress.com

Academic Press
An imprint of Elsevier Science
200 Wheeler Road, Burlington, Massachusetts 01803, USA
www.academicpressbooks.com

Library of Congress Catalog Card Number: 2003103058

International Standard Book Number: 0-12-243851-5

PRINTED IN THE UNITED STATES OF AMERICA
03 04 05 06 9 8 7 6 5 4 3 2 1

Contents

Foreword

Perhaps it is best to confess right away my personal perspective on Forewords: in other words, what a Foreword is and isn't. It definitely is not a "book review"; instead it should be a welcoming invitation for the reader to the contents of the book and its special character, avoiding as much as possible the opprobrium of being dubbed an "advertising copy," a gushing and uncritical paean of praise. It is also more fun to write a Foreword, especially in this case because I have had the good fortune of knowing the two authors for over three decades. I met Ted Evans when he came to his first meeting of the American Geophysical Union in the heady early days of global plate tectonics. I remember having animated discussions about the reality of the natural submicrometer-sized magnetite grains in basalts, whose presence he had to infer from single domain-like magnetic properties. They were too small to be studied by an optical microscope; however, in a couple of years he and M.L. Wayman took transmission electron micrographs to prove the existence of natural single domain magnetite, a much sought after but also much missed natural magnetic carrier. In Friedrich Heller's case, what caught my attention was his work on the basalts used to construct Hadrian's Wall near the boundary between Scotland and England. Through a laboratory study of acquisition of viscous magnetization in a known field, Heller and Markert could "date" the placement of the basalts in the wall from their natural (viscous) magnetization. It thrilled me that in both cases purely magnetic measurements could lead to applications in mineralogy and archaeology. It is only natural then that, some thirty years later, these two authors and long-standing friends of each other have joined forces to write the first authoritative "how to" book on Environmental Magnetism.

As Evans and Heller recall in their preface, this interdisciplinary field of scholarly inquiry was born only in 1986, with the publications of the eponymous monographs by R. Thompson and F. Oldfield. And yet, while this latter book was a collection of novel applications of magnetism to lake and fluvial sediments, the present volume is more of a consolidated description, from basics to applied, of a mature discipline. In some ways, another recent monograph, *Quaternary Climates, Environments and Magnetism,* edited by B. Maher and R. Thompson, is the true descendant of Thompson and Oldfield's *Environmental Magnetism.* The present volume could be used as a textbook for beginning graduate students with a background in college physics, as well as for specialists in biology, archaeology, or atmospheric pollution, or for others who are curious about the strengths and weaknesses of environmental magnetism as a tool of choice. I cannot help mentioning, tongue in cheek, that while perusing this volume, the readers will also gather a great vocabulary, with the likes of cardiomagnetism, pneumomagnetism, malacology, magnetoclimatology and phreatomagmatism. (I think I know which of my two friends is responsible for including these terminologies, a "hazard" of interdisciplinary research.) One other related volume comes to mind, D. Dunlop and Ö. Özdemir's *Rock Magnetism: Fundamentals and Frontiers*. Readers with a background in physics, whose taste may be whetted by this volume, would do well to consult Dunlop and Özdemir for explanations with greater depth and subtlety than the present volume, whose emphasis lies in clarifying the interdisciplinary connection.

And, speaking of such connections, it is necessary once more to emphasize the broad sweep of the topics covered in this book. From pedologists to geomorphologists, isotope geochemists to microbiologists, all will have an opportunity to truly appreciate what environmental magnetic techniques can or cannot do, and *why*. In the future, when one more marine geologist asks me whether sediment magnetic susceptibility is directly or inversely proportional to paleotemperature, I will enjoy saying, "Neither; why don't you look up Evans and Heller?" Somehow the importance of first constructing an intelligent model of the natural process before interpreting environmental magnetic parameters has not been communicated well enough to the geoscience community. Evans and Heller have done a wonderful job of providing examples to do just that, and do it well.

Colleagues, raise your glasses with your fluid of choice to this timely, comprehensive, and comprehensible work!

Subir K. Banerjee
Institute for Rock Magnetism
University of Minnesota—Twin Cities
U.S.A.

Preface

Environmental magnetism is a relatively new science. It essentially grew out of numerous interdisciplinary studies involving sediments in British lakes, but soon expanded to include sediments in other natural archives that also retain records of past global changes. Prominent among these are marine sediments, windblown deposits on land, and the thin layer of soil covering much of the continents. The materials residing in these various settings are of two main types: one transported in from elsewhere, the other created *in situ*. Material flux takes place in the hydrosphere, the atmosphere, and the cryosphere, the most important agents being rivers, ocean currents, ground water, wind, rain, snow, glaciers, ice sheets, and icebergs. We will be looking at examples of all of these.

For the most part, the transported material itself exists in granular or particulate form, typically in the size range 10^{-4}–10^{-5} m. Depending on the ambient conditions, these mineral grains may suffer some chemical change (such as oxidation) during transport and deposition, but by and large they are passive and inert. However, once they are in place, many chemical changes may occur. Indeed, some grains may entirely disappear while others may be created. This is particularly so in soils (the pedosphere), which harbor a complex interplay of chemical, physical, and biological activity. Whatever the particular history of a given geological repository, experience shows that magnetic measurements can be of great value in our attempts to understand the environmental conditions that prevailed in the past. This is because magnetic minerals—particularly iron oxides—occur more or less universally, iron being one of the most common elements in the Earth's crust. They may be present in minor amounts (usually less than 1%), but they are easily, rapidly, and nondestructively detected.

Early developments along these lines were brought together in the seminal 1986 textbook, *Environmental Magnetism*, by Roy Thompson and Frank Oldfield that marked the real birth of the subject. From this promising beginning, the subject has matured into a full-fledged scientific discipline practiced throughout the world. By the mid-1990s, the level of activity was such that an updated review was provided by Verosub and Roberts (1995). Furthermore, specialized laboratories were coming on stream, entire conferences were being devoted to the latest advances, and the relevant literature was growing exponentially. At the start of a new millennium, the time seemed ripe for a new, state-of-the-art summary and analysis.

Anyone setting out to cover such a broad subject must endeavor to strike a balance between the underlying principles (which embrace physics, chemistry, biology, mineralogy, geography, geology, and geophysics) and the major applications (which range from archeology to zoology). Our goal has been to provide sufficient groundwork to allow advanced undergraduates, graduate students, and interested professionals (all of diverse backgrounds) to grasp the essential aspects of magnetism, mineralogy, and the many processes by which the observed magnetic signals are encoded in the various natural archives. The latter half of the book then introduces a wide selection of real examples chosen to reflect the diversity of topics that lend themselves to enviromagnetic analysis. In addition to various aspects of past global change (e.g., ice ages, Milankovitch theory, paleoprecipitation), these cover the assessment of material flux by various agents (e.g., wind, ground water, ocean currents) in different environments, magnetism in the biosphere (e.g., magnetotactic bacteria, cardiomagnetism, homing pigeons), pollution monitoring (e.g., soil contamination, sewage outfall, pneumomagnetism), and archeology (e.g., magnetic mapping, speleomagnetism, hominid evolution). Finally, we close by stepping back, as it were, and taking an overview of the Earth's magnetic environment in order to place the whole subject into its planetary perspective.

The exponential increase in publications that was occurring when we set out to write this book has continued unabated, and we have been compelled to be selective. Even so, the bibliography contains in excess of 600 entries, three quarters of which were published in the years since Thompson and Oldfield's book appeared.

We are grateful to Frank Cynar, who first invited us to embark on this project and who has been a constant source of encouragement and guidance throughout. Likewise, we are indebted to the entire production team: Angela Dooley, Jennifer Helé, Kelly Mabie, and Nancy Zachor—without their skill and dedication, our efforts would never have come to fruition.

We thank the many friends and colleagues who have helped us by providing data, photographs, figures, and other information: Geoff Bartington, Cathy Batt, Teresa Bingham-Müller, Ulrich Bleil, Jan Bloemendal, Mark Dekkers, Ramon Egli, Brooks Ellwood, Jörg Fassbinder, Fabio Florindo, Maja Haag, Paul Hesse, Kalevi Kalliomäki, Karen Kohfeld, Kurt Konhauser, Masuru Kono, Carlo Laj, Jean-Louis Le Mouël, Neil Linford, Derek Lovley, Tadeusz Magiera, Jim Marvin, Adrian Muxworthy, Clare Peters, Nikolai Petersen, Chris Pike, Andrew Roberts, Joe Rosenbaum, Robert Scholger, Simo Spassov, Joe Stoner, Gerhard Stroink, Matsuori Torii, Piotr Tucholka, Hojatollah Vali, and Marianne Vincken.

We are grateful to Beat Geyer, Gerry Hoye, and Dean Rokosh for much help with the art work, always willingly, efficiently, and cheerfully carried out.

We thank our families for their constant understanding and moral support, particularly during the more difficult times. It is a special pleasure to record the joy and inspiration that the presence of little Andreas has provided.

The whole undertaking would never have been successfully concluded without the patience, encouragement, and unfailing support of Anita and Barbara, to whom we express our heartfelt gratitude.

Michael E. Evans and Friedrich Heller
Edmonton and Zürich, February 2003

1

INTRODUCTION

1.1 PROSPECTUS

Our environment — be it local or global — is in need of care and attention. This brute fact has now forcefully registered itself in the minds of all people bent on survival — that is, most of us. A clear demonstration is provided by unprecedented attempts to reach international agreements — the Montreal Protocol, the Rio Summit, the Kyoto Accord. It is also the driving force behind an enormous range of scientific inquiry aimed at providing a better understanding of the complex interplay of factors which constitute what is now referred to as earth systems science, involving atmosphere, hydrosphere, biospheres, and lithosphere. Indeed, it is quite legitimate — perhaps even necessary — to extend the field of inquiry even further. The lithosphere is no more than a mosaic of slabs at the mercy of viscous upwelling and downwelling currents deeper in the Earth, in a region called the asthenosphere. At even greater depths is the liquid core, wherein complex motions generate the geomagnetic field, which, in turn, is responsible — through its interaction with the solar wind — for the magnetosphere. And so on...

This book is concerned with one tiny aspect of this vast interconnected web of scientific effort, namely the occurrence and uses of magnetic materials in the natural and cultural environment. At first sight, it is perhaps surprising that magnetism has become a useful topic in environmental studies. There are several reasons, the two most fundamental being that, first, *all* substances exhibit some form of magnetic behavior and, second, iron is one of the commonest elements in the Earth's crust. The former follows from the basic nature of matter, the latter from a cosmic accident. There are more practical considerations, however. With modern equipment, it is experimentally easy to detect useful magnetic signals from environmental materials, such as soils and various sediments, even if the magnetic component makes up less than a thousandth of the whole sample. Magnetism thus provides a tracer of environmental conditions. To make use of this tracer, however, knowledge of the magnetic substances involved and of their relevant magnetic properties is required. Furthermore, some understanding of the techniques used is necessary if the possibilities —

and limitations—of the subject are to be properly appreciated. All this groundwork occupies Chapters 2, 3, and 4.

The rest of the book is devoted to a discussion of the many applications of magnetic measurements in various environmental settings on land, in lakes, in the ocean, and even in various biological organisms (including humans). Once sequestered in a suitable host, magnetic particles constitute a natural archive of conditions existing in former times. If we can learn how to interpret such records, we have the possibility of investigating not only the present but also the past. Chapter 5 is concerned with the two central aspects of this goal, namely how the information was captured in the environmental record and how we can succeed in decoding what is there. An important aspect relates to the time dimension (Chapter 6). The next five chapters then discuss specific topics in which environmental magnetism is involved: paleoclimate (Chapter 7), mass transport (Chapter 8), biomagnetism (Chapter 9), pollution (Chapter 10), and archeology (Chapter 11). Finally, Chapter 12 gives a brief planetary perspective of our magnetic environment.

In order to explain the basic concepts, we often consider simplified situations, but a number of case histories are also brought into the discussion to guard against straying too far from harsh reality. These are chosen on pedagogic grounds, for the force with which they illustrate the point in question and not for their overall significance in the research spectrum. Hence they are not necessarily the first, nor the fullest, nor even the best-known examples. To redress the balance, we provide an extensive bibliography. There is now a vast corpus of published data from sites representing all environmental settings in all parts of the globe, with the result that even the bibliography is inevitably selective.

1.2 AN EXAMPLE

Before taking the plunge—wrestling with experimental details, digesting basic magnetic data, appreciating the significance of the case histories, and generally coming to terms with the subject as a whole—let us pause to consider an instructive example. We choose one with which we are personally familiar and which vividly illustrates the interconnectedness of the many topics impinging on environmental magnetism (Heller and Evans, 1995).

In parts of China, there exists a thick blanket of windblown dust that has accumulated over the last few million years and now stands at thicknesses commonly exceeding a hundred meters (Fig. 1.1). For millennia, this *huangtu* (yellow earth) has been the substrate on which civilizations have prospered, providing both the means of agricultural production and the raw material for domestic and artistic ceramics (including the celebrated terracotta army of the emperor Qin Shi Huang). In recent years, this material has attracted a great deal of scientific interest for another reason: stratigraphic fluctuations in the magnetic minerals it contains provide evidence of the waxing and waning of ice ages. Broadly speaking, sediments formed during cold, dry (glacial) times are about half as magnetic as their warm, moist (interglacial) counterparts. The magnetic minerals are essentially behaving like a combined geological

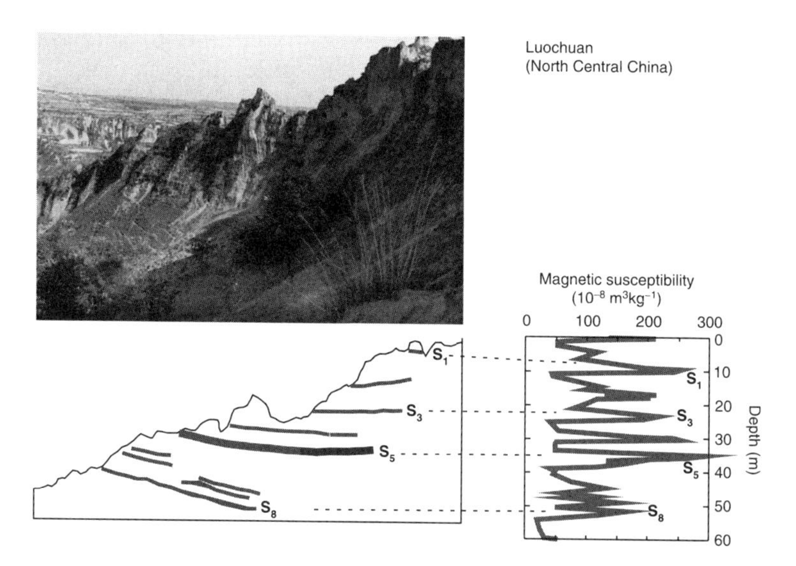

Figure 1.1 The famous sedimentary section at Luochuan, China. The alternating yellow and brown strata provide a visible manifestation of past climatic changes. During cold, dry glacial periods, windblown dust accumulates. When conditions become warmer and wetter, interglacial soils are formed, turning the yellow pristine dust a rich brown color. This process is dramatically reflected in the magnetic susceptibility variations shown in the depth profile at the lower right. Some of the prominent soils are indicated (strictly speaking, they should be referred to as buried fossil soils or paleosols). See color plate.

thermometer and rain gauge. If we were to succeed in calibrating it, actual quantitative estimates of ancient temperatures and precipitation would be forthcoming. In the meantime, we speak of the magnetism as a *paleoclimatic proxy*.

The ice ages themselves are driven by very small changes in the Earth's motion in space caused by gravitational attraction between the planets of the solar system. Detailed astronomical calculations show that the Earth's orbital parameters vary with certain specific periodicities (measured in tens to hundreds of thousands of years), and spectral analysis has shown that these periodicities can be identified in the magnetic profiles. Furthermore, the deposition of dust in China is strongly influenced by the Asian monsoonal atmospheric circulation system, which is one of two major systems controlling climate change in the northern hemisphere (the other being the North Atlantic air–ocean system). Magnetic data help demonstrate that the monsoon system itself has intensified over the last million years due to uplift of the Tibetan Plateau, which itself is driven by forces in the asthenosphere. Another crucial magnetic contribution concerns the thorny problem of how all these events can be properly arranged in geological time. Until about 20 years ago, the actual time span covered by these Chinese sediments was poorly known. Then it was discovered that, as well as carrying a paleoclimatic signal, these thick sequences of dust carry a record of the times when the Earth's magnetic field as a whole flipped polarity, magnetic north becoming magnetic south and vice versa. Because the times when these inversions took place are accurately known, they provide a suitable clock. Thus, we see how the magnetism of the *huangtu* brings together such diverse topics as celestial mechanics, climatic variability, atmospheric circulation, plate tectonics, and the dynamics of the Earth's core. Not bad for a pile of dust!

1.3 SCOPE OF THE SUBJECT

Because all matter consists of atoms with circulating charged particles, everything in and around us is, strictly speaking, magnetic. However, for the moment, we need only note that environmentally important minerals, for our purposes, fall in the broad subset of materials exhibiting properties like those of iron—*ferromagnetism*. Pure elemental iron is found in meteorites and on the moon's surface, but it is extremely rare in terrestrial samples: there is too much oxygen around. Consequently, we need to consider certain compounds of iron, such as the iron oxide magnetite (Fe_3O_4).

How do such minerals get into environmental circulation in the first place? There are many sources. (1) They can be formed naturally as a small part of many igneous rocks, such as basalt. After erosional breakdown, these grains are released and eventually find their way into river catchments, from which they may be delivered to the sea or into lakes. In both cases, sediments are formed. (2) If geological circumstances change, these sediments may in time be eroded and subsequently redeposited. (3) Alternatively, mineral grains may find themselves in arid environments from which they may be entrained into the atmosphere and then deposited downwind—perhaps repeatedly. (4) Volcanic eruptions may produce ash clouds that deliver mineral particles directly to the atmosphere. (5) An entirely different source is biological, particularly from the so-called magnetotactic bacteria. These fascinating organisms create pure magnetite particles some tens of nanometers in diameter, which they use for navigational purposes. After death, the organic parts decay but the magnetic particles remain. (6) Complex chemical and biological processes involved in soil development are another important source of magnetic minerals in the environment. (7) Human activity also adds magnetic material to the environment as a result of the burning of fossil fuels and industrial activities such as steel production. This list is illustrative rather than exhaustive. It serves to indicate the wide variety of pathways that characterize environmental magnetism, and it makes clear the convenience of dealing with the available data in terms of particular environmental settings. Here is a quick preview of some of the topics covered in detail later:

- **Lakes** have long been appreciated as repositories of magnetic paleoenvironmental information (Thompson *et al.*, 1975). They are, however, often limited to relatively short times in the past—the last 10,000 years or so. On the other hand, this can provide high time resolution so that even historical events such as deforestation can be identified in the magnetic record. Some lake studies have managed to penetrate deeper into the past, as in the case of Lac du Bouchet in France (Thouveny *et al.*, 1994) and Lake Baikal in Siberia (Peck *et al.*, 1994). Magnetic data from these two investigations provide important proxies for climatic change over the past 140,000 years and 5 million years, respectively.

• **Marine sediments** have become an extremely important archive of mineral magnetism related to several diverse aspects of environmental variability. To illustrate the richness of this natural archive, consider the following examples. Bloemendal and deMenocal (1989) describe how cyclic variations in the magnetic content of sediments in the western Arabian Sea monitor the amount of dust blown from Africa and Arabia by monsoon winds. Furthermore, these variations are strongly correlated with fluctuations in the solar energy falling on the northern hemisphere calculated from astronomical theory (Berger, 1988). A second example, from the southern hemisphere, is reported by Lean and McCave (1998), who demonstrate a convincing correlation between magnetic properties of samples from a Tasman Sea core and the well-known climatically driven fluctuations in oxygen isotopes found in shells of marine microorganisms. By means of electron microscopy, they go on to show that the magnetic signal is due to bacterially formed magnetite, the abundance of which — in the open ocean — is climatically controlled. In an exciting development, Barthès *et al.* (1999) demonstrate how magnetic measurements made directly in marine boreholes can provide detailed chronological control, one of the most irksome problems in the whole of the Earth sciences. In addition to this high-resolution magnetic chronostratigraphy, their North Sea well (originally drilled for hydrocarbon exploration) yields a magnetic record of variations in northern hemisphere ice cover.

• **Loess** is the correct scientific word for the windblown dust (huangtu) of China discussed in the preceding example. In addition to the famous Chinese occurrences, such deposits occur in many places around the globe. Indeed the word itself comes from an old German word (Löβ, essentially meaning "loose," referring to the unconsolidated nature of the material) first used to describe similar deposits in the Rhine Valley. A discontinuous belt of loess stretches from western Europe through central Asia to China. Significant amounts are also found in the Americas — in Alaska, in the Mississippi Valley, and in the pampa of Argentina. In fact, it was the Alaskan loess that provided the first land-based evidence of the long-period cyclicity in climate variability caused by orbital forcing (Begét and Hawkins, 1989).

• **Soils** exhibit a wide variety of magnetic behavior and have been intensively, and extensively, studied. Mullins (1977) and Maher (1998) provide comprehensive reviews. Early work by Le Borgne (1955) indicated that topsoil often displays greatly enhanced magnetism compared with the bedrock on which it formed. In some cases, this results from fire, but other situations are found in which the normal soil-forming processes *(pedogenesis)* produce new magnetic material. This topic has been of immense importance in working out past climatic changes as recorded by buried fossil soils *(paleosols)*.

• **Biomagnetism** is a relatively new area that is being rapidly explored by environmentalists. Magnetic minerals produced by various organisms, particularly bacteria (Bazylinski and Moskowitz, 1997), are widespread and can provide an important source of magnetic information. Living populations of magnetotactic bacteria have been found in soils, in lake sediments, and in the deep ocean. There is even a suggestion that fossil bacterial magnetite has been found in a meteorite from Mars.

• **Pollution** of our environment produces widespread, and easily detected, magnetic signals (Flanders, 1994). In particular, the burning of ordinary bituminous coal produces ash that sometimes contains more than 10% Fe_3O_4. The production of steel is also a potent source of Fe_3O_4, which can be carried by winds to distances of several tens of kilometers from the source. The ubiquitous nature of magnetic particulates is demonstrated by their documented presence on tree trunks, leaves, and buildings: even the humble dust ball lurking under your furniture is a highly efficient collector, with more than a million particles per gram having been reported.

• **Archeology** is yet another subject to which environmental magnetism has been successfully applied. Enhanced magnetism of the soil on some archeological sites allows the detection of buried structures (Becker and Fassbinder, 1999). In some cases, it provides vital information concerning the evolution of long-occupied sites, as at the Cahokia Mounds State Historic Site in southwestern Illinois (Dalan and Banerjee, 1998). Furthermore, magnetic fingerprinting of certain building materials allows their source to be pinpointed; granite used for columns in the Roman Forum, for example, has been traced to individual quarries in the Eastern Desert of Egypt (Williams-Thorpe *et al.*, 1996). Even the pigment in mural paintings at Pompeii carries a record of the geomagnetic field that existed at the time they were executed (Zanella *et al.*, 2000).

2

BASIC MAGNETISM

2.1 DIAMAGNETISM, PARAMAGNETISM, FERROMAGNETISM

The classical theories of both diamagnetism and paramagnetism first appeared in 1905 in a paper by Paul Langevin (1872–1946). Diamagnetism is a property of all materials. It arises from the interaction of an applied magnetic field and the motion of electrons orbiting the nucleus. Because electrons carry charge, they experience a sideways Lorentz force (Hendrik Lorentz, 1853–1928) when moving through a magnetic field. The outcome can be appreciated from a simple case involving an electron traveling clockwise in a circular orbit centered at the origin and lying in the xy plane, with an external magnetic field applied in the $+x$ direction. For half the orbit ($x > 0$), the Lorentz force will parallel $-z$; for the other half, it will parallel $+z$. A torque therefore arises parallel to the y axis, and this causes the orbit to precess—like a gyroscope—around the field direction. This so-called *Larmor precession* (Joseph Larmor, 1857–1942) gives rise to a magnetic moment in the opposite direction to the applied field. For our purpose, the effect is so small that it can almost always be neglected. It is typically a hundred times smaller than paramagnetism and a hundred thousand times smaller than ferromagnetism. Quartz (SiO_2) and many other minerals that occur naturally in sediments, rocks, and soils are diamagnetic, as is water. There are certain cases, then, where the diamagnetic signal from these substances may become appreciable. One example is the weakly magnetized water-saturated sediment sometimes encountered in lake studies. Another occurs in the laboratory when samples are heated for various experiments. For this purpose, quartz sample holders are often used. Now diamagnetism is independent of temperature, whereas paramagnetism and ferromagnetism decrease markedly as the sample is heated. At high temperatures, therefore, the diamagnetism of the sample holder itself may complicate the experimental results.

In the context of environmental magnetism, paramagnetism is much more important than diamagnetism. It arises by virtue of the fact that the electron behaves as though it were spinning about its own axis as well as orbiting the nucleus. It therefore possesses a spin magnetic moment in addition to its orbital magnetic moment. The total magnetic moment of an atom is given by the vector sum of all the electronic moments. If the spin and orbital magnetic moments of an atom are oriented in such a way as to cancel one another out, the atom has zero magnetic moment. This leads to diamagnetic behavior. If, on the other hand, the cancellation is only partial, the atom has a permanent magnetic moment. This leads to paramagnetism. For example, sodium has one unpaired electron in its $3s$ subshell. Such atoms will tend to be aligned by an external magnetic field, but thermal energy will always prevent perfect alignment. Consequently, the resulting magnetization decreases as temperature increases, the balance being a matter of statistics. Many minerals of interest to environmental studies are paramagnetic, although they generally turn out to produce "noise" rather than "signal." Nevertheless, it is important to monitor possible paramagnetic contributions to the net magnetization of a sample in order to isolate properly the ferromagnetic component, which is usually where environmental information is encoded.

Ferromagnetism is much stronger than diamagnetism and paramagnetism. It is particularly associated with the elements iron (hence the name), nickel, and cobalt but also occurs in many natural minerals such as certain very important iron oxides described in Chapter 3. Because of its unfilled $3d$ subshell, the iron atom possesses a fundamental magnetic moment of 4 Bohr magnetons ($4\mu_B$, see Box 2.1) (Niels Bohr, 1885–1962). In the crystal lattice of ferromagnetic materials, adjacent atoms are sufficiently close together that some of the electron orbitals overlap and a strong interaction arises. This so-called *exchange coupling* means that, rather than being directed at random, the magnetic moments of all the atoms in the lattice are aligned,

Box 2.1 Bohr Magneton

All electrons behave like microscopic magnets with a fundamental quantity of magnetic moment called the Bohr magneton, μ_B. Its magnitude is given by $eh/4\pi m$, e and m being the electron charge and mass and h being Planck's constant; substituting the appropriate values for these fundamental quantities leads to $\mu_B = 9.27 \times 10^{-24}$ Am^2. Each electron subshell in an atom can accept a maximum number of electrons arranged with their magnetic moments aligned in either of two antiparallel directions. A filled subshell has an even number of electrons and therefore has zero magnetic moment. We are particularly interested in the element iron. Its 26 electrons are arranged like this: $1s^2 2s^2 2p^6 3s^2 3p^6 3d^6 4s^2$. All the subshells are full except for $3d$, which is four electrons short of the full d-subshell complement. The six electrons in the $3d$ subshell provide a net magnetic moment of $4\mu_B$ because they are aligned five in one direction and only one in the opposite direction, following a basic requirement of quantum mechanics known as Hund's rule.

↑↓	↑↓	↑↑↑↓↓↓	↑↓	↑↑↑↓↓↓	↑↑↑↑↑↓	↑↓
$1s$	$2s$	$2p$	$3s$	$3p$	$3d$	$4s$

giving rise to a strong magnetization. This arrangement is usually depicted as a regular array of arrows, all the same length and all parallel. This is ferromagnetism in its simplest form, but exchange coupling can give rise to other configurations.

In antiferromagnetism, the atomic magnets all have the same strength but neighboring atoms have oppositely directed moments. Although possessing strong exchange coupling, such materials have zero net magnetization. In some cases, however, a weak magnetization can arise from lattice defects and vacancies or from situations in which the atomic moments are slightly tilted out of perfect antiparallelism (spin canting).

There is yet another important way in which exchange coupling acts, giving rise to the phenomenon of ferrimagnetism. Here, the crystal lattice contains two kinds of sites with cations in two different coordination states. The outcome, in terms of our mental picture, is that two types of arrow are required, one longer than the other. As in antiferromagnetism, the two sets are opposed, but a strong magnetization can obviously arise if the two types are sufficiently unequal. This point will be discussed further in Chapter 3 in the context of specific minerals of interest.

2.2 MAGNETIC SUSCEPTIBILITY

Suppose a suitable piece of a material in which we are interested is placed in a uniform magnetic field (H) and thereby acquires a magnetization per unit volume of M (Fig. 2.1). Its magnetic susceptibility (κ) is defined as the magnetization acquired per unit field,

$$\kappa = M/H \tag{2.1}$$

In SI units, both M and H are measured in A/m, so κ is dimensionless. Strictly speaking, κ is called the volume susceptibility: to obtain what is called the mass susceptibility, we divide by the density (ρ),

$$\chi = \kappa/\rho \tag{2.2}$$

Because κ is dimensionless, χ has units of reciprocal density, m^3/kg.

In some situations, it is more convenient to introduce the *magnetic moment* of the entire body. This is simply given by the product Mv, where v is the total volume, the resulting units being Am^2.

In diamagnetic materials, the precessing electrons give rise to values of χ on the order of 10^{-8} m^3/kg. Water is one of the strongest, with $\chi = -0.90 \times 10^{-8}$ m^3/kg, many common rock-forming silicates, such as quartz and calcite, having values about half as large.

Paramagnetic materials have strongly temperature-dependent susceptibilities described by Curie's law (Pierre Curie, 1859–1906),

$$\kappa = C/T \tag{2.3}$$

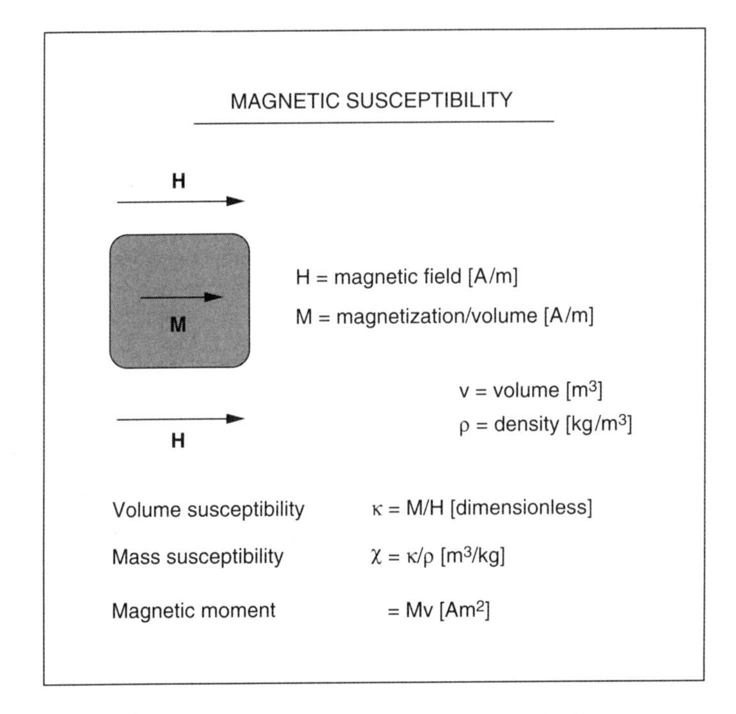

Figure 2.1 Definition of magnetic susceptibility and related parameters.

where T is absolute temperature and C is Curie's constant (see Box 2.2). At room temperature, the thermal energy tending to disrupt alignment is thousands of times greater than the magnetic energy trying to align the atomic moments, m. For n atoms, the result is that the net magnetization is a small fraction of the total maximum, nm. This can be approached only at very low temperatures or by the application of extremely high fields. At room temperature, one needs fields on the order of 10^9 A/m, whereas a typical laboratory electromagnet can reach only about 10^6 A/m. This means that, for all practical purposes, if we experimentally determine a graph of M versus H for a paramagnet, it will be restricted to a region near the origin where the relationship is linear, the slope being equal to the susceptibility. The mass susceptibilities (χ) of common rock-forming silicates, such as fayalite or biotite and the iron sulfide pyrite, are typically about 5×10^{-7} m^3/kg (within a factor of 2).

In ferromagnetic materials, the relationship between M and H is more complicated (and consequently more interesting) than those for diamagnets and paramagnets. One important difference is that it is relatively easy to achieve saturation, where all the atomic moments are aligned. In some cases, this occurs in fields that are well within the range of laboratory electromagnets. Normal practice, therefore, is to measure at low fields (less than $\sim 10^3$ A/m) near the origin of the M-H graph. The

Box 2.2 Curie's Law

Paramagnetism arises from the tendency of atomic magnetic moments, m, to be aligned by an external magnetic field, H, all the time opposed by the disrupting effect of thermal energy. At any given temperature, the balance of thermal and magnetic energies leads to a statistical alignment such that the probability of finding an atomic moment at an angle θ to the magnetic field depends exponentially on the ratio of the two energies, that is, $e^{mH \cos \theta / kT}$ (where k is Boltzmann's constant and T is absolute temperature). The weakness of the alignment can readily be checked by substituting typical values. Considering atoms with a magnetic moment of 1 Bohr magneton in typical laboratory fields at room temperature leads to $mH/kT(=\alpha)$ in the range 10^{-3} to 10^{-4}. Thermal perturbations vastly outweigh magnetic alignment. The actual magnetization is given by

$$M = nm(\coth(\alpha) - 1/\alpha) = nmL(\alpha)$$

where n is the total number of atoms and $L(\alpha)$ is called the Langevin function. For small α, $L(\alpha)$ is approximately equal to $\alpha/3$, and Curie's law is obtained:

$$\kappa = M/H = nm^2/3kT = C/T$$

slope then gives the *low-field susceptibility* (alternatively called the *initial susceptibility*—but note that the word "initial" is often omitted). A much more important consideration, however, is the inevitable tendency of strongly magnetic objects to demagnetize themselves (see Box 2.3). The result is that the measured susceptibility is given by

$$\kappa = \kappa_i/(1 + N\kappa_i) \tag{2.4}$$

where κ_i is the actual intrinsic susceptibility that would be measured in the absence of a demagnetizing field. Experimentally, this can be arranged by using a ring-shaped sample, called a *Rowland ring* after its inventor Henry Rowland (1848–1901). The demagnetizing factor, N, is simply determined by the shape of the sample. For a sphere, it is 1/3. For such a sample, as κ_i increases by an order of magnitude from 10 to 100, κ changes by only 26% (from 2.31 to 2.91, in fact). In the limit, κ approaches $1/N$, and the measured susceptibility is completely controlled by the shape of the sample. This is clearly illustrated in Fig. 2.2: a material with an intrinsic susceptibility of 100, for example, suffers a reduction of 97% as the sample shape varies from a long rod to a sphere.

We have discussed the phenomenon of demagnetization in terms of bulk material, but the same arguments hold for typical environmental samples. Now, however, the control is exerted by the shape of the individual magnetic mineral grains inside the sample, not the overall shape of the sample itself. Of course, the grains inside

Box 2.3 Demagnetizing Factor

Consider an elongated sample situated in an external field, H, applied parallel to the sample's long axis. It becomes magnetized as shown in the inset diagram, with magnetic poles at each end. These poles produce a field inside the sample, H_d, which is opposed to H.

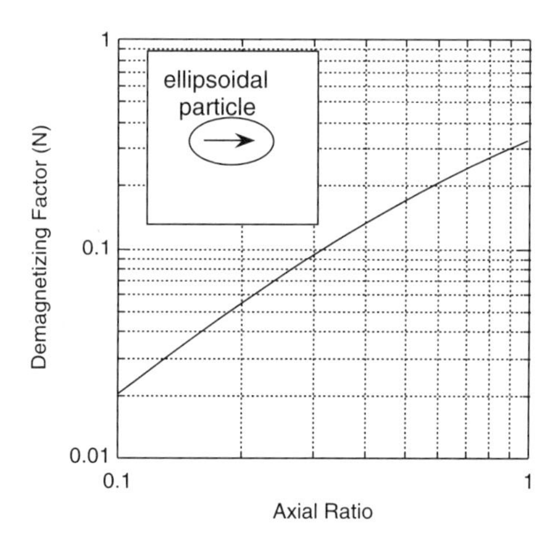

This *demagnetizing field* depends on the shape of the sample and its magnetization (M), that is, $H_d = NM$, where N is the so-called *demagnetizing factor*. Thus,

$$H_{\text{internal}} = H - H_d = H - NM = H - N(\kappa_i H_{\text{internal}})$$

where κ_i is the intrinsic susceptibility of the material. The susceptibility actually observed is

$$\kappa = M/H = [\kappa_i H_{\text{internal}}]/[H_{\text{internal}}(1 + N\kappa_i)] = \kappa_i/(1 + N\kappa_i).$$

If the sample is long and thin, the poles are far apart, N approaches zero, and the effect is negligible. The simplest case to deal with mathematically is the ellipsoid of revolution. For a prolate ellipsoid (wherein one axis is longer than the other two), N is only about 0.02 for an axial ratio of 10:1. However, when the sample is more equidimensional, the demagnetizing effect cannot be neglected. For example, in the case of a sphere, $N = \frac{1}{3}$ (see the accompanying graph).

the sample will not generally be aligned, so some form of spatial averaging will take place. Specific minerals of interest will be discussed in detail in Chapter 3. For the moment, we consider the useful example of a population of roughly equidimensional grains of magnetite (Fe_3O_4): it is found experimentally that the susceptibility of most well-characterized samples falls in the range 3.1 ± 0.4 SI (Heider *et al.*, 1996), which corresponds to 5.2×10^{-4} m^3/kg $< \chi <$ 6.7×10^{-4} m^3/kg. Recall that this is approximately a thousand times greater than that of most relevant paramagnetic materials and a hundred thousand times greater than most diamagnetic values.

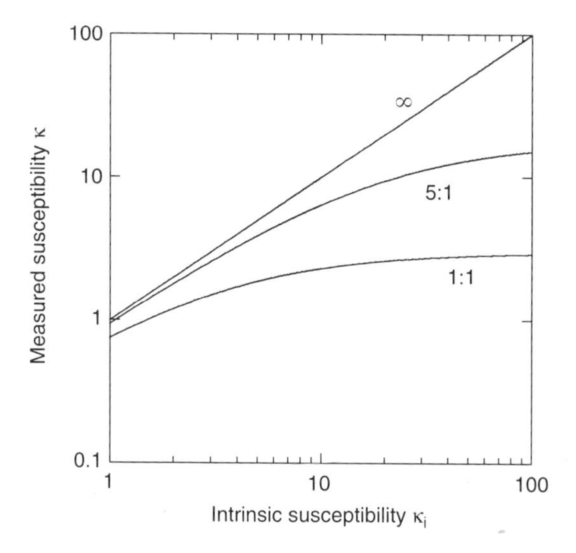

Figure 2.2 This plot shows the drastic effect of demagnetization on measured susceptibility (κ) for materials of high intrinsic susceptibility (κ_i). For an infinitely long rod there is no reduction, whereas for a sphere (axial ratio 1:1) susceptibility is reduced by more than a factor of 30.

2.3 MAGNETIC HYSTERESIS

In the previous section, discussion centered on magnetic susceptibility, which measures the ability of a substance to acquire magnetization while the external magnetic field (H) is being applied. This is referred to as the *induced magnetization*. For diamagnets and paramagnets, when the external field is removed, the magnetization disappears. But for ferromagnets, this is not so. This feature is usually investigated by first applying a strong field so that the magnetization (M) is saturated (Fig. 2.3). As H is then decreased to zero, M does not fall to the origin. This is the phenomenon of *magnetic hysteresis*: it leaves the sample with a permanent magnetization, or *magnetic remanence*. If the field is now increased in the negative direction, M gradually falls to zero and then reverses and eventually saturates again. Repeated cycling of H traces out a *hysteresis loop*.

It is useful to identify and name certain key points on such a loop, as indicated in Figure 2.3. After application of a sufficiently high field, the sample acquires its *saturation magnetization* (M_s). Removal of this field leaves the sample with its *saturation remanence* (M_{rs}), but if the original field was insufficient to achieve saturation, we speak only of the sample's *remanence* (M_r). Application of a reversed field to M_{rs} eventually leads to the point where the overall magnetization, M, equals zero. The field necessary to achieve this is called the *coercive force* (H_c). [This quantity is not really a force (which would be measured in newtons), but the picturesque old-fashioned term is still universally applied—it has the merit of conjuring up the

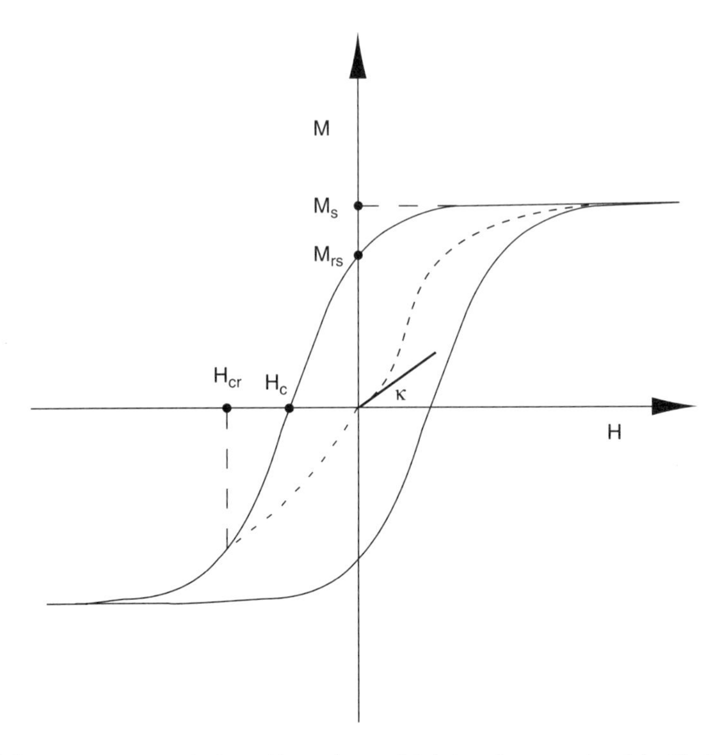

Figure 2.3 Magnetic hysteresis. Several key points are labeled on the axes and explained in the text. The initial susceptibility (κ) is given by the slope of the *M–H* curve in low fields. H_c is known as the coercive force, whereas the field necessary to reduce M_{rs} to zero is called the coercivity of remanence, H_{cr}.

picture of an unwilling sample yielding under the action of an external agent.] To arrive at the point where the sample has zero remanence after the removal of the field (i.e., to get to the origin of the *M–H* graph), a somewhat stronger negative field is required. This is called the *coercivity of remanence* (H_{cr}). These four key elements of the hysteresis loop (M_s, M_{rs}, H_c, and H_{cr}) turn out to be extremely useful diagnostic tools. A few typical hysteresis loops are shown in Fig. 2.4. In Chapter 4, we will see how they are applied to environmental problems. However, let us not overlook the great technological importance of hysteresis. Two remanence points ($+M_r$ and $-M_r$) provide the two states necessary for a binary system (1 and 0), from which it is a short step to magnetic recording, the basis of all modern computer hard drives.

2.4 GRAIN SIZE EFFECTS

If you were to look inside a magnetized ferromagnet, you would discover that it is divided into small regions in which the magnetization is uniform but that the magnetization vector within each region differs from that of its neighbors. This is why $M_{rs} < M_s$ (see Fig. 2.3). These regions are called *magnetic domains* (Fig. 2.5).

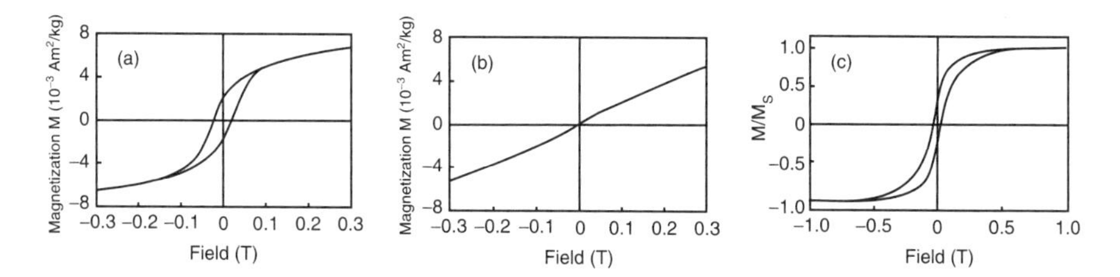

Figure 2.4 (a and b) Examples of hysteresis curves from the central equatorial Atlantic (Frederichs *et al.*, 1999). The hysteresis loop of the sample in (a) is relatively wide open at low coercivity. Its "rectangular" shape indicates the presence of single-domain particles of magnetite. In the sample in (b), the ferrimagnetic content is greatly diminished. The "sigmoid"-shaped loop hardly opens and implies the presence of a coarser grained magnetite mineral fraction. (c) Mixtures of minerals with different coercivities may produce constricted hysteresis loops that are narrow in the middle section but wider above and below this region. Hence they are called wasp-waisted. The sample in (c) is a Pleistocene lacustrine sediment from Butte valley in northern California. On the basis of additional rock magnetic investigations, Roberts *et al.* (1995) ascribe its wasp-waistedness to the simultaneous occurrence of superparamagnetic and single-domain magnetite. Because the hysteresis loop is open at applied fields above 0.4 T, they even do not exclude a contribution from high-coercivity minerals such as hematite or goethite. a and b, © Springer-Verlag, with permission of the publishers and the authors. c, © American Geophysical Union. Reproduced by permission of American Geophysical Union.

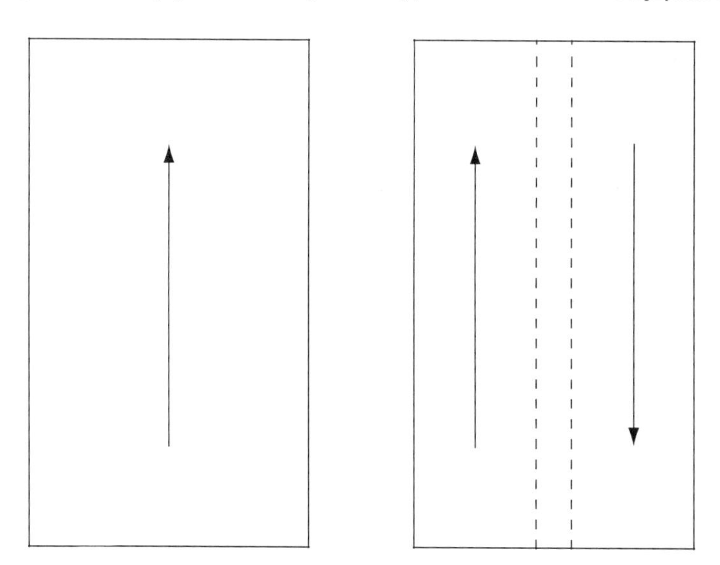

Figure 2.5 Schematic representation of magnetic domains. In the two-domain particle, the dashed lines represent the domain wall, in which the individual atomic moments gradually rotate from the direction in one domain to that in its neighbor.

They arise from the minimization of the overall energy budget of the sample, as explained in Box 2.4.

Mineral grains containing many domains are called multidomain (MD) particles; those containing only one are referred to as single-domain (SD) particles. The boundary between the two types is not sharp — there is a significant middle ground consisting of grains containing only a few domains. Strictly speaking, such grains are

Box 2.4 Magnetic Energy Budget

Consider a spherical particle of the common magnetic mineral magnetite (Fe_3O_4). If it is small enough, it will be uniformly magnetized—all its atomic magnetic moments will be parallel. In this magnetically polarized state, the north and south poles on the surface give rise to what is called *magnetostatic energy*, E_M, given by $v(\mu_0 N M_s^2/2)$. Here, v is the particle's volume, M_s its saturation magnetization (= 480 kA/m), and N its demagnetization factor (= 1/3, see Box 2.3); μ_0 is the permeability constant [defined in (2.8)]. If the particle is now divided into two equally sized, oppositely polarized, hemispheres (called *domains*), the magnetostatic energy is approximately halved. But to do this a price must be paid. The boundary between the two regions—called a *domain wall*—costs $\sim 10^{-3}$ J/m^2. This is because the wall has finite thickness within which the atomic magnetic vectors gradually rotate from the direction in one domain to that of its neighbor. Extra energy is involved because the crystalline magnetite has "easy" and "hard" directions of magnetization—it is anisotropic. To minimize the overall energy, the domains themselves are magnetized along crystallographic easy directions. The magnetic vectors in the wall must therefore be forced out of such directions, a process that requires energy. The critical size for single-domain (SD) behavior can be found by comparing the total energies of the two configurations and substituting the appropriate numerical values. Give it a try; you will find that below $\sim$50 nm, magnetite particles will be SD.

MD, but they possess many of the properties of assemblages of true SD grains. Stacey (1963) first realized the importance of grains of this kind, for which he coined the term pseudo–single-domain (PSD) particles. In nature, geological processes lead to a wide distribution of grain sizes with the result that all three categories are found in environmental investigations.

There is a fourth size-dependent property that is particularly important to us, namely the property of *superparamagnetism*. It arises from the time stability of remanence. This is best understood by considering the behavior of a hypothetical assemblage of identical SD particles. Unless they are at a temperature of absolute zero, thermal energy causes random fluctuations of the individual magnetic moments associated with each and every particle. A finite chance exists that some of the moments flip completely through 180°, leading to a progressive decrease in the net magnetization of the whole sample. Superficially, it is rather like the spontaneous decay of radioactive substances, but the underlying physics is entirely different, of course. Both processes lead to an exponential decrease with time, with the decay rate being described in terms of a characteristic time. In the case of radioactivity, it is common practice to quote the half-life, but for thermodynamic phenomena such as the decay of magnetism, the standard procedure is to define a *relaxation time* (τ), such that

$$M_t = M_0 e^{-t/\tau} \tag{2.5}$$

where M_0 is the initial remanent magnetization at time zero and M_t is its decreased value at time t. As Louis Néel (1904–2000) pointed out, the relaxation time itself is given by

$$\tau = fe^{E_1/E_2} \tag{2.6}$$

where f is a frequency factor on the order of $10^9\ s^{-1}$, E_1 is the potential energy barrier opposing each 180° magnetization flip, and E_2 is the thermal energy (Néel, 1955). The behavior of the whole ensemble of grains thus represents a constant struggle between alignment (due to E_1) and its disruption (due to E_2). The thermal energy equals kT, where k is Boltzmann's constant (Ludwig Boltzmann, 1844–1906) and T is the absolute temperature. The potential energy barrier equals Kv, where K is a coefficient arising from grain anisotropy (crystalline and/or shape) and v is the grain's volume. The end result is that τ depends extremely strongly on the ratio v/T. If the grain size is sufficiently small, τ can diminish to a matter of seconds or even less. The material is still ferromagnetic but the remanence is disappearing before your very eyes—the assemblage is said to be superparamagnetic (SP, for short). Substitution of typical numerical values for equidimensional magnetite shows that, at room temperature, the relaxation time increases from less than a minute for 28-nm grains to more than a billion years for 37-nm grains. This leads naturally to the notion of a *critical diameter* above which remanence can be considered stable. Alternatively, in some situations (e.g., fired archeological pottery; see Chapters 6 and 11) it is convenient to speak of a *blocking temperature* below which the remanence is stable. A magnetization acquired by cooling from an elevated temperature is called a thermoremanent magnetization (universally abbreviated to TRM; see later).

The actual dimensions of grains falling in the various categories (MD, PSD, SD, SP) are very much a function of the mineral in question. In magnetite, direct microscopic observations indicate that two-domain patterns (definitely PSD) persist up to $\sim 10^{-6}$ m, whereas to accommodate about 10 domains, a grain of some 10^{-4} m may be required (Dunlop and Özdemir, 1997). These are all small sizes—bear in mind that the distance between atoms in solid iron is $\sim 3 \times 10^{-10}$ m and the wavelength of visible light is $\sim 5 \times 10^{-7}$ m. One useful way of illustrating domain behavior is to map out the various fields on a plot of grain size versus grain shape (Evans and McElhinny, 1969; Butler and Banerjee, 1975). This is done for magnetite in Fig. 2.6, to which has been added the modifications suggested by three-dimensional micromagnetic calculations (Fabian *et al.*, 1996). These more recent calculations indicate that equidimensional grains as large as 140 nm may act as single domains.

Regardless of the precise locations of the boundaries separating the different size-dependent behaviors, it is both practicable and useful to identify the distinct magnetic properties of MD, PSD, SD, and SP assemblages. For this exercise, several diagnostic tests—discussed in Sections 2.6 and 2.8—are available. Because the dominant grain size present is controlled by the original process of formation and the subsequent history, such tests often provide useful environmental information concerning the origin and evolution of a particular deposit.

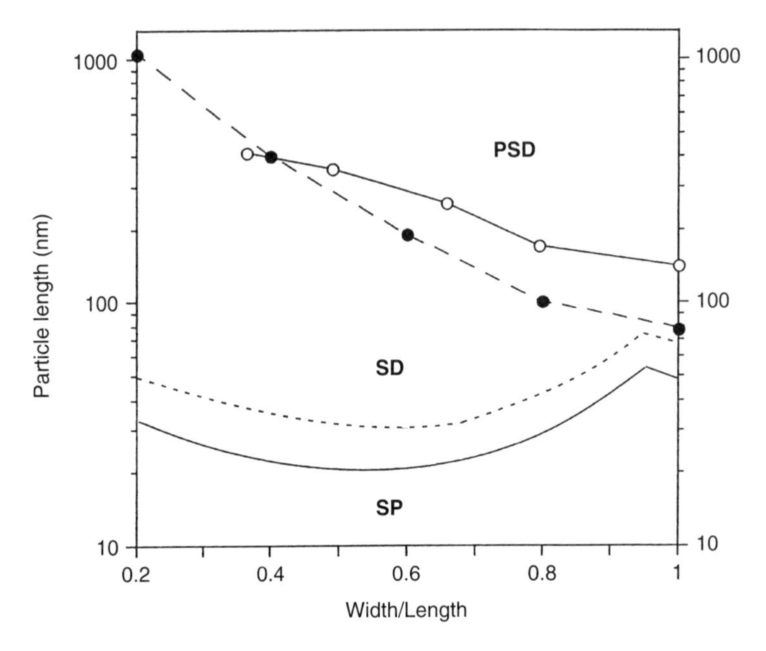

Figure 2.6 Size–shape regions for various domain states in magnetite. The lower three curves are from Butler and Banerjee (1975) and the uppermost one is from Fabian *et al.* (1996). The lowermost curve represents a relaxation time of 100 seconds. For axial ratios less than ~0.95, this curve is calculated on the basis of shape anisotropy, but the small bend near the right-hand axis results from the importance of magnetocrystalline anisotropy in near-equidimensional particles. The lower dashed curve (short dashes) is similar to the curve below it but is calculated for a relaxation time of 4.5 billion years (the age of the Earth). The upper dashed curve (long dashes) was calculated from a simple energy balance model, whereas the solid line with the open circles results from a full three-dimensional micromagnetic calculation. The superparamagnetic (SP), single-domain (SD), and pseudo–single-domain (PSD) fields are indicated.

2.5 SUMMARY OF MAGNETIC PARAMETERS AND TERMINOLOGY

For convenience, the most important magnetic quantities and the SI units in which they are measured are gathered together in Table 2.1 (see also Appendix 1). For more details on magnetic units in general, see Payne (1981). It is also useful to summarize here (see Table 2.2) some unavoidable jargon that will arise in later chapters. As we saw previously, while a sample is being held in a field, it will have an *induced magnetization*. When the field is removed, the sample may retain a *remanent magnetization* (or *remanence*, for short). The remanence could arise in a number of ways, each of which is given a name (not to confuse the student, but to provide useful information, usually to indicate that the manner in which it became magnetized is known).

When a natural sample is first collected and before any laboratory experiments have been conducted on it, one speaks of its *natural remanent magnetization* (NRM). This is a neutral term reflecting our ignorance concerning the sample's history.

Table 2.1 Common Magnetic Quantities

Volume susceptibility	κ	dimensionless
Mass susceptibility	χ	$m^3\ kg^{-1}$
Magnetizing field	H	Am^{-1}
Magnetic induction	B	T
Magnetization	M	Am^{-1}
Magnetic moment	Mv	Am^2
Saturation magnetization	M_s	Am^2kg^{-1} (mass normalized)
Saturation remanence	M_{rs}	Am^2kg^{-1} (mass normalized)
Coercive force	H_c or B_c	Am^{-1}or T
Coercivity of remanence	H_{cr} or B_{cr}	Am^{-1}or T

Table 2.2 Common Types of Remanent Magnetization

Natural remanent magnetization	NRM
Thermoremanent magnetization	TRM
Isothermal remanent magnetization	IRM
Saturation IRM	SIRM
Anhysteretic remanent magnetization	ARM
Depositional remanent magnetization	DRM
Chemical remanent magnetization	CRM

A remanence acquired by cooling from an elevated temperature (in a volcanic lava flow, for example) is a *thermoremanent magnetization* (TRM). A remanence acquired by exposure to a field at ambient temperature is an *isothermal remanent magnetization* (IRM). This can arise in nature (in a lightning strike, for example) but more often refers to laboratory procedures where a sample has been exposed to a known field (it is equivalent to the quantity M_r described in Section 2.3). If the field used to impart an IRM is sufficient to achieve saturation, we speak of saturation isothermal remanence (SIRM), which is equivalent to M_{rs} (see Fig. 2.3). Be warned, however, that the acronym SIRM is often used to represent the remanence acquired by a sample after exposure to what happens to be the highest field available to a particular investigator. This is usually on the order of 1 T and may, or may not, actually reach true saturation. The coercivity spectrum obtained by incremental IRM acquisition is a powerful—and popular—laboratory technique.

For completeness, we also include here certain terms that will be discussed in greater detail as they arise later in the book. A widely used experimental procedure involves magnetizing a sample by means of a small bias field in the presence of an alternating magnetic field that is smoothly reduced to zero from a predetermined maximum: this is *anhysteretic remanent magnetization* (ARM) (see Fig. 4.12). The alternating field plays a role not unlike that provided by thermal agitations in TRM but avoids the danger of unwanted chemical changes caused by heat.

In Chapter 5, the terms *detrital (or depositional) remanent magnetization* (DRM) (see Box 5.1) and *chemical remanent magnetization* (CRM) (see Box 5.2) will be used in connection with paleomagnetism. They provide two other mechanisms (in addition to TRM) by which geological formations can acquire, and retain, a record of past changes in the geomagnetic field.

2.6 ENVIROMAGNETIC PARAMETERS

The items listed in Table 2.1 are fundamental parameters that arise in any discussion of the properties of magnetic materials—in physics, chemistry, and engineering, for example. Those in Table 2.2 are rather more specialized, being restricted mostly to geophysics and geology. There is yet a third group (see Table 2.3) that is absolutely essential to us in our pursuit of environmental magnetism. The parameters involved—and certain combinations of them—will crop up time and time again throughout this book so it is worth gathering them together at the outset. They have been introduced by various authors with specific purposes in mind, and their use will become clear when actual examples arise throughout the book. Rather than

Table 2.3 Selected Enviromagnetic Parameters

χ_{lf}	Low-field susceptibility
χ_{hifi}	High-field susceptibility
χ_{ferri}	Ferrimagnetic susceptibility
χ_{fd}	Frequency-dependent susceptibility
χ_{ARM}	Anhysteretic remanent susceptibility
Bivariate ratios:	
S	S-ratio (= "soft" IRM/"hard" IRM)
SIRM/κ_{lf}	Granulometry indicator
ARM/SIRM	Granulometry indicator
$M_{\mathrm{rs}}/M_{\mathrm{s}}$	Magnetization ratio
$B_{\mathrm{cr}}/B_{\mathrm{c}}$	Coercivity ratio
Bivariate plots:	
$M_{\mathrm{rs}}/M_{\mathrm{s}}$ vs $B_{\mathrm{cr}}/B_{\mathrm{c}}$	Day plot
κ_{ARM} vs κ_{lf}	King plot
H_{u} vs $H_{\mathrm{c'}}$	FORC diagram

make the list exhaustive (not to mention exhausting), we have chosen a representative cross section to portray the current state of the art. This should allow the reader to appreciate the rationale behind other combinations currently in use as well as those yet to be devised.

2.6.1 Susceptibility

First, let us consider the various forms in which the all-important parameter susceptibility is useful. As we saw previously, in its mass-normalized form this is usually given the symbol χ. In some instances, this will appear as χ_{lf} to stress that it has been measured in a low magnetic field (typically $< 1\,mT$) as opposed to χ_{hifi}, the susceptibility given by the slope of the magnetization curve at high fields, beyond closure of the hysteresis loop (i.e., above $\sim 100\,mT$; see Fig. 2.4). Subtracting χ_{hifi} from χ_{lf} yields the ferrimagnetic susceptibility, χ_{ferri}. This is because χ_{hifi} measures the contribution of the paramagnetic and antiferromagnetic minerals present: when these are subtracted, we are left with the ferrimagnetic component that saturates in relatively low fields (typically $< \sim 200\,mT$). Another extremely important susceptibility parameter is its frequency dependence, χ_{fd}. This is the difference in susceptibility observed when the apparatus being used is driven at two different frequencies. It is particularly useful for detecting the presence of very small, superparamagnetic particles (see Chapter 4). [Note that some authors label the two frequencies as lf (low frequency) and hf (high frequency), which leads to χ_{lf} and χ_{hf}. To prevent confusion, we reserve lf for low field, not low frequency. We avoid hf altogether. Where necessary, we indicate — as a subscript — the actual measuring frequency used.] On another point of nomenclature, it should be noted that all these susceptibility quantities have their corresponding volumetric susceptibility counterparts, denoted κ instead of χ.

2.6.2 ARM Susceptibility

The ARM susceptibility is the mass-normalized ARM (in Am^2/kg) per unit bias field (H, in A/m). It turns out to be a useful parameter in its own right and also as one factor in certain widely used ratios. Moreover, division by H represents an essential normalization if different experimenters use different bias fields. Its most useful property is that it preferentially responds to SD particles because, gram for gram, these acquire more remanence than particles containing domain walls that allow lower magnetostatic energy configurations to be achieved. For example, Maher (1988) compiles results for a series of essentially pure magnetite powders of known grain size, giving χ_{ARM} values of $\sim 8 \times 10^{-3}\,m^3kg^{-1}$ for particles with a mean diameter of 0.05 microns, but only $8 \times 10^{-4}\,m^3kg^{-1}$ for 1-micron particles. (Recall that 1 micron $= 10^{-6}$ m.)

2.6.3 *S*-Ratio

The main purpose of the so-called *S*-ratio is to provide a measure of the relative amounts of high-coercivity ("hard") remanence to low-coercivity ("soft") remanence. In many cases, this provides a fair estimate of the relative importance of antiferromagnetics (such as hard hematite) versus ferrimagnetics (such as soft magnetite). The procedure is to saturate a sample in the forward direction (SIRM) and then expose it to a backfield (typically equal to 0.3 T). The *S*-ratio is obtained by dividing the "backwards" remanence by the SIRM. Values close to unity indicate that the remanence is dominated by soft ferrimagnets (e.g., see Fig. 4.18). (Note that some authors retain the algebraic [negative] sign for the backward IRM.)

2.6.4 ARM/SIRM and SIRM/κ_{lf}

These ratios are widely employed as grain size indicators for magnetite (e.g., see Fig. 4.21). Small particles yield higher values because they are more efficient at acquiring remanence, particularly ARM (e.g., see Maher, 1988; Dunlop and Xu, 1993; Dunlop, 1995). Broadly speaking, it is found experimentally that SIRM as a function of grain diameter follows a power law over a very wide range of grain sizes (from $\sim$0.04 to $\sim$400 μm). On the other hand, ARM follows two separate power laws above and below $\sim$1 μm. For smaller grains, the slope is steeper, so that samples containing a higher fraction of SD-PSD particles will yield higher ARM/SIRM ratios. For the SIRM/κ_{lf} ratio, the observed size dependence of the numerator, coupled with the size independence of the denominator (Heider *et al.*, 1996), again leads to higher values where smaller particles are more abundant.

It has emerged that the SIRM/κ_{lf} ratio is also useful for indicating the presence of the iron sulfide greigite (Roberts *et al.*, 1996; see also Chapter 3).

2.6.5 M_{rs}/M_s and B_{cr}/B_c and the Day Plot

These two ratios are sometimes used separately (e.g., see Fig. 4.18) but are particularly useful when used simultaneously on a graph of M_{rs}/M_s versus B_{cr}/B_c — sometimes referred to as a Day plot (Day *et al.*, 1977). For the most part, this type of analysis is valid only if other evidence points to magnetite as the dominant magnetic mineral present. This is because most of the experimental data available refer to this mineral. Nevertheless, this restriction is not too severe because magnetite is, in fact, a commonly occurring mineral. Moreover, it is strongly magnetic and will often dominate the magnetic properties of a sample even when present in relatively small amounts. The ratio M_{rs}/M_s is ≥ 0.5 for single-domain particles (Dunlop and Özdemir, 1997, p. 320) and decreases as particle size increases into the PSD and MD fields. This is because the presence of domain walls allows each particle to take up a remanence configuration that minimizes its magnetostatic energy (see Box 2.4), which, for the whole assemblage of particles, leads to a much reduced value of M_{rs}. How far it will be reduced can be understood from the following argument. The slope of the hysteresis loop near the origin is close to $1/N$ (where N is the demagnetizing factor, $\approx 1/3$), which means that $|M_{rs}| \approx 3H_c$. The coercive force (H_c) in MD grains depends on the strength of domain wall pinning, which, in turn, depends on the level of internal stress within the particle. According to Dunlop and Özdemir (1997), it is not likely to exceed 10 mT ($\approx$ 8 kA/m) for MD magnetite. Finally, therefore,

$M_{rs}/M_s \leq 3 \times (8\ \text{kA/m})/(480\ \text{kA/m}) = 0.05$ (see Table 3.1 for the M_s value for magnetite). In practice, therefore, SD behavior is classified as $M_{rs}/M_s \geq 0.5$ and MD behavior as $M_{rs}/M_s \leq 0.05$. Between these two values, PSD behavior is indicated. Whereas the remanence ratio M_{rs}/M_s can never exceed unity, the coercivity ratio B_{cr}/B_c can never be less than unity (see Fig. 2.3). Calculating its exact value, however, is not as straightforward as was the case for the remanence ratio. For MD particles, theory predicts a value given by $(1 + N\kappa_i)$, where κ_i is the intrinsic susceptibility of the material. Taking a typical value of $\kappa_i = 10$, Dunlop and Özdemir (1997, p. 318) obtain a lower limit of ~4 for MD coercivity ratios. The corresponding upper limit for SD behavior is very poorly constrained. Empirically, it seems that it cannot exceed ~2, so most practitioners have rather arbitrarily accepted a value of 1.5. Values between 1.5 and 4 are taken to indicate the presence of PSD particles.

The so-called Day plot therefore consists of rectangular zones with SD behavior defined by $M_{rs}/M_s \geq 0.5$ and $B_{cr}/B_c \leq 1.5$, MD behavior defined by $M_{rs}/M_s \leq 0.05$ and $B_{cr}/B_c \geq 4.0$, with the PSD zone sandwiched in between. Figures 4.15 and 4.19a are good examples of how the Day plot is commonly used. Dunlop (2002a,b) has undertaken a thorough reassessment of the way in which these kinds of data can be interpreted. His penetrating analysis provides a new road map allowing a more subtle means of navigating the Day plot. In particular, he shows how various mixtures of superparamagnetic (SP), single-domain (SD), pseudo–single-domain (PSD), and multidomain (MD) particles can sometimes be unraveled. This is particularly useful because it has been widely found that there is a strong tendency for enviromagnetic materials to yield values falling in the PSD zone when, in fact, they actually contain mixtures of grains in different domain states. The detail is averaged out—instead of a nice black-and-white zebra, all we get is a fuzzy gray horse.

Dunlop's approach is to use theoretical hysteresis curves combined with experimental results obtained from samples of known composition and grain size to redefine the standard rectangular zones and then to construct "mixing curves" obtained by combining various proportions of pairs of end members (e.g., SD + MD, SD + SP). He suggests that the previously accepted boundaries be adjusted so that the MD zone is now defined by $M_{rs}/M_s \leq 0.02$ and $B_{cr}/B_c \geq 5.0$ (compared with the earlier values of 0.05 and 4.0, respectively). He retains the well-established threshold of $M_{rs}/M_s \geq 0.5$ for SD behavior but favors $B_{cr}/B_c \leq 2.0$ (rather than 1.5). Finally, he points out that the addition of SP particles leads to the identification of a new field with approximate limits $0.1 \leq M_{rs}/M_s \leq 0.5$ and B_{cr}/B_c ratios as high as 100. This is particularly important because SP particles have often been completely misplaced on the Day plot or else omitted from it altogether.

Figure 2.7 presents a good example of a binary mixture interpretation applied to lake sediment data. Most of the points fall in the PSD box, but there is a clear trend that strongly suggests that each sample actually contains a mixture of SD and MD particles (with the SD fraction ranging from ~10 to ~85%). The tendency for the most SD-rich samples to plot to the right of the mixing curve suggests that a third, SP, component may also be present.

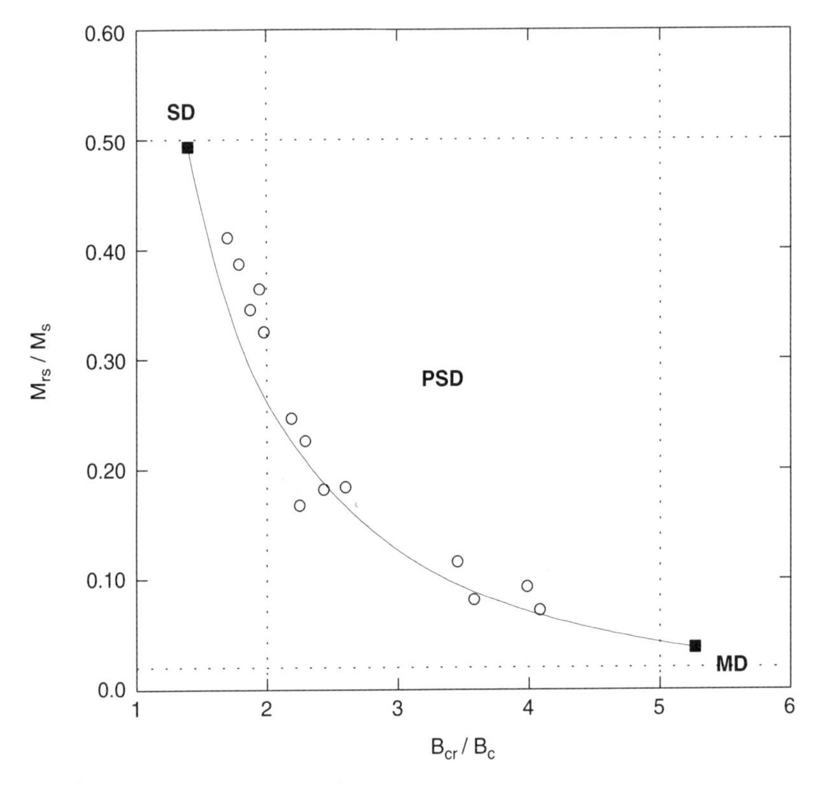

Figure 2.7 Day plot illustrating the effect of a binary mixture of single-domain (SD) and multi-domain (MD) particles. The samples involved are lake sediments from Minnesota. (Modified from Dunlop, 2002b.)

2.6.6 κ_{ARM}/κ_{lf} and the King Plot

If a sample's dominant magnetic mineral is magnetite, this dimensionless ratio provides a means of assessing grain sizes (King *et al.*, 1982). This is because both parameters increase linearly with increasing magnetite concentration, but, as pointed out before, smaller grains are relatively more efficient at acquiring remanence. Thus, if the two parameters are plotted on a graph (with κ_{ARM} as the ordinate), smaller grains yield steeper slopes, as indicated in Fig. 2.8 (see also Figs. 4.16 and 9.10). Such a graph is often referred to informally as a King plot. (It is able to use κ rather than χ because both parameters are measured on the same samples, so that most investigators omit the superfluous step of normalizing by mass or volume.) Notice that the slope changes relatively slowly as a function of size for large grains and much more rapidly for smaller grains. This means that what we might call the "resolving power" of this procedure is much greater for SD and PSD distributions than for samples dominated by MD assemblages.

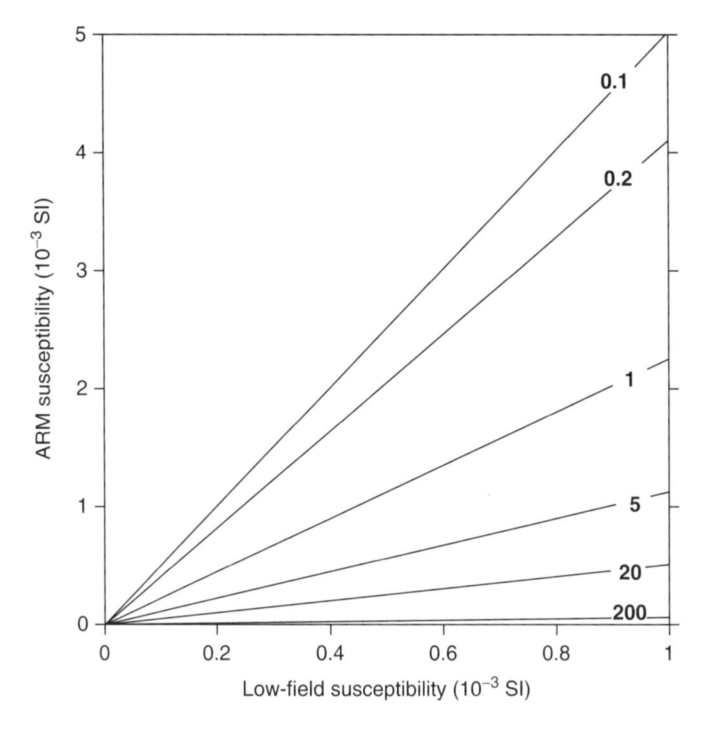

Figure 2.8 Relationship between ARM susceptibility (κ_{ARM}) and low-field susceptibility (κ_{lf}) for magnetite particles of different size (given in microns). (Adapted from King *et al.*, 1982.) © Elsevier Science, with permission of the publishers.

2.6.7 H_c, H_u and FORC Diagrams

The analysis of hysteresis properties has been extended by measuring the $M(H)$ curve not only in one major hysteresis cycle between a large positive and a large negative field. Saturating in a positive field and then reversing the field to a number of negative field values and subsequently returning to positive saturation produces a number of curves which have been named first-order reversal curves (FORCs) (Mayergoyz, 1986). They are generally transformed for better visualization into contour plots known as FORC diagrams (Pike *et al.*, 1999; Roberts *et al.*, 2000).

As Pike and Marvin (2001) explain, FORC measurements start out by saturating a sample in a strong positive field. Then the field is changed to a negative field H_r (Fig. 2.9a). A FORC is measured when reverting back from H_r to full positive saturation. The magnetization at the applied field H_a on the FORC with reversal field H_r is denoted by $M(H_r, H_a)$, where $H_a > H_r$. The difference between successive FORCs arises from irreversible magnetization changes that occur between successive reversal fields (Fig. 2.9b). The FORC distribution is defined as the mixed second derivative:

$$\rho(H_r, H_a) \equiv -\frac{\partial^2 M(H_r, H_a)}{\partial H_r \partial H_a}, \tag{2.7}$$

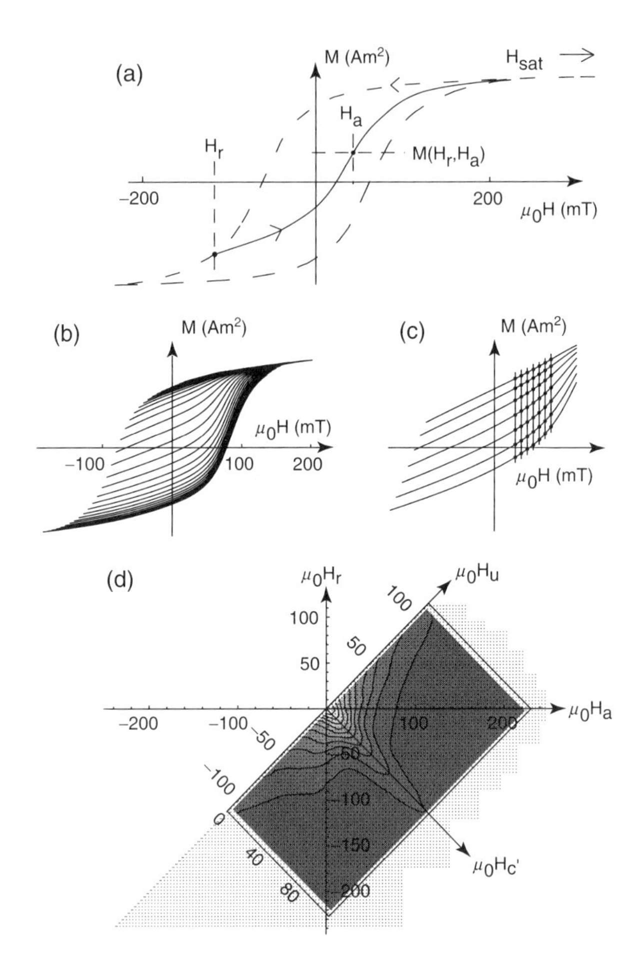

Figure 2.9 Illustration of how FORC diagrams are constructed. (a) After positive saturation the field is reversed to H_r and then increased again to saturation along the initial field direction. The magnetization at H_a is denoted by $M(H_r, H_a)$. The dashed line represents the major hysteresis loop. (b) A set of 33 consecutive FORCs for a typical floppy disk sample. (c) A subset of seven consecutive FORCs of the floppy disk sample measured at equal field increments. The data points (filled circles) therefore plot on an evenly spaced grid in the $\{H_r, H_a\}$ coordinate system. (d) Example of an $\{H_r, H_a\}$ plot with the $\{H_{c'}, H_u\}$ plot superimposed. A local square grid (in which each row of the grid points ranges from H_r to H_a) evaluates the data density $\rho\{H_r, H_a\}$. The number of grid points used around each data point determines the degree of smoothing of the FORC distribution. The particular FORC distribution illustrated is based on a set of 99 FORCs. It characterizes an MD magnetite–bearing deep sea sediment sample. (Adapted from Roberts *et al.*, 2000, and Pike and Marvin, 2001.) © American Geophysical Union. Modified by permission of American Geophysical Union and the authors.

which is well defined for $H_a > H_r$. When plotting a FORC distribution, it is convenient to change coordinates from $\{H_r, H_a\}$ to $\{H_u = (H_a + H_r)/2, H_{c'} = (H_a - H_r)/2\}$ (Fig. 2.9d).

FORC diagrams represent microcoercivity $H_{c'}$ along the horizontal axis, while magnetic interactions cause vertical spread along H_u. Thus, noninteracting SD

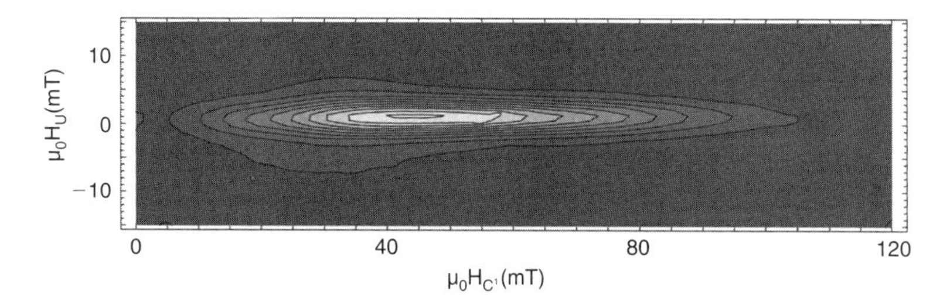

Figure 2.10 Example of a FORC diagram of weakly interacting SD titanomagnetite from the Yucca Mountain ash flow (southern Nevada). Note that contours center around a well-defined coercivity maximum of ~45 mT and do not reach the ordinate. Vertical spread is minimal. (Adapted from Pike and Marvin, 2001, with permission of the authors.)

particles produce horizontally elongated contour lines on a FORC diagram, peaking at the appropriate $H_{c'}$ with little vertical spread (Fig. 2.10). They have no contours close to the ordinate. Thermal relaxation of SP and (small) SD particles yields maximum density of vertical contour lines near $H_{c'} = 0$ (Pike *et al.*, 2001a). MD grains seem to have vertical contour lines centered on their $H_{c'}$ with a contour density spread over a comparatively large H_u interval because the domains within MD particles interact with each other (Pike *et al.*, 2001b). Often they form contour line patches that are shaped like acute triangles in various attitudes (Roberts *et al.*, 2000).

Loess/paleosol samples from Moravia show distinctly different FORC distributions (van Oorschot *et al.*, 2002). The paleosol sample (Fig. 2.11, upper panel) is dominated by well-dispersed fine-grained SD magnetite grains that have a wide range of coercivities (up to 50 mT) with very little vertical spread (within ± 2 mT) indicating virtually no interaction. Increased contour density close to the ordinate might indicate the presence of SP material. A few triangular contours point to the subordinate presence of MD grains. The loess sample (Fig. 2.11, lower panel) also contains SD magnetite, which is centered more to the left, indicating slightly coarser grain size. The vertical distribution is wider (most contours within ± 4 mT), and more contours of MD-like triangular shape are observed. Thus, MD contributions seem to be more significant in the loess sample. SP grains cannot be discerned in the weakly magnetic loess sample.

At present, FORC diagrams are able to recognize magnetic interactions qualitatively and to identify SP, SD, and MD particles of magnetic minerals that may constitute a complex magnetic rock mineralogy. According to Pike *et al.* (2001b), quantitative tools for interpreting and modeling FORC diagrams can be expected to improve these capabilities in the near future.

2.7 MAGNETIC UNITS

So far, so good. The required magnetic parameters have been successfully introduced. But because we will need to discuss real data resulting from actual laboratory

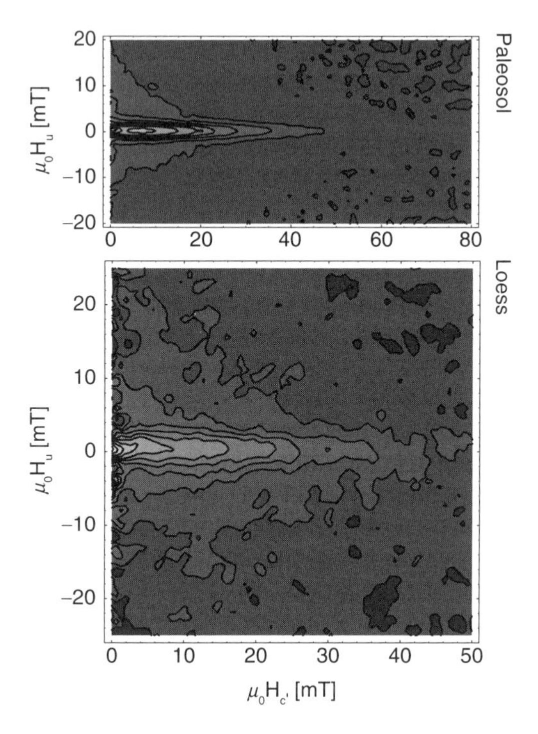

Figure 2.11 FORC diagram of a paleosol and a loess sample from Moravia (from van Oorschot *et al.*, 2002). The maximum field for both diagrams was 500 mT, and 106 FORCs were measured. Further explanation is given in the text. (From van Oorschot *et al.*, 2002.) © Blackwell Publishing, with permission of the publishers and the authors.

investigations, we must now take a brief detour concerning the matter of units. Over the last 200 years or so, several measurement systems have been devised leading to different units being used to measure the same physical quantities: kilometers versus miles, pounds versus kilograms, and joules versus calories are familiar examples. Nowhere has the confusion been more troublesome than in the treatment of magnetism. At least four systems have been in use at various times, and the modern reader requires conversion tables to use the older literature (e.g., see Appendix 1). No purpose would be served in dwelling here on the fundamental reasons behind this complexity (for a particularly lucid discussion, see Feynman *et al.*, 1964). Our sole purpose is to introduce the system universally employed by enviromagnetic practitioners.

In this book, we stick to the Système Internationale (SI, for short), which is now taught in all high schools. Even so, a complication arises because there are two kinds of magnetic field, H and B. The H field we have already seen in Fig. 2.1. It is measured in A/m. In the absence of matter (i.e., in a vacuum), the two fields are related by

$$B = \mu_0 H \tag{2.8}$$

where μ_0 is the so-called *permeability constant*. In SI, it has the value $4\pi \times 10^{-7}$ Vs/Am, which means that B not only differs from H in size but also is measured in different units, namely Vs/m^2, or tesla (T) (Nikola Tesla, 1856–1943). Human nature being what it is, even the experts often do not distinguish between B and H. Indeed, for many purposes it is enough simply to refer to the "magnetic field" (you will find many examples throughout this book!). (Perhaps this laziness can be excused—after all, it is very common to give one's weight in kilograms rather than the correct SI unit of force, the newton.) Strictly speaking, of course, B and H should not be mixed up: some authors therefore refer to H as the *magnetizing field* and B as the *magnetic induction* or *flux density*. The Earth's magnetic B field has a strength of $\sim 5 \times 10^{-5}$ T, and a typical magnet for holding notes on your refrigerator door has a field of $\sim 10^{-2}$ T. Because the tesla is a large unit, it is common to give laboratory fields in millitesla (1 mT $= 10^{-3}$ T).

2.8 PUTTING IT ALL TOGETHER

The various enviromagnetic parameters (and their combinations) discussed here are generally employed for the purpose of answering three broad questions:

- Composition (i.e., which magnetic minerals are present?)
- Concentration (i.e., how much of each one is present?)
- Granulometry (i.e., what are the dominant grain sizes present?)

Variations in each of these offer useful information concerning environmental change, several examples of which are described in Chapter 4.

As far as composition is concerned, the most useful parameter mentioned so far is the *S*-ratio. But there are other important diagnostic tests, such as the Curie point and the so-called Verwey and Morin transitions. These are covered in Chapter 3, where the main features of the important environmental magnetic minerals are summarized. Several nonmagnetic techniques are also of great value in this context, including X-ray diffraction, Mössbauer spectroscopy, and microscopy (both optical and electron).

Concentration-dependent parameters include χ, SIRM, and M_s. These increase monotonically with the amount of magnetic material present. They can therefore signal increases and decreases of magnetic influx into an area, perhaps as a result of changes in climate (see Chapter 7), subsurface fluid flow (see Chapter 8), biological activity (see Chapter 9), or industrial pollution (see Chapter 10). However, most parameters, other than M_s, are also dependent on grain size. This difficulty provides the motivation for the use of certain biparametric ratios that attempt to take account of variations in the total amount of magnetic material present. Successful removal of concentration dependence then emphasizes the role of grain size.

The ratio of ARM susceptibility to low-field susceptibility is one of the most widely used concentration-independent parameters. As pointed out earlier, for magnetite, χ_{ARM} is strongly size dependent whereas χ_{lf} lies close to 6×10^{-4} m^3/kg

(= 3.1 SI) over a very wide range of sizes, from 0.01 μm all the way up to 6 mm (Heider *et al.*, 1996). Only as the superparamagnetic range is entered does systematic change occur (when the driving frequency of the measuring instrument becomes comparable to the relaxation times of the magnetic particles in the sample; see Eq. (2.6) and the description of susceptibility instruments in Chapter 4). At this point the susceptibility increases abruptly by an order of magnitude. For this reason, Maher (1988) recommends an experimental sequence in which the frequency dependence of the material under investigation is measured first, to check for the presence of grains near the SD/SP threshold. Then χ_{ARM} is determined and normalized to remove the concentration effect. This is usually done using χ_{lf} as the denominator (corresponding to the King plot), although Maher herself prefers to divide by SIRM.

3

ENVIROMAGNETIC MINERALS

3.1 INTRODUCTION

Any data retrieval methodology involves three steps: input, storage, and output. Storage requires some kind of physical material able to capture input variations and retain them for later output — the arrangement of pigment on paper, for example. No ink, no storage. In computer technology, data are encoded magnetically on floppy disks and hard drives. No magnetic particles, no information highway. Enviromagnetic studies depend on information stored in natural archives by virtue of the magnetic grains they contain. If we are to decipher environmental changes correctly, we must come to grips with the whole process. Thus, the purpose of this chapter is to introduce the magnetic minerals responsible for the information "storage": later chapters will deal with "input" and "output." Several comprehensive textbooks are available that deal with the magnetic properties of naturally occurring minerals, a subject that has, over the years, been variously referred to as *rock magnetism, mineral magnetism*, and *petromagnetism* (Nagata, 1961; Stacey and Banerjee, 1974; O'Reilly, 1984; Dunlop and Özdemir, 1997). In a review article, Rancourt (2001) has stressed the ubiquitous nature of ultrafine magnetic particles, or environmental nanomaterials, as he calls them.

The list of minerals relevant to the geosciences runs into thousands, and to these must be added numerous biominerals manufactured by organisms to make shells and other body parts. Iron, the fourth most abundant element in the Earth's crust [5% by weight, after oxygen (47%), silicon (28%), and aluminum (8%)], is a common constituent of many of these. However, our task is greatly simplified by the fact that very few naturally occurring minerals exhibit the magnetic properties we seek. Conditions on Earth are such that iron is almost always combined with other elements, particularly oxygen — witness the universal tendency of steel to rust. Consequently, we can focus our attention on a handful of iron oxides, iron oxyhydroxides, and iron sulfides, for which some relevant data are given in Table 3.1 (see also www.geo.umn.edu/orgs/irm/bestiary).

Table 3.1 Properties of Common Magnetic Minerals

Mineral	Formula	M_s (kA/m)	T_c (°C)
Magnetite	Fe_3O_4	480	580
Hematite	α-Fe_2O_3	~2.5	675
Maghemite	γ-Fe_2O_3	380	590–675
Goethite	α-FeOOH	~2	120
Pyrrhotite	Fe_7S_8	~80	320
Greigite	Fe_3S_4	~125	~330

This book is concerned with magnetism in various environmental settings here on Earth, but we will have occasion to touch upon extraterrestrial magnetism in some instances. It is therefore worth remembering that iron is a common element throughout the universe and occurs in various forms in the solar system. For example, iron oxides are prominent in the soil and surface dust of Mars, hence its sobriquet — the red planet. Furthermore, at least some of the oxides present there are magnetic, as was first demonstrated in 1976 by one of the early experiments employing simple permanent magnets on the *Viking* landers (for a compilation of papers on all aspects of the earlier exploration of Mars, see Kieffer *et al.*, 1992). Subsequently, in 1997, an array of permanent magnets on the *Pathfinder* lander positively identified magnetic iron oxide as an important component of the particulates suspended in the Martian atmosphere (Gunnlaugsson, 2000; see also http://mars.sgi.com). Other known extraterrestrial occurrences of magnetic iron minerals are meteorites and the moon. In both these cases, the most prominent contributors are iron–nickel alloys and/or pure metallic iron. As carriers of magnetic remanence, these provide important information concerning magnetic fields during the early evolution of the solar system, as described in Chapter 12. Furthermore, it was the investigation of single-domain (~5–10 nm) iron particles in the *Apollo 11* lunar dust that prompted Stephenson (1971a,b) to emphasize the importance of measuring the frequency dependence of magnetic susceptibility, a technique that is now routine in environmental magnetism (see Chapter 4). Finally, an entirely different (and so far unique) occurrence of extraterrestrial iron oxide is that in the famous meteorite ALH84001, which was found in Antarctica but is known on chemical grounds to have come from Mars. There has been much debate concerning the possibility that the tiny (10 to 100 nm) iron oxide crystals it contains represent an early form of microbial life, a topic to which we return in Chapter 9.

3.2 IRON OXIDES

Three minerals — magnetite, hematite, and maghemite — dominate our discussion of magnetism in, and on, the Earth's crust.

3.2.1 Magnetite

This is a good place to start. As Dunlop and Özdemir (1997) point out in their comprehensive monograph, magnetite is "the single most important magnetic mineral on earth." It occurs in igneous, sedimentary, and metamorphic rocks; it is common in the unconsolidated deposits typically involved in environmental studies; and it is widely manufactured by certain bacteria that use it for navigational purposes (see Chapter 9). It is also an important source of iron ore, exemplified by the great deposits of northern Sweden.

Magnetite (Fe_3O_4) is a dense, shiny black mineral that is totally opaque in microscope thin sections. Crystallographically, it is cubic with spinel structure. Its oxygen atoms thus form a face-centered cubic framework; that is, there is an O^{2-} ion at each corner and in the center of each face of the cube that constitutes the basic building block of the crystal lattice. That is a total of 14 anions (8 corners plus 6 faces), but sharing with the neighboring cubes reduces this to 4. Those at the 6 face centers are each shared 50:50 with an adjacent cube and are thus equivalent to 3 atoms; those at the 8 corners are each shared with seven neighboring cubes and are thus collectively equivalent to a single atom. Such a framework possesses two kinds of interstitial spaces [tetrahedral (known as A sites) and octahedral (known as B sites)] in which the cations are lodged (see Fig. 3.1). These constitute two sublattices having antiparallel, but unequal, magnetic moments. Magnetite is therefore ferrimagnetic. All the cations in the A sublattice are Fe^{3+}, but the B sublattice, which has twice as many occupied sites, contains equal numbers of Fe^{3+} and Fe^{2+} (this overall arrangement of the cations constitutes what is called an *inverse spinel*). The net result is that the trivalent moments cancel out, leaving an overall moment of $4\,\mu_B$ arising from the divalent ions (you can check this out very quickly by extending the information given in Box 2.1).

Taking into account the volume of the unit cell, the *spontaneous magnetization* works out to be 480 kA/m, which makes Fe_3O_4 the most magnetic naturally occurring mineral. It is for this reason that it was historically exploited — in the form of so-called *lodestones* — to construct primitive compasses. Its directional properties were

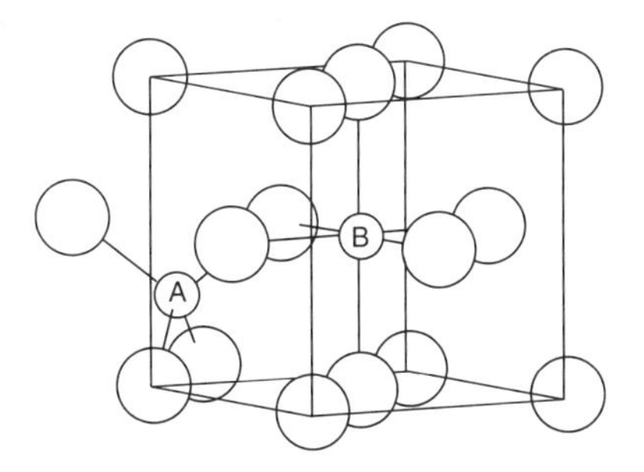

Figure 3.1 Tetrahedral and octahedral cation sites in the crystal structure of magnetite (Fe_3O_4). The large circles represent oxygen anions arranged in a face-centered cubic framework. The iron cations reside in the interstitial spaces, of which there are two kinds — at the center of tetrahedra (A sites) and octahedra (B sites). The diagram illustrates only one of each of these, but the complete structure contains twice as many tetrahedra as octahedra. However, only one eighth of the tetrahedral sites and one half of the octahedral sites are occupied (the former entirely with Fe^{3+} ions, the latter with a 50:50 mix of Fe^{3+} and Fe^{2+} ions). Because these two sublattices are ferrimagnetically coupled, the magnetic moments of the trivalent ions cancel out, leaving only the divalent ions to account for the overall magnetization of magnetite.

known in China some 2000 years ago. In Europe, an Italian scholar by the delightful (and singularly appropriate) name of Peter the Wayfarer (Petrus Peregrinus) wrote a detailed discourse in 1269 describing his extremely insightful experiments on spherical pieces of lodestone. By 1600, William Gilbert—physician to Elizabeth I—had extended and perfected this kind of work to the point where he was able to publish what is often referred to as the first modern scientific treatise, *De Magnete.*

Two important temperatures characterize magnetite—the *Curie point* and the *Verwey transition.* The first of these [named in honor of Pierre Curie (1859–1906), see also Box 2.2] occurs at 580°C, the temperature at which thermal energy overcomes the exchange coupling and the ferrimagnetism is lost (Fig. 3.2). The second [named after its discoverer, E. J. W. Verwey (1905–1981)] occurs at about −150°C and marks a change in the crystallographic distribution of the iron cations such that the previously cubic framework is slightly distorted to monoclinic symmetry. This is a subtle effect, but it alters the crystalline anisotropy (see Box 2.4) that, in many cases, can result in abrupt changes in magnetic (and other) properties. Both the Curie point and the Verwey transition provide excellent diagnostic tests.

Although magnetite occurs widely, it is also common to find a whole range of variants in which iron is replaced by titanium. This gives rise to a solid-solution series known as the *titanomagnetites.* Magnetite (Fe_3O_4) thus appears as one end member of a whole spectrum, the other end being represented by the mineral *ulvöspinel*

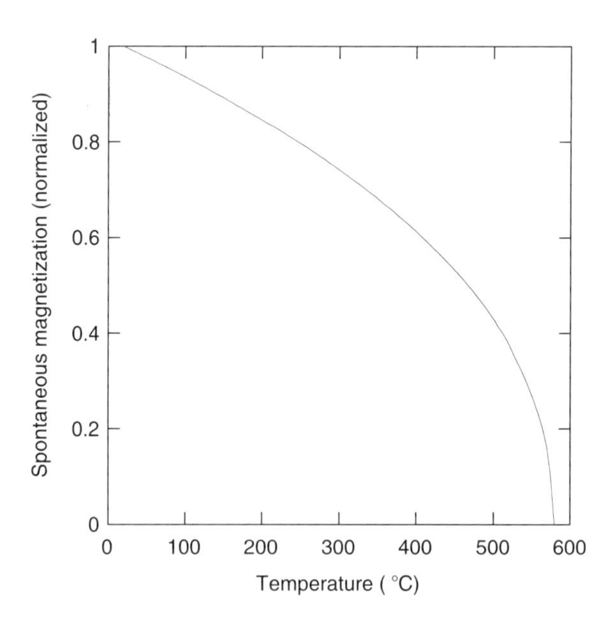

Figure 3.2 Variation of spontaneous magnetization of magnetite between room temperature (RT) and the Curie point ($T_C = 580$°C). The curve shown [$M_T/M_{RT} = ((T_C - T)/(T_C - RT))^{0.43}$] is a best fit to various experimental data (see Dunlop and Özdemir, 1997).

Box 3.1 Ternary Diagrams

In a three-component system, any particular composition can be plotted graphically on a diagram wherein each vertex of an equilateral triangle represents 100% (and the entire side opposite that vertex represents 0%) of one of the constituent elements, as shown in the diagram (left).

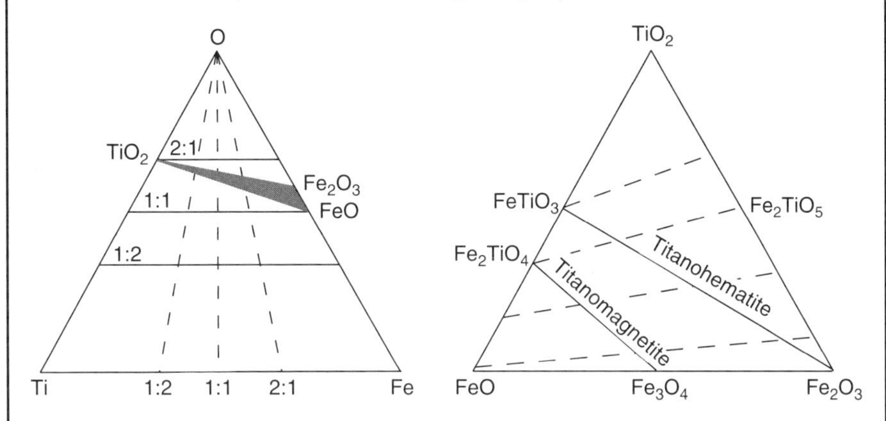

At points inside the triangle, the diagram works because the sum of the perpendicular distances from any point to the three sides is equal to the height of the triangle (i.e., 100%). (To get the idea, try plotting $FeTiO_3$.) Horizontal lines indicate constant ratios of (Fe + Ti)/O: lines converging on the oxygen vertex indicate constant Fe/Ti. The latter are very useful as they illustrate the effects of oxidation, an extremely important process in environmental magnetic studies. The drawback with the diagram is that the minerals of interest fall in the restricted range shown by the shading. It is normal practice, therefore, to construct the ternary diagram so as to emphasize this compositional field, with vertices at TiO_2, FeO, and Fe_2O_3. This is done in the diagram on the right, on which the key minerals and solid solution series are labeled. The latter fall along lines sloping down to the right, which are equivalent to horizontal lines in the other diagram. On the other hand, oxidation is indicated by lines sloping up to the right.

(Fe_2TiO_4). The general formula for the titanomagnetites is therefore written as $Fe_{3-x}Ti_xO_4$ ($0 \leq x \leq 1$). Because there are three chemical elements involved, it is very useful to follow the normal practice of representing the composition on a triangular—or ternary—diagram, as explained in Box 3.1.

The Ti^{4+} cations are located on the octahedral sites. For each titanium atom inserted, one remaining trivalent Fe ion must become divalent in order to maintain charge neutrality. At $x = 1$, there are no trivalent ions left: ulvöspinel thus has only Fe^{2+} ions on its occupied A sites and a 50:50 mix of Fe^{2+} and Ti^{4+} on its occupied B sites. The Ti^{4+} cation has zero magnetic moment because it has no unpaired electrons. This leads to a steady decrease in the spontaneous magnetization as the amount of titanium substituted into the lattice increases. As we saw before, for $x = 0$ (magnetite), the spontaneous magnetization is $4\,\mu_B$ per formula unit. For $x = 1$ (ulvöspinel), this falls to zero because the lattice now has equal numbers of divalent iron ions on the two opposing sublattices. In other words, ulvöspinel is antiferromagnetic. Increasing titanium content also causes the crystal lattice to expand from 8.396 Å for magnetite to 8.54 Å for ulvöspinel (Fig. 3.3). X-ray diffraction measurements can therefore be used as a means of identifying the composition of the titanomagnetite grains present in a sample, although this usually requires that the grains first be extracted. A more popular diagnostic test is based on the almost linear decrease in the Curie point that takes place as titanium is added to the lattice (Fig. 3.4).

Magnetite itself and the titanomagnetites in general are formed initially in a variety of igneous rocks. For example, the widespread basaltic lavas that carry the marine magnetic anomalies so crucial to plate tectonics typically contain crystals of $Fe_{2.4}Ti_{0.6}O_4$ (often referred to as TM60). On the continents, a whole range of iron oxide compositions is found in various igneous products. As a result of weathering

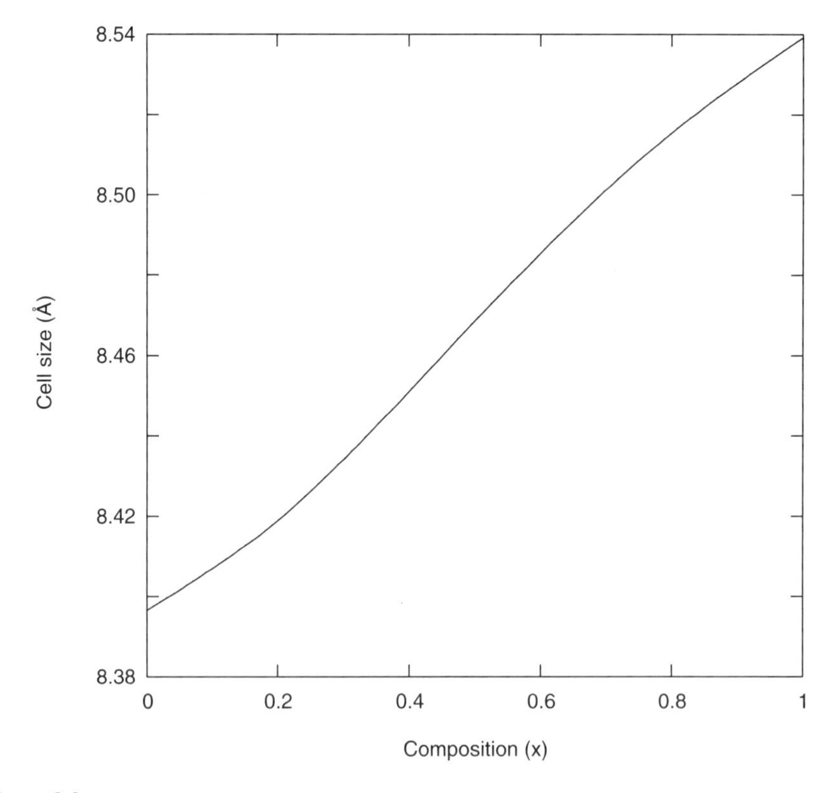

Figure 3.3 The unit cell parameter dependence on Ti content (x) for the titanomagnetite solid solution series. The curve shown is based on experimental data from several authors (for a summary, see O'Reilly, 1984).

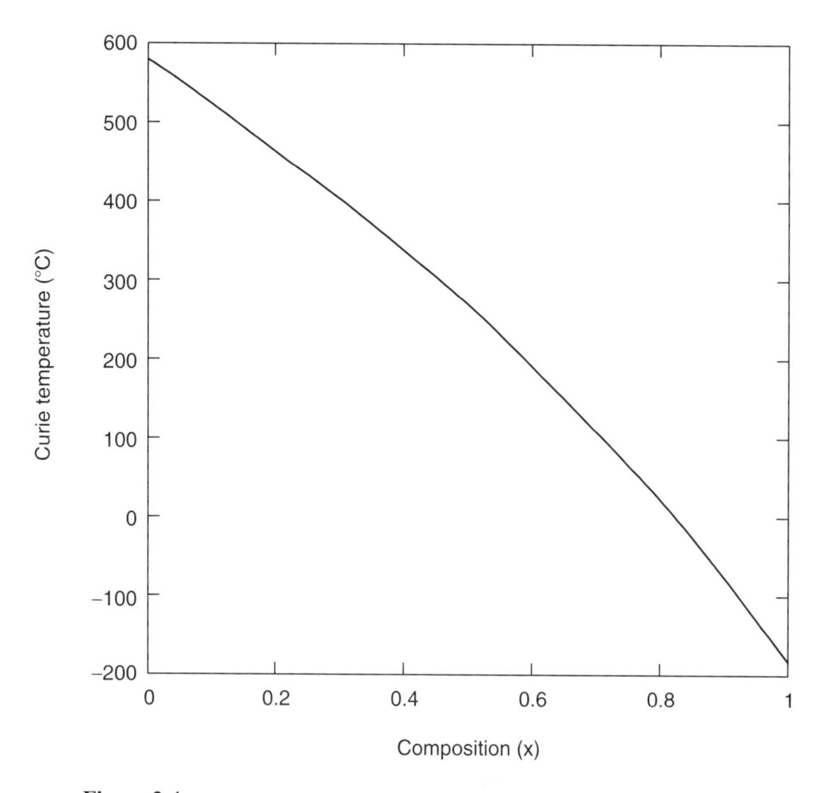

Figure 3.4 Curie point dependence on Ti content (x) in the titanomagnetites.

and erosion, the mineral grains involved eventually find their way into a variety of sedimentary environments, where they provide the magnetic records we seek. However, there are modes of occurrence other than as discrete, individual grains. The most important of these are magnetite/ilmenite intergrowths. These arise by partial oxidation of a general titanomagnetite that moves the composition off the magnetite/ulvöspinel line (see Box 3.1). If cooling is slow enough, *exsolution* may then take place, leading to a separation into two distinct phases, usually close to pure magnetite and pure ilmenite (Burton, 1991; Lindsley, 1991). The two phases are intergrown in an intimate microstructure in which the ilmenite forms lamellae on the [111] crystal planes with the magnetite taking up the space between. In this way, magnetite regions with various morphologies can arise, the effective size of which is much smaller than the overall grain size. Haggerty (1991) has assembled an impressive collection of optical photomicrographs covering the whole range of textures found in the iron–titanium minerals. Electron microscope images of magnetite/ilmenite intergrowths can be found in Davis and Evans (1976). Another (but very rare) possibility involves no oxidation, the exsolution leading to an orthogonal [100] framework of ulvöspinel interspersed with small cubes of magnetite (Nickel, 1958; Evans and Wayman, 1974). Finally, several examples have been described in which tiny magnetite grains occur as inclusions precipitated inside common silicates (Evans *et al.*, 1968; Davis, 1981;

Bogue *et al.*, 1995). In terms of environmental magnetism, one important implication is that this type of magnetite is protected from the various (bio)geochemical reactions that its "naked" counterparts may suffer during such processes as soil formation and burial diagenesis (Maher and Hounslow, 1999; see also Chapter 5).

3.2.2 Hematite

This mineral occurs widely in nature, being particularly common in soils and sediments of environmental significance. It is also responsible for the magnetization carried by "red beds"—red sandstones and shales that provide a major source of data in classic paleomagnetism. Many iron ores are hematitic, most notably the great deposits mined in the Lake Superior region.

Hematite possesses hexagonal crystal structure in which alternate planes contain trivalent iron ions magnetized in (almost) opposite directions (Fig. 3.5). The slight departure from antiparallelism—called *spin canting*—is crucial. It turns hematite from an antiferromagnetic mineral into a weakly ferromagnetic one with a spontaneous magnetization of about 2.5 kA/m and a Curie point of 675°C (Fig. 3.6). Thus, although being about 200 times weaker than magnetite, hematite is thermally more stable. The higher Curie point is useful for identification purposes, particularly in cases where magnetite and hematite coexist. On cooling from the Curie point, the magnetization of hematite rises sharply to a plateau that is maintained down to about −15°C. Here, it passes through the *Morin transition* (Morin, 1950), where the spin canting, and hence the weak ferromagnetism, is lost.

It is again found that iron can be replaced by titanium giving rise to a second solid solution series, the *titanohematites* (see the ternary diagram in Box 3.1). At one end is hematite (α-Fe_2O_3), at the other $FeTiO_3$ *(ilmenite)*. The general formula is therefore written as $Fe_{2-y}Ti_yO_3(0 \leq y \leq 1)$. For every Ti^{4+} cation added, one of the remaining Fe^{3+} ions must be converted to Fe^{2+}. Once again, the substitution of titanium leads to an almost linear decrease in Curie temperature (Fig. 3.7). As with the titanomagnetites, compositions rich in titanium ($y > \sim 0.7$ for the titanohematites, $x > \sim 0.8$ for the titanomagnetites) lead to Curie points below room temperature.

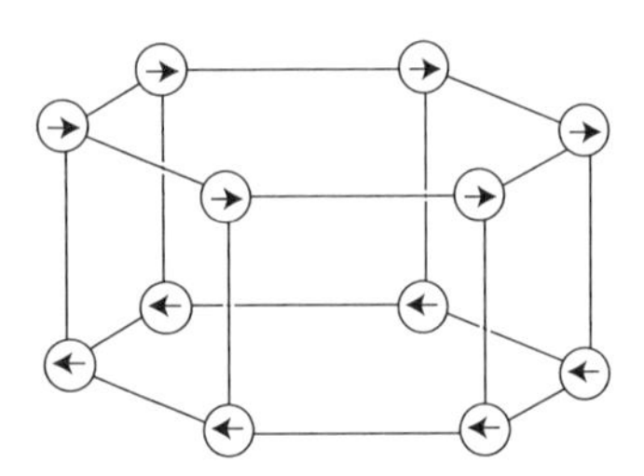

Figure 3.5 Simplified representation of the spatial arrangement of the iron cations (all trivalent) in hematite. The arrows indicate how basal [0001] planes of cations are ferromagnetically coupled within planes and antiferromagnetically coupled between planes. The antiparallelism is not exact, however, and this *spin canting* leads to a net magnetization in the basal plane. Imagine the top layer of arrows rotated slightly clockwise and the lower layer rotated slightly counterclockwise (viewed from above). The result is a net magnetization pointing toward you.

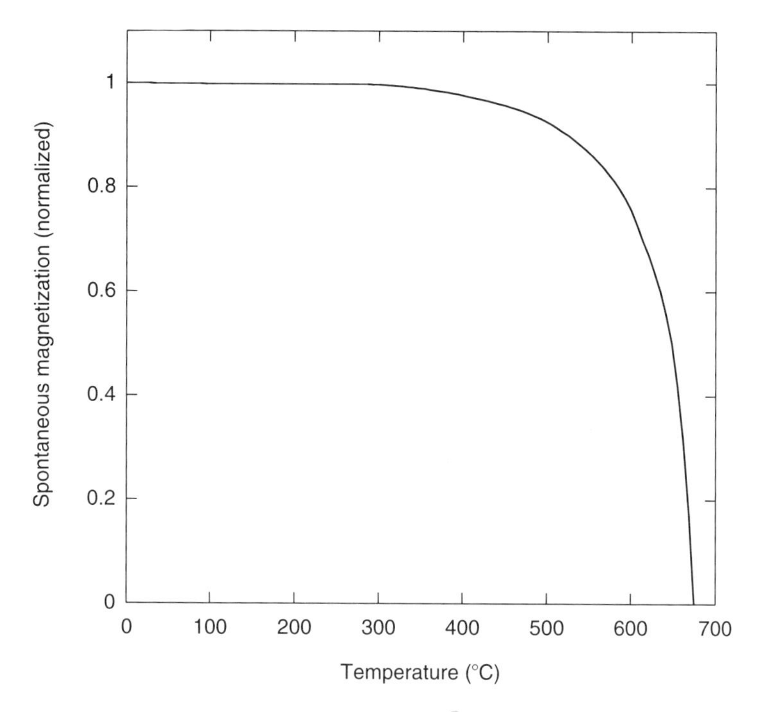

Figure 3.6 Variation of spontaneous magnetization of hematite between room temperature (RT) and the Curie point ($T_C = 675°C$).

The spin canted arrangement and its weak (sometimes called *parasitic*) ferromagnetism persist until y reaches ~ 0.45, beyond which the cations become partially ordered and the titanohematites become ferrimagnetic. As a result, the spontaneous magnetization rises rapidly to a maximum (at $Fe_{1.3}Ti_{0.7}O_3$) of $2.8\,\mu_B$ ($0.7 \times 4\,\mu_B$) before falling off again to very low values at $y = \sim 0.95$. Although theoretically very important, this ferrimagnetic behavior of the titanohematites is of limited significance for environmental magnetists. One reason for this is that at compositions beyond about $y = 0.7$, the observed Curie points are below room temperature. More important, however, is the fact that most intermediate compositions are unstable at ordinary temperatures and exsolve into Ti-rich and Ti-poor phases that form intergrown microstructures similar to those described earlier for the titanomagnetites. In classical paleomagnetism, such microstructures are often responsible for the property of *self-reversal*, wherein a geological formation containing such grains acquires a remanence in the opposite direction to the prevailing ambient magnetic field. In the early days, when the geomagnetic polarity timescale was emerging (see Chapter 6), rocks of this type caused no end of difficulty. Fortunately, it turns out that they are generally quite rare.

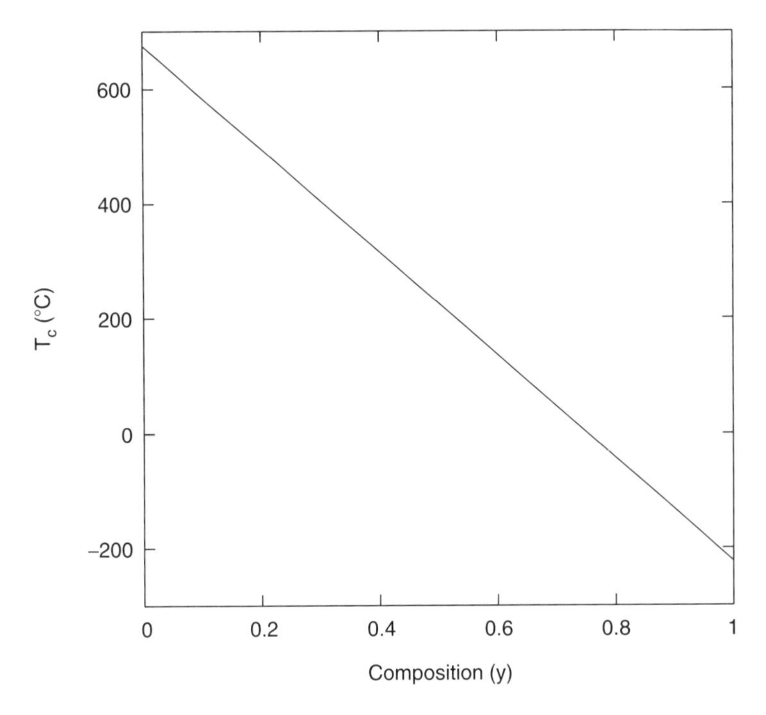

Figure 3.7 Curie point dependence on Ti content (y) in the titanohematites.

3.2.3 Maghemite

Before the advent of the compact disc, maghemite was commercially significant as the storage element used in tape recording. For our purposes, however, it is important in environmental studies because it occurs widely in soils. Its chemical formula is identical to that of hematite, and both minerals therefore occupy the same position on the ternary diagram (see Box 3.1). However, they do not share the same crystal structure or magnetic properties. To prevent confusion, a prefix is introduced, hematite being designated as α-Fe_2O_3, maghemite as γ-Fe_2O_3. Maghemite is simply the fully oxidized form of magnetite: it has a cubic crystal structure with a unit cell edge of 8.337 Å, somewhat smaller than that of magnetite (see Fig. 3.3). The oxidation process involves the divalent iron ions. Two thirds of them have their valence state changed from Fe^{2+} to Fe^{3+}, and the remaining one third are removed from the lattice entirely. The sites from which atoms are removed remain vacant, and such structures are therefore said to be *cation deficient*. The valence change and the loss of cations collectively result in a decrease of the room-temperature spontaneous magnetization to 380 kA/m (from 480 kA/m for magnetite). The Curie temperature (~645°C) is difficult to determine experimentally because maghemite is metastable; at elevated temperatures it suffers an irreversible crystallographic change to hematite with a consequent dramatic loss of magnetization. Indeed, this characteristic behavior is often more useful for identification purposes than the Curie point itself. Even here, great care is required in interpreting experimental data because the temperature at which the conversion takes place (the so-called *inversion temperature*) is very variable—values anywhere from 250 to 900°C have been reported!—and seems to depend on grain size and the presence of impurities (see Dunlop and Özdemir, 1997).

Referring again to the ternary diagram shown in Box 3.1, we can now anticipate the existence of a whole field of compositions lying between the two solid-solution series represented by the ulvöspinel–magnetite and the ilmenite–hematite joins. Minerals represented by points in this field are called *titanomaghemites*. Any given composition can be arrived at by oxidation from the appropriate position in the titanomagnetite series, as indicated by the dashed lines.

3.3 IRON OXYHYDROXIDES

Weathering of bedrock produces a wide variety of products, among which are numerous hydrous iron oxides. Of these, only goethite (α-FeOOH) is magnetically significant in its own right. Some of the others, such as ferrihydrite ($5Fe_2O_3 \cdot 9H_2O$, also known as limonite) and lepidocrocite (γ-FeOOH), are noteworthy in that they may undergo chemical changes to produce hematite and magnetite (see Chapter 5), which may be magnetically important in soils (Schwertmann, 1988a,b; Zergenyi *et al.*, 2000) and in the red cement of certain rock types of paleomagnetic significance (Hedley, 1968).

Goethite is hexagonal and antiferromagnetic—but not perfectly so. It also possesses a weak ferromagnetism whose origin is poorly understood. It is thought to be a *defect moment* due to unbalanced numbers of atomic moments. The corresponding Curie point is about 120°C and the spontaneous magnetization is $\sim 2\,kA/m$, slightly less than that of hematite. Although it has been the subject of a great many laboratory investigations (often involving synthetically derived powders), it is only recently that goethite has been systematically sought in natural environments. France and Oldfield (2000) studied a number of soils and recent sediments with a view to assessing the significance of goethite for environmental magnetists. They selected samples representing different settings from a variety of sites around the world—soils from China and Portugal, lateritic weathering products from Indonesia, river sediments from the United States, lake sediments from England, and turbidite sediments from the deep Atlantic Ocean—and concluded that goethite is much more widespread than has often been thought.

3.4 IRON SULFIDES

The elements iron and sulfur combine in various ratios to form a number of distinct minerals. *Troilite* (FeS) is common in meteorites and lunar samples but does not occur on Earth. *Pyrite* (FeS_2), on the other hand, is very common but paramagnetic.

Between these compositional bounds, there are two naturally occurring minerals that are important for environmental magnetism: *pyrrhotite* and *greigite*. Pyrrhotite actually crystallizes in several forms, the most common being Fe_7S_8, which is monoclinic and ferrimagnetic, and Fe_9S_{10}, which is hexagonal and antiferromagnetic. In the former, the ferrimagnetism arises from the fact that alternate planes (within which atomic spins are ferromagnetically coupled) have oppositely directed magnetic moments. This is rather similar to the magnetic structure of hematite except that now there is no spin canting; the ferrimagnetism is due to the fact that magnetic moments of the different planes are not all equal. The result is an overall magnetization of $\sim$80 kA/m, with a Curie point of 320°C.

In practice, Fe_7S_8 and Fe_9S_{10} are often found in close association with each other, so the magnetic properties of natural samples can be quite variable. Although hexagonal pyrrhotite (Fe_9S_{10}) is antiferromagnetic at room temperature, it undergoes a crystallographic transition (called the λ transition by some authors, the γ transition by others) at about 200°C, where it becomes ferrimagnetic, with a subsequent Curie point at $\sim$265°C. During heating, therefore, samples containing Fe_9S_{10} exhibit a rapid increase in magnetization above 200°C. This provides an excellent means of assessing the relative amount of Fe_9S_{10} present in natural samples (Schwarz, 1975). Pyrrhotite is a common minor constituent of igneous, metamorphic, and sedimentary rocks as well as sulfide ores. In the mining district of Sudbury, Canada, it occurs in intimate association with $(Fe, Ni)_9S_8$ (pentlandite), which constitutes the world's most important source of nickel.

Greigite (Fe_3S_4) is a cubic mineral with spontaneous magnetization and Curie point similar to those of pyrrhotite ($\sim$125 kA/m, $\sim$330°C). Until recently, it was thought to be rather rare in nature, but it is now known to occur widely in many sedimentary environments. For example, Roberts (1995) found it in samples of Plio-Pleistocene marine sediments from Taiwan, Miocene coal measures from the Czech Republic, Cretaceous marine sediments from Alaska, and Mio-Pliocene lacustrine sediments from California. Sagnotti and Winkler (1999) describe a variety of cases from sites in Italy and also provide a summary of relevant work by others, including an extensive bibliography. Greigite is particularly associated with the sulfate reducing, anoxic conditions under which many lacustrine and marine sediments form, where it represents an intermediate step in the chemical pathway leading to pyrite. Greigite also occurs as magnetosomes originating from magnetotactic bacteria living in sulfur-rich habitats (see Chapter 9).

The enviromagnetic significance of the iron sulfides is discussed in detail by Snowball and Torii (1999).

3.5 IRON CARBONATE

Siderite ($FeCO_3$) is a paramagnetic iron mineral that is common in carbonate sediments. It often forms by direct precipitation from water, either marine or lacustrine. In the latter, it sometimes provides small-scale commercial deposits known as bog iron ore. It also occurs in hydrothermal veins and as concretionary nodules in clays. Ellwood *et al.* (1988) have described siderite in carbonate rocks from the United States and the Czech Republic and also from anoxic deep-sea sediments from the South Atlantic. They propose that it results from the metabolic activity of certain bacteria (see Chapter 9 for more on this general topic). For our purposes, the importance of siderite really arises from its oxidation products — magnetite, maghemite, and hematite. These certainly affect paleomagnetic investigations because they often carry a chemical remanent magnetization (CRM, see Box 5.2). Once created, these magnetic minerals will play a significant role in environmental magnetism, particularly in carbonate-rich environments. However, it may no longer be possible to prove that they formed from preexisting siderite.

3.6 SOME EXAMPLES

In this section we illustrate the occurrence and means of identification of several magnetic mineral species of importance in typical enviromagnetic investigations. The examples chosen are selected to show a variety of natural settings and experimental approaches. They certainly do not represent all the possibilities. The offering is more of a smorgasbord than a four-course meal. The menu is given in Table 3.2.

In this selection, the diagnostic criteria of importance are the Curie point (T_C), the Verwey and Morin crystallographic transitions (VT and MT, in magnetite and hematite, respectively), the coercivity spectra obtained from room-temperature incremental isothermal remanent magnetization (IRM) experiments [see Fig. 2.3, Section 2.5, and Chapter 4], and the occurrence of known chemical transformations (CTs) that take place during laboratory heating experiments. Other procedures are available (see Chapter 4), but we emphasize these because they emerge naturally from laboratory experiments carried out more or less routinely by environmental magnetists.

Table 3.2 Selected Examples of Magnetic Mineral Occurrences

Where	What	How	Who	Figure
(a) Siberia	Magnetite	IRM	Chlachula *et al.*, 1998	3.8
(b) Germany	Magnetite	VT, T_C	Fassbinder and Stanjek, 1993	3.9a
(c) Portugal	Goethite	IRM	France and Oldfield, 2000	3.9b
(d) England	Hematite	IRM, MT	France and Oldfield, 2000	3.9c,d
(e) China	Maghemite	CT	Evans and Heller, 1994	3.10
(f) Taiwan	Pyrrhotite	T_C	Torii *et al.*, 1996	3.11a
(g) Italy	Greigite	CT	Sagnotti and Winkler, 1999	3.11b

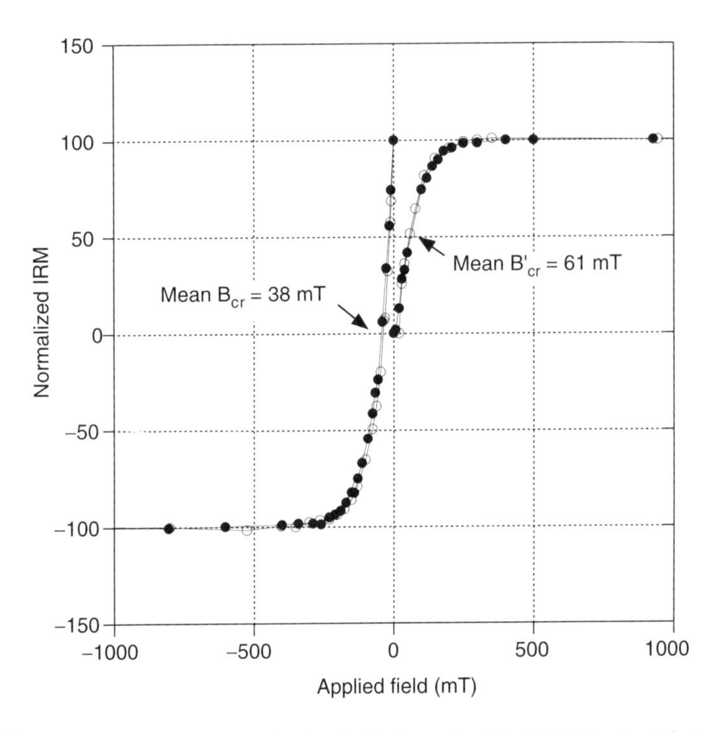

Figure 3.8 Isothermal remanent magnetization (in the forward and backfield directions) for two samples of loess from Siberia. B'_{cr} is the remanent acquisition coercive force = the applied field at which 50% of the eventual saturation IRM is achieved in the forward direction and B_{cr} is the remanent coercive force = the field at which the IRM of a previously saturated sample is reduced to zero in the backfield direction. These quantities are used in the text. (Modified from Chlachula *et al.*, 1998.)

1. At Kurtak in southern Siberia, Chlachula *et al.* (1998) have studied a 34-m-thick loess sequence spanning the last ~150,000 years (see further discussion in Chapter 7). IRM acquisition experiments yielded low coercivities typical of a "soft" magnetic material such as magnetite (Fig. 3.8): most of the coercive force spectrum lies below 100–150 mT. Dankers (1981) has determined reference curves for samples of magnetite, titanomagnetite, and hematite. He uses the parameters B'_{cr} (the remanent acquisition coercive force = the applied field at which 50% of the eventual saturation IRM is achieved in the forward direction) and B_{cr} (the remanent coercive force = the field at which the IRM of a previously saturated sample is reduced to zero in the backfield direction) to distinguish the three types of material he investigated. For magnetite, titanomagnetite, and hematite, he obtains $B'_{cr}/B_{cr} = 1.6 \pm 0.2, 1.2 \pm 0.2$, and unity, respectively. The results for the two Kurtak samples shown in Figure 3.8 yield a mean ratio compatible with magnetite, namely $B'_{cr}/B_{cr} = 1.61$. The trouble with results of this kind, however, is the fact that maghemite is also magnetically soft and exhibits coercivities similar to those of magnetite. To secure a firm identification, therefore, it is advisable to carry out thermal experiments also, as in the next example.

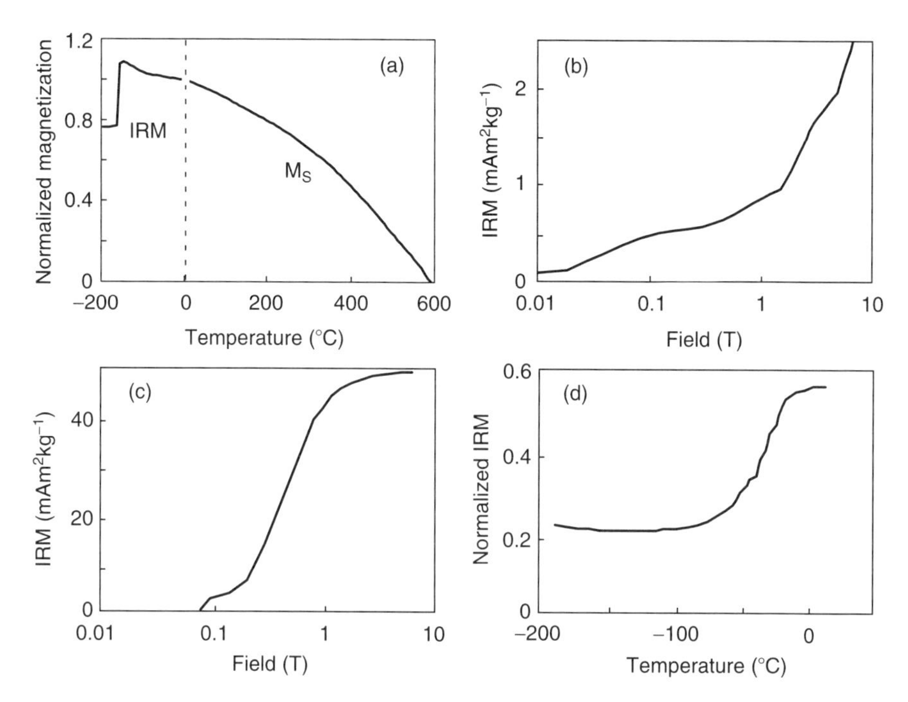

Figure 3.9 (a) Heating and cooling experiments on a magnetic extract from the remains of a Neolithic wooden post at an archeological site in Germany. The Curie point and Verwey transition of magnetite are clearly shown. (Compiled from Fassbinder and Stanjek, 1993.) (b) Acquisition of isothermal remanent magnetization (in fields of 10, 20, 30, 40, 60, 80, 100, 150, 200, 300, 400, 600, 800, 1000, 1500, 2000, 3000, 4000, 5000, 6000, and 7000 mT) for a Portuguese soil sample containing goethite. (Redrawn from France and Oldfield, 2000.) (c and d) Magnetic data for hematite-rich sediments from a small lake in northern England. (Redrawn from France and Oldfield, 2000.) (c) Acquisition of isothermal remanent magnetization (using the same field values as Fig. 3.9b). (d) Change in IRM as a function of temperature showing a clear Morin transition. The ordinate represents the magnetization normalized to the room-temperature magnitude of the "hard" IRM (i.e., $IRM_{7T} - IRM_{0.4T}$). a, © Polish Academy of Sciences, with permission of the publishers. b, c, and d, © American Geophysical Union. Modified by permission of American Geophysical Union.

2. This example comes from the fascinating case described by Fassbinder and Stanjek (1993), who investigated magnetic extracts from the remains of wood excavated at a German archeological site (see further discussion in Chapter 11). Heating and cooling experiments yielded clear evidence of both the Curie point and Verwey transition so characteristic of magnetite (Fig. 3.9a). Another convincing example involving these two diagnostic temperatures is described by Torii *et al.* (2001), who studied sand samples from the Taklimakan Desert, a potential source region for the thick loess deposits in China (see further discussion in Chapter 7). It is also worth noting that very clear Verwey transitions and magnetite Curie points have been reported by Matasova *et al.* (2001) on the Siberian material discussed in section 1.

3. France and Oldfield (2000) report magnetic results for several samples that show strong evidence of the presence of goethite. A particularly good example is the IRM acquisition curve for a Portuguese soil sample (Fig. 3.9b). This has a coercivity

spectrum that extends beyond fields at which hematite typically saturates (4–5 T) and even beyond the highest field available experimentally (7 T). Such high coercivities are particularly diagnostic of goethite.

4. Urswick Tarn is a small lake in northwest England that is known to have received hematite inwash as a result of nearby mining activity during the 19th century. Sediment samples from the lake bottom studied by France and Oldfield (2000) exhibit IRM acquisition behavior typical of hematite (Fig. 3.9c), saturating in fields of a few tesla. Furthermore, low-temperature experiments show a clear Morin transition (Fig. 3.9d).

5. The identification of maghemite presents several difficulties to the environmental magnetist as it is easily confused with magnetite. Both minerals are magnetically strong and soft. Methods relying on coercivities are generally ineffective. But when heating is attempted (to seek the Curie point), maghemite very commonly converts to hematite as described before. This chemical transformation offers diagnostic possibilities, as shown in Fig. 3.10. This is an example taken from the Chinese loess (Evans and Heller, 1994). On heating, it exhibits a clear magnetite Curie point, but there is also an inflection between 300 and 400°C. This can be attributed to the presence of maghemite that transforms to weakly magnetic hematite and thus leads to a decrease in the overall magnetization when the sample is cooled (see Box 7.1 for a quantitative analysis of these results).

6. The identification of the sulfide minerals pyrrhotite and greigite by purely magnetic means is a tricky business. In most cases, other evidence such as X-ray

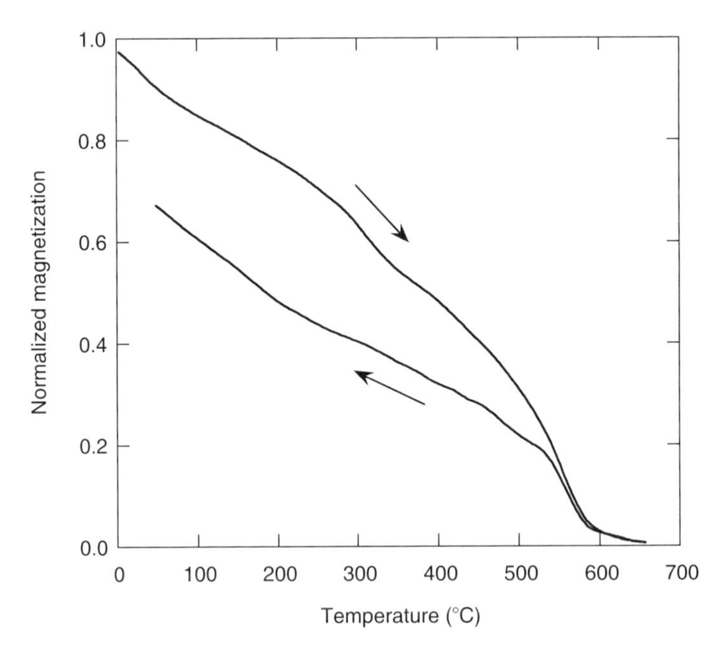

Figure 3.10 Thermomagnetic heating and cooling curves for a sample of Chinese loess. (Redrawn from Evans and Heller, 1994.)

diffraction analysis and optical microscopy already points to the presence of iron sulfides. Indeed, one may anticipate them on general grounds if the depositional setting is anoxic and pyrite is present. As far as magnetism is concerned, the problem then boils down to discriminating between pyrrhotite- and greigite-bearing sediments. Torii *et al.* (1996) have addressed this problem. They point out that one possibility relies on the fact that monoclinic pyrrhotite undergoes a crystallographic transition at 34 K that affects its magnetic properties, rather like the Verwey transition in magnetite (see Dekkers *et al.*, 1989). However, the examples that Torii and his coauthors report indicate that this is a very subtle effect not likely to constitute a routine test. More promise is offered by their modification of a technique due to van Velzen and Zijderveld (1992) for detecting chemical alteration during heating experiments in the laboratory. This is because monoclinic pyrrhotite is chemically stable up to its Curie point, whereas greigite is very prone to significant alteration to pyrite and marcasite (cubic and orthorhombic FeS_2, respectively) (Roberts, 1995). To compare the thermal behavior of pyrrhotite and greigite, this section should therefore be read in unison with section 7 following.

During stepwise thermal demagnetization, a decrease in magnetism between one temperature and the next could arise from progressive thermal unblocking [see Eq. (2.6)] or from chemical conversion to a nonmagnetic (or less magnetic) mineral. To find out which it is, van Velzen and Zijderveld (1992) suggest that the sample be initially given a saturation IRM (SIRM) and then remagnetized in the same field after each thermal step. If unblocking alone is taking place, each successive IRM will restore the magnetization back to its original level. On the other hand, if "magnetizable" material has been irreversibly lost, only partial restoration will be observed.

7. Following the argument set out in section 6, the results of the van Velzen and Zijderveld (1992) technique applied to greigite- and pyrrhotite-bearing samples are compared in Figure 3.11. Notice how the reintroduced SIRM essentially brings the pyrrhotite curve back to the initial value after each heating step (Fig. 3.11a). This implies that the lower curve is dominated by unblocking with little chemical alteration. The rapid fall beyond 320°C can therefore be interpreted as the monoclinic pyrrhotite Curie point. The greigite sample, on the other hand, shows the characteristic behavior of chemical alteration: beyond ~200°C the successive SIRMs fall far short of bringing the magnetization back to its starting value (Fig. 3.11b).

3.7 ROOM-TEMPERATURE BIPLOTS

For the introductory discussions in Section 3.6 concerning the identification of the magnetic minerals present in a sample, we restricted attention to single parameters (such as the Curie point) or simple ratios (such as the coercivity ratio B'_{cr}/B_{cr}). However, reality is often more complicated and calls for more refined procedures. One common response has been to combine various parameters (either singly or as ratios) into two-dimensional graphs that (hopefully!) separate the various magnetic components into distinct parameter spaces. This, of course, is the intent of the Day plot (see Section 2.6). There, however, the aim is to specify the domain state of

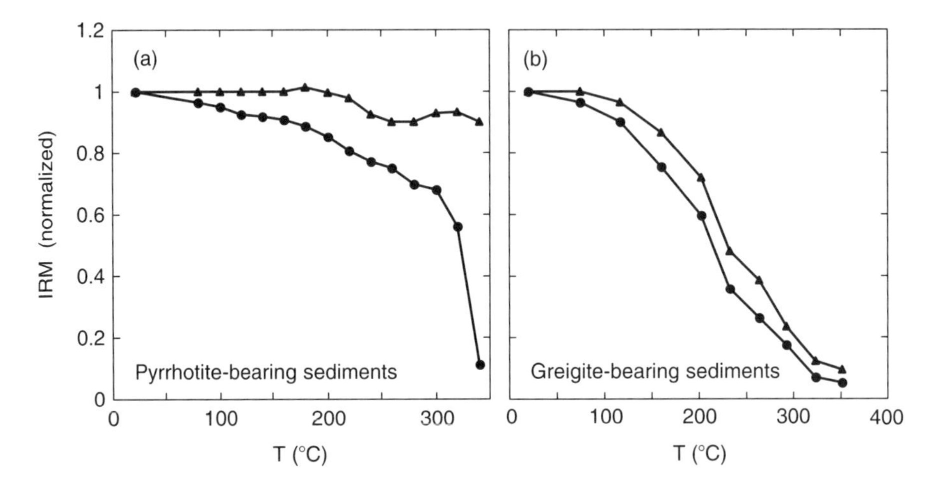

Figure 3.11 Stepwise thermal demagnetization of saturation isothermal remanent magnetization (SIRM), using the van Velzen and Zijderveld (1992) method (see text for detailed description). (a) Pyrrhotite-bearing sediment from Taiwan. (Modified from Torii *et al.*, 1996.) (b) Greigite-bearing sediment from Italy. (Modified from Sagnotti and Winkler, 1999.) In both graphs, circles indicate measurements after each thermal demagnetization step and triangles represent SIRM remagnetization prior to the next heating step. a, © American Geophysical Union. Modified by permission of American Geophysical Union. b, © Elsevier Science, with permission of the publishers.

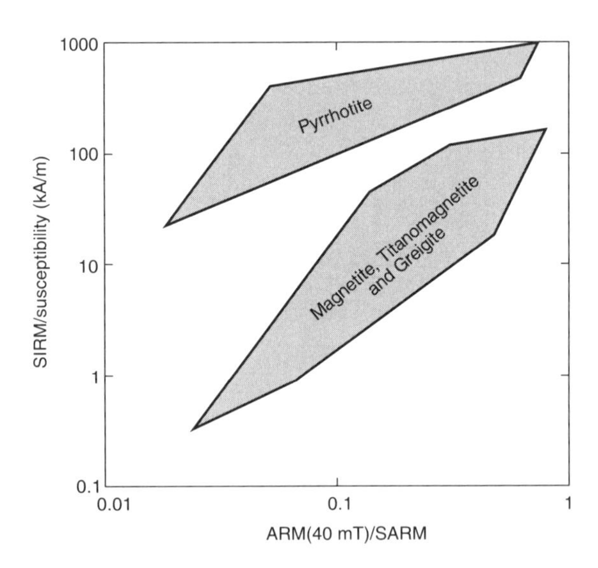

Figure 3.12 Biplot of saturation isothermal remanence/susceptibility (SIRM/χ) versus anhysteretic remanent magnetization after 40-mT alternating-field demagnetization/total anhysteretic remanent magnetization [ARM(40 mT)/SARM]. (Note that the acronym SARM is used for its conceptual similarity to SIRM; there is no implication that it represents a saturation of any kind.) (Redrawn from Peters and Thompson, 1998a.) © Elsevier Science, with permission of the publishers.

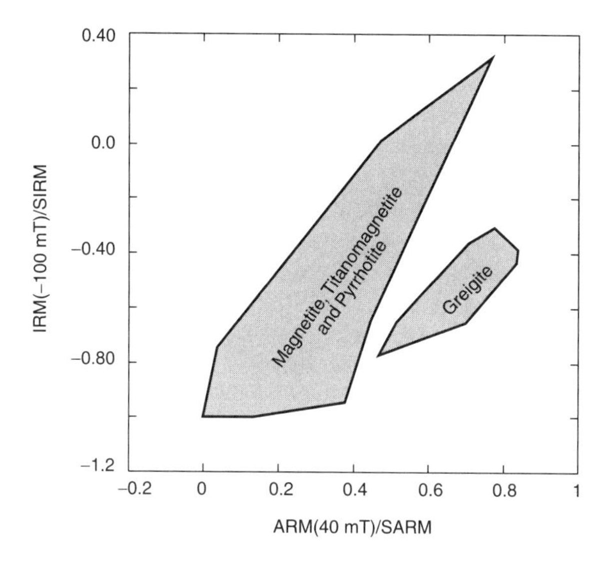

Figure 3.13 Biplot of saturation isothermal remanence after 100-mT "backfield" demagnetization/saturation isothermal remanence [IRM(−100 mT)/SIRM] versus anhysteretic remanent magnetization after 40-mT alternating-field demagnetization/total anhysteretic remanent magnetization [ARM(40 mT)/SARM]. (Note that the acronym SARM is used for its conceptual similarity to SIRM; there is no implication that it represents a saturation of any kind.) (Redrawn from Peters and Thompson, 1998a.) © Elsevier Science, with permission of the publishers.

already identified (or presumed) magnetite. Peters and Thompson (1998a) have extended earlier suggestions as to how such biplots can be employed to furnish compositional information. Their goal is to distinguish the common magnetic minerals from one another using combinations of simple room-temperature measurements without resorting to expensive, time-consuming, mineral separation techniques or heating the sample and risking chemical alteration. Using data from 56 natural samples representing "various sediment, soil and rock types," they found that combinations of IRM, ARM, and susceptibility measurements were able to distinguish magnetite (and/or titanomagnetite) from pyrrhotite (Fig. 3.12) and greigite (Fig. 3.13). Semiempirical graphs of this kind serve as useful first steps in the qualitative identification of the main magnetic minerals present in the material under investigation. However, they should be used with caution. Further work may extend the parameter fields occupied by the various minerals, and, of course, real samples may well contain mixtures. Nevertheless, the individual parameters involved (susceptibility, IRM, ARM) are widely measured as a matter of routine in enviromagnetic work, so it is always worthwhile to calculate and plot the results in several ways as a guide to further analysis.

4

MEASUREMENT AND TECHNIQUES

4.1 INTRODUCTION

Magnetic minerals with different mineral magnetic properties provide natural archives of environmental processes that may be found in archeological surveys, in soil studies, in suspended or deposited sediments of rivers, lakes, and seas, or in emission products of urban pollution, to mention only a few settings of interest to us. This chapter aims first at presenting some techniques for measuring the mineral magnetic parameters that have been introduced in earlier chapters. Among these, the most relevant are low-field susceptibility, natural remanent magnetization and laboratory remanences such as the widely used ARM and IRM, saturation magnetization, coercive force and coercivity of remanence, and their dependence on temperature and applied field.

We will then describe and illustrate how these measured parameters have been recorded in the environment and what their mutual relations can tell us. How can they be utilized — possibly in connection with nonmagnetic methods — to identify type and quantity of enviromagnetic minerals? How can the grain size distributions, which are so indicative of many environmental processes, be evaluated? How can mixtures of different enviromineral types be separated?

Finally, we will turn to methods of interpretation that have been developed for the investigation of environmental change in marine, lacustrine, or eolian sediment sections and for modeling input source parameters and transport mechanisms. The application of some useful statistical methods will be illustrated.

4.2 MEASUREMENT OF MAGNETIC PARAMETERS

4.2.1 Low-Field or Initial Susceptibility

Magnetic susceptibility may be taken as a measure of how "magnetizable" a sample is. It helps to identify the type of material and the amount of iron-bearing minerals present. In environmental studies, it is a very convenient parameter because virtually all materials can be measured and the measurement is simple and fast (typically a few seconds). The measurements are also nondestructive and can be made in the laboratory or in the field. Low-field susceptibility therefore is often ideal in reconnaissance studies where a large sample set is needed in order to find representative samples for other expensive or time-consuming analyses (Dearing, 1999).

Low-field susceptibility measurements usually rely on two main types of alternating current (AC) susceptibility instruments: bridge circuits or induction coils tuned to resonance. Both instrument types are commonly used in rock magnetic laboratories worldwide because they are commercially available, a bridge-type instrument being produced, for instance, by AGICO Inc. (http://www.agico.com) and inductor coil instruments offered, for instance, by Bartington Instruments Ltd. (http://www.bartington.com).

The operating principle of susceptibility-measuring instruments follows the extension of Eq. (2.8) when material is present in addition to free space. The magnetic induction field B is then described by the following equation:

$$B = \mu_0(H + M) \tag{4.1}$$

where μ_0 is the permeability of free space, H is the applied field strength, and M is the induced magnetization of the sample. M is related to H by the sample's susceptibility, $\kappa = M/H$ [see (2.1)].

Dividing (4.1) through by H and solving for κ, we get:

$$\mu_0\kappa = \mu_r - \mu_0 \tag{4.2}$$

where μ_r is the relative permeability of the sample ($= B/H$, which is dimensionless). Thus κ is related to μ_r, which is closely associated with the characteristics of AC circuits containing inductive elements. Hence Eq. (4.2) forms a basis of measuring low-field susceptibility. For a more extensive discussion refer to Collinson's (1983) informative textbook.

The AC bridges for magnetic susceptibility measurement follow the classic technique developed by Charles Wheatstone (1802–1875) in 1843 in order to determine accurately an unknown resistance. Among the modern successors, transformer bridges have been very successful because they can be designed for extremely high sensitivity (Jelínek, 1973; Fuller, 1987; Jelínek and Pokorny, 1997). One arm of the bridge network consists of an air-cored pickup coil L where a sample may be inserted, which changes the inductance of the coil, and a second equalizing coil L′ with electrical properties identical to those of the sample coil (Fig. 4.1). The other two arms are formed by the secondary winding of the transformer, which is fed by a generator through the primary winding. The out-of-balance voltage of the system, which is tuned to resonance with the capacitance C_1, is measured at point P. Some modern instruments are designed in such a way that only rough balancing of inductance and resistivity of the circuit is required prior to measurement. Alternatively, they may be fully compensated automatically when under computer control.

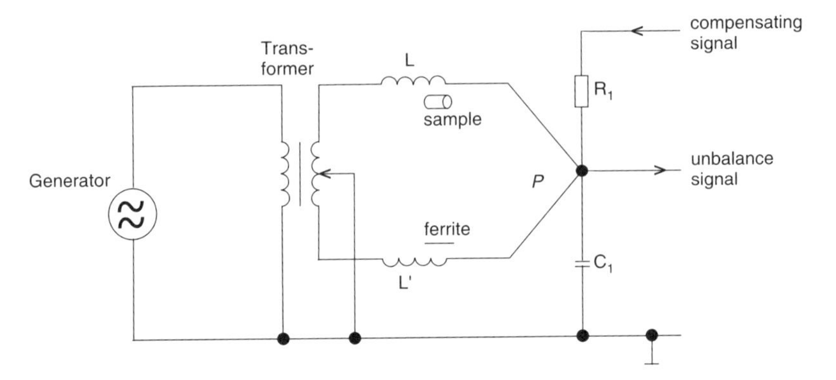

Figure 4.1 Schematic circuit diagram for a modern transformer susceptibility bridge. (From Jelínek and Pokorny, 1997.) © Elsevier Science, with permission of the publishers.

During a measurement session before sample insertion, any residual unbalanced signal that may be caused by temperature drift of the coils is automatically compensated by manual or computer-controlled triggering. After insertion of the sample, the output signal at point P due to the inductance change of the sample coil is approximately proportional to the susceptibility of the sample.

Sample shape and size are determined by the physical size and the region of high homogeneity of the measurement coil field. Cylindrical 1-inch samples of favorable length to avoid shape anisotropy (Scriba and Heller, 1978) or cubic samples up to some 10 cm^3 are commonly used for susceptibility measurement. The sensitivity achieved in the best fully automatic instruments is about 1.2×10^{-8} SI units (Jelínek and Pokorny, 1997).

The other widely used susceptibility-measuring system employs wound inductor coils built into a series resonant network. The probe consists of a very high thermal stability oscillator for which a wound inductor is the principal frequency-determining component. When the inductor loop contains only air, the value of μ_0 determines the frequency of oscillation. When the inductor is placed within the influence of the sample to be measured, the value of μ_r modulates the frequency of oscillation. Thus, it is this frequency modulation that depends on the relative permeability μ_r of the specimen. The meter to which the sensor is connected digitizes the μ_0- and μ_r-dependent frequency values with a resolution of better than one part in a million. They are a measure of the magnetic susceptibility.

The value of μ_0 is constant, but the variable of interest, the specimen permeability, may be relatively small. Therefore any thermally induced sensor drift needs to be eliminated by occasionally obtaining a new "air" or μ_0 value before measuring the actual sample. By appropriate calibration, the oscillation frequencies can be converted into magnetic susceptibility values, which may be calculated by an on-line computer. Sample size and shape follow considerations similar to those valid for susceptibility bridges. Sensitivity of the most sensitive commercially available sensors is about 2×10^{-6} SI units.

One advantage of the loop sensors is their wide versatility for different indoor and outdoor applications in environmental research because the coils can be built easily in different shapes. Open sensors of variable internal diameter may be used for measurement of environmental materials such as lake sediment cores placed in long tubes and successively advanced through the loop. Similarly, coils of variable diameter (of the order of 1–10 cm) may be used to scan surface areas of interest, for instance, soil profiles or archeological excavation sites, with high spatial resolution.

The oscillation frequency of loop sensors can be modified as well, for instance, by a factor of 10 (as in the Bartington device, which normally operates at $\sim$0.5 kHz but also provides the option of $\sim$5 kHz). This causes the magnetization of materials of very fine grain size near the SD–SP boundary to be blocked as soon as the operation frequency becomes comparable to the relaxation frequency of grains [see Eq. (2.6)]. Even more sophisticated instruments allow frequency-dependent susceptibility measurements over a wide range of frequencies (e.g., from 10 to 10,000 Hz) so that the spectrum of grain sizes below the SD–SP boundary can be evaluated fully, especially in combination with low-temperature measurements (Worm and Jackson, 1999).

Because the temperature dependence of low-field susceptibility is very diagnostic not only of grain size of ferromagnetic minerals but also of various magnetic transitions that appear in the temperature curves as spikes or steps (such as the isotropy point near the Verwey transition in magnetite, the Morin transition in hematite, or the loss of ferromagnetic properties at the Curie temperature), nonmagnetic furnaces and cryostats have been constructed for susceptibility meters that provide the possibility of temperature-dependent measurement, usually at least between 77 and 1000 K. Some instruments are available with helium cryostats that allow susceptibility measurement down to very close to absolute zero temperature.

Although bulk values of low-field susceptibility provide a wealth of useful information about the concentration, composition, and grain size of magnetic materials present in environmental materials, susceptibility should generally be regarded as a tensor parameter; that is, susceptibility in general is anisotropic. Principal susceptibilities and principal directions of this symmetric second-order tensor can be determined by measuring a sample in different directions. Because the directional differences in a sample are usually quite small, very sensitive measuring equipment is required such as the instruments just described. A minimum of six independent directional susceptibility data is required to determine the full tensor. Its orientation and magnitude are usually given by the directions and magnitudes of the three principal axes of the susceptibility ellipsoid. Evaluation of the degree and orientation of susceptibility anisotropy has been described in great detail, for instance, by Granar (1958) and Jelínek (1977). A comprehensive summary can be found in the textbook of Tarling and Hrouda (1993).

4.2.2 Remanent Magnetization

If the external field is switched off, ferromagnetic substances usually still have a magnetic moment, called remanence or remanent magnetization (see Chapter 2). Many instruments measuring remanent magnetization at high sensitivity have been developed because of the outstanding importance of the natural remanent magnetization (NRM) of rocks in the field of paleomagnetism (Collinson, 1983). Nowadays mainly two major instrument types are in use: spinner magnetometers, in which either pickup coils or so-called fluxgate probes sense the remanence as a rotating sample induces flux in these detectors, and cryogenic magnetometers, which are based on SQUID technology.

Using spinner magnetometers, the B field of a rotating sample is measured continuously by a stationary sensor. A pickup coil experiences an induced electromagnetic force due to Faraday's law that results in an AC output voltage whose amplitude and frequency depend on the rotation speed, the magnitude of the remanence component perpendicular to the rotation axis, and the coil properties. The fluxgate sensor measures the B field of the rotating sample directly. Hence, the induced signal amplitude does not depend on rotation frequency but depends only on the magnitude of the remanence component in the plane of measurement.

Figure 4.2 illustrates the operating principles of a fluxgate sensor. Physically, it consists of a thin piece of highly permeable μ-metal that may be shaped as a slab or a ring depending on the final measurement configuration. The μ-metal is packed into

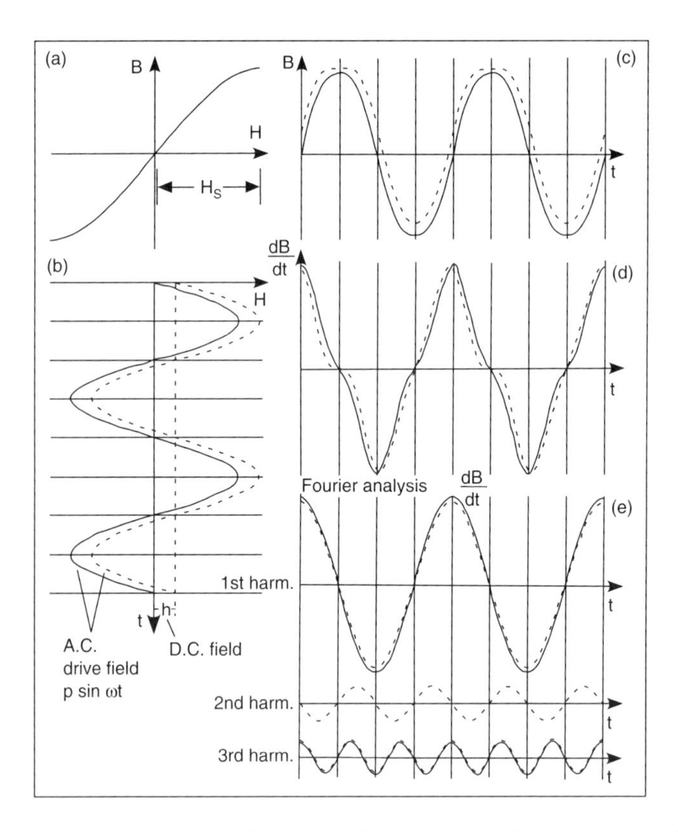

Figure 4.2 Principles of fluxgate operation. (a) $B(H)$ in μ-metal core of fluxgate sensor (see also Box 4.1); (b) time dependence of applied fields ($H(t)$, $h(t)$); (c) flux versus time $B(t)$ in the core; (d) output dB/dt from secondary coil; (e) simplified Fourier spectrum of (d). (Modified from Kertz, 1969.) © Spektrum Akademischer Verlag, with permission of the publishers.

Box 4.1 Mathematical Derivation of Fluxgate Operation Principles

The $B(H)$ curve in Fig. 4.2a can be expressed by a third-order polynomial of:

$$B(H) = 3H - H^3 \quad \text{if} \quad |H| \leq 1 \equiv H_s$$

that is, $H = 1$ corresponds to the saturation field H_s.

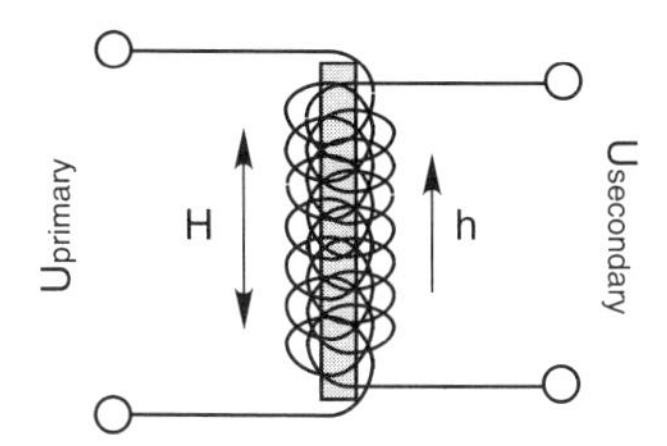

The fluxgate sensor consists of a μ-metal die (gray) surrounded by two induction coils, the primary coil producing the AC drive field H and the secondary measuring the output voltage $U_{\text{secondary}}$, which is proportional to dB/dt. The h is the external field to be measured. It may be simply static or may be caused by a slowly rotating rock sample.

The fluxgate sensor is driven in the primary coil by an alternating field H with a direct current field h superposed (Fig. 4.2b):

$$H(t) = h + p \sin \omega t$$

with p the amplitude and ω the frequency of the AC field, then the time dependence of B can be determined using de Moivre's formulas for trigonometric functions:

$$B(t) = h\left(3 - h^2 - \frac{3}{2}p^2\right) + 3p\left(1 - h^2 - \frac{1}{4}p^2\right) \sin \omega t + \frac{3}{2}hp^2 \cos 2\omega t + \frac{1}{4}p^3 \sin 3\omega t$$

The derivative measured in the secondary coil then results in:

$$\frac{dB}{dt} = 3p\left(1 - h^2 - \frac{1}{4}p^2\right)\omega \cos \omega t + 3hp\omega \sin 2\omega t + \frac{3}{4}p^3 \omega \cos 3\omega t$$

Conclusion: The direct field is proportional to the amplitude of the second harmonic, which exists only if a direct field is present.

two coils, a primary coil driven by AC (typical frequency ~10 kHz) and a secondary coil that outputs the flux changes dB/dt induced by the primary current (see Box 4.1). Without additional external DC field h, the flux B in the μ-metal follows the full line in Figure 4.2c; the output is presented by its full line derivative in Figure 4.2d. If a DC field h, caused, for instance, by a nearby sample, is present, the dashed flux lines are

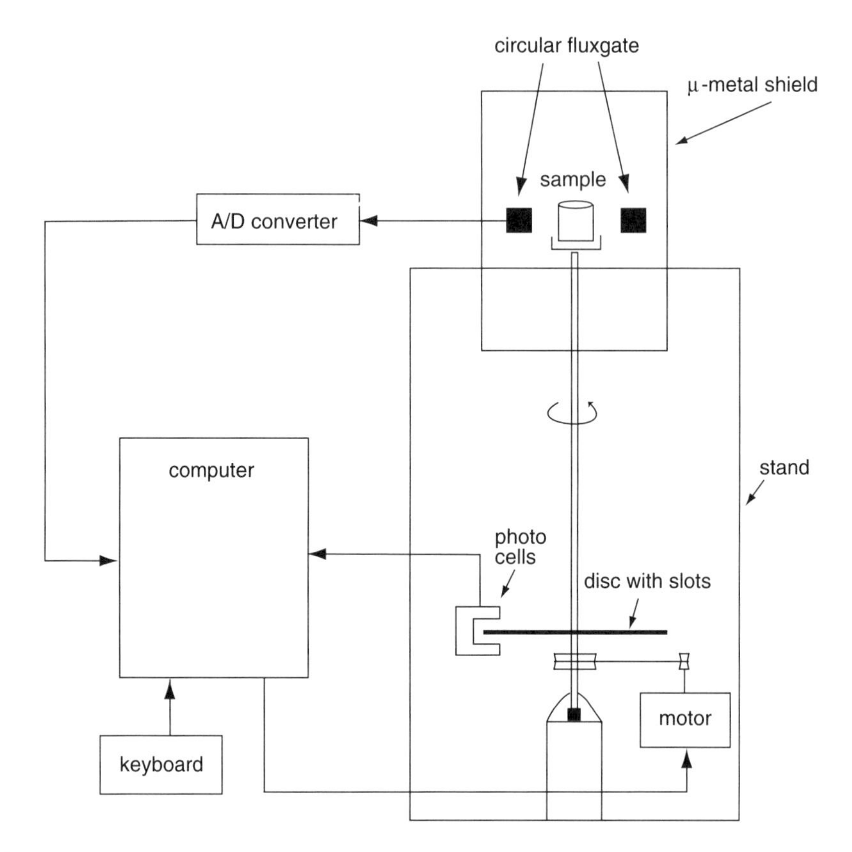

Figure 4.3 Schematic diagram of a computer-controlled fluxgate spinner magnetometer (Molyneux, 1971). The sample is located on top of a vertically rotating shaft in field-free space generated by a μ-metal shield. © Blackwell Publishing, with permission of the publishers.

observed, which are asymmetrical with respect to time. Thus, if the waveform of the output signal dB/dt is analyzed by the Fast Fourier Transform technique, a second harmonic is observed in this case only. Its amplitude is proportional to the magnitude of the DC field h (Fig. 4.2e).

Using a fluxgate spinner magnetometer (e.g., the commercially available Molspin magnetometer; http://www.molspin.com/), the sample may be placed on top of a vertical rotating shaft in the center of a ring-shaped fluxgate sensor (Fig. 4.3). At the bottom of the shaft a phase-determining disk with 128 slots is mounted which triggers about every 3° a reading of the flux entering the fluxgate. If the instrument is computer controlled, both signals — flux and slot position — are transmitted to the computer, which — after appropriate calibration — determines the phase and amplitude of the magnetization in the measuring plane by means of fast Fourier analysis. In order to increase the signal-to-noise ratio, the number of revolutions may be set to preselected values for stacking the readings and thus obtaining improved sensitivity. The three-dimensional remanence vector is calculated after this procedure has been repeated at least three times in three orthogonal measurement planes. The vector

Box 4.2 Sample Orientation

Although enviromagnetic studies often do not require exact knowledge of the orientation of samples in the field, it is often advisable to collect oriented samples. The natural remanent magnetization (NRM) of samples in a lake core may give useful information about paleosecular variation and hence provide age information for the sediment column. Alternatively, susceptibility anisotropy of soils

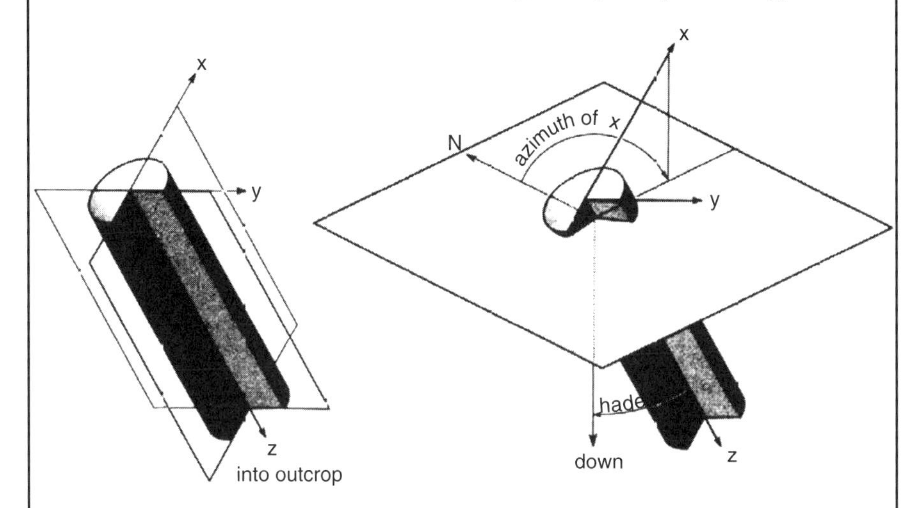

may contain important information about soil-forming processes. The orientation of the samples should be known accurately for both these examples. Depending on the mechanical properties of the substrate to be studied, oriented samples may be extracted using a drilling machine or push-in sampling digging tubes or just plastic boxes pressed into the sediment.

The azimuth with respect to geographic north is measured using a magnetic or a sun compass, whereas the dip with respect to the horizontal or the hade with respect to the vertical of a cylindrical or cubical sample may be determined simply with the inclinometer mounted in most compasses. The orientation is marked on the sample and defines a coordinate system that provides an orthogonal reference frame for the directional magnetic properties of the samples (remanent magnetization, anisotropy of susceptibility).

direction is expressed by declination and inclination with respect to a predetermined coordinate system marked on the specimen (Box 4.2).

If x, y, z are the remanence components measured in three orthogonal directions of a sample, then:

Declination DEC $= \tan^{-1}(y/x)$
Inclination INC $= \tan^{-1}(z/(x^2 + y^2)^{1/2})$
Magnetization $M = (x^2 + y^2 + z^2)^{1/2}$

The measurement time depends on the strength of the remanent magnetization so that integration times may vary. Typically, a sample will be fully measured within 2 to 20 minutes. The sensitivity limit of fluxgate magnetometers is in the range of 10^{-4} to 10^{-5} Am^{-1} for 1-inch cylindrical samples (2.5 cm diameter, 2.25 cm height $\rightarrow$ 11.4 cm^3 volume).

Cryogenic magnetometers have become very popular for extremely sensitive and also rapid measurement of remanent magnetization. Their basic measurement element is the superconducting quantum interference device (SQUID). The SQUID is basically a ring that, upon cooling through the transition temperature critical to achieve superconductivity, excludes the external magnetic field by spontaneously generating currents that inhibit further flux penetration by the field. This is called the Meissner effect (Fritz Walther Meissner, 1882–1974) and makes the ring an ideal diamagnet (Fig. 4.4a). When the external field B is large enough to cause the current circulating in the SQUID to exceed a critical value I_c, the device becomes resistive (Fig. 4.4b), magnetic flux momentarily enters the ring, and the current drops by quantized values of $\Delta I = n\phi_0$. Here $n\phi_0$ denotes multiples of the flux quantum ϕ_0. Because the current has now dropped below the critical current (Fig. 4.4c), the superconducting state is achieved again. With further increase of the external field, the switching from superconductivity to resistivity and back again—sometimes called resistivity switching—will occur as often as the external field strength B requires flux quantum jumps. The number of jumps can then be used to measure the field strength, a procedure called flux counting.

The sensitivity of the device may be greatly increased by reducing the cross section of the ring at a certain point. This is the principle of the Josephson weak link (Fig. 4.4). A very small external flux change equivalent to one ϕ_0 can now easily produce a high current density at the weak link so that resistivity switching is initiated by extremely small fields. The geometric dimensions may be chosen so that the switching causes the circulating current to even change polarity (Fig. 4.4d).

Suppose now that a sinusoidal external AC field is applied to the sensor which is large enough to exceed the critical current I_c; the SQUID current will follow the external field B_{ext} as shown by full lines in Figure 4.5a. Each time a flux quantum enters or exits the sensor, a voltage spike can be detected in a suitably placed pickup coil (Fig. 4.5b). The frequency spectrum of these spikes produces voltage components (EMF in Fig. 4.5b) at the resonance frequency of the circuit, which, depending on the sense of the polarity change, are either positive or negative. The phase of the voltage spikes induced in the pickup coil with respect to the drive phase is modulated in the presence of additional DC fields as shown by dashed lines in Fig. 4.5a and b. This modulation of the frequency spectrum of the voltage spikes caused by the radio-frequency (RF) drive field can be utilized for measuring the magnitude of the applied DC field. The effect is in a way similar to the frequency modulation that was seen when discussing the operation of fluxgate magnetometers. Modern SQUID magnetometers run on drive frequencies of 20–30 MHz, and circulating currents of the order of 1 μA correspond to a change of one flux quantum (Goree and Fuller, 1976; Collinson, 1983). A more extensive treatment of SQUID operation may be found in Fuller (1987).

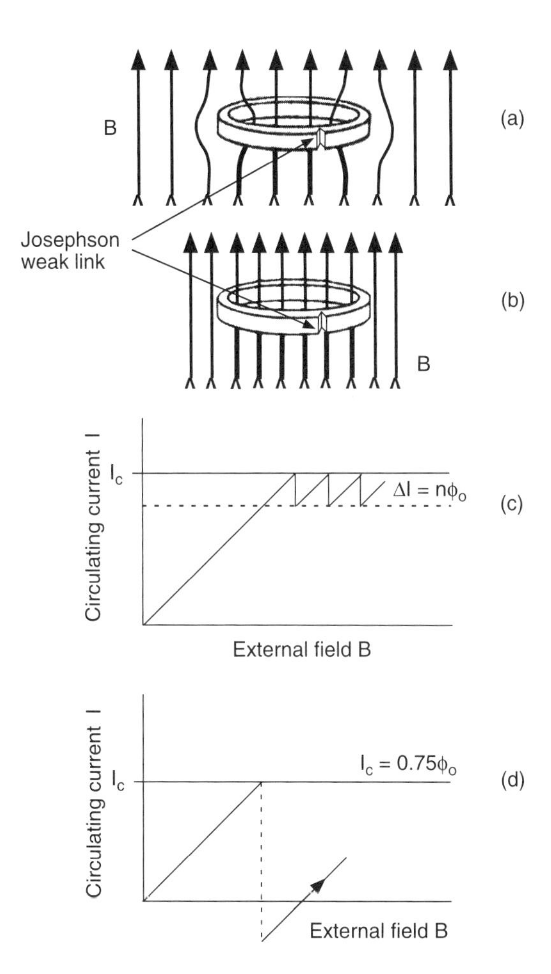

Figure 4.4 Schematic diagram and operation of a superconducting quantum interference device (SQUID) (a) in superconducting state and (b) becoming resistive when an applied field causes the circulating current to reach the critical current value I_c. (c) Quantization of current–field relation with increasing field B. (d) Critical current equivalent to $0.75\phi_0$ causes the circulating current to change polarity. Flux quantum $\phi_0 = h/2e = 2.07 \times 10^{-15}$ weber with h = Planck's constant and e the charge of an electron. A Josephson weak link has been indicated schematically. (Modified from Goree and Fuller, 1976.) © American Geophysical Union. Modified by permission of American Geophysical Union.

As Fuller (1987) points out, the flux of a sample can be measured by means of three basic arrangements:

1. The sample may be inserted directly into the SQUID. This provides optimum flux linking but requires measurement of small samples at liquid helium temperature. It is not a very practical procedure for most environmental applications.

2. The sample may be kept at room temperature outside the magnetometer but placed as near as possible to the SQUID sensor. As illustrated in Fig. 4.6, this can be achieved by the use of thin-walled dewars, as in the so-called high-temperature

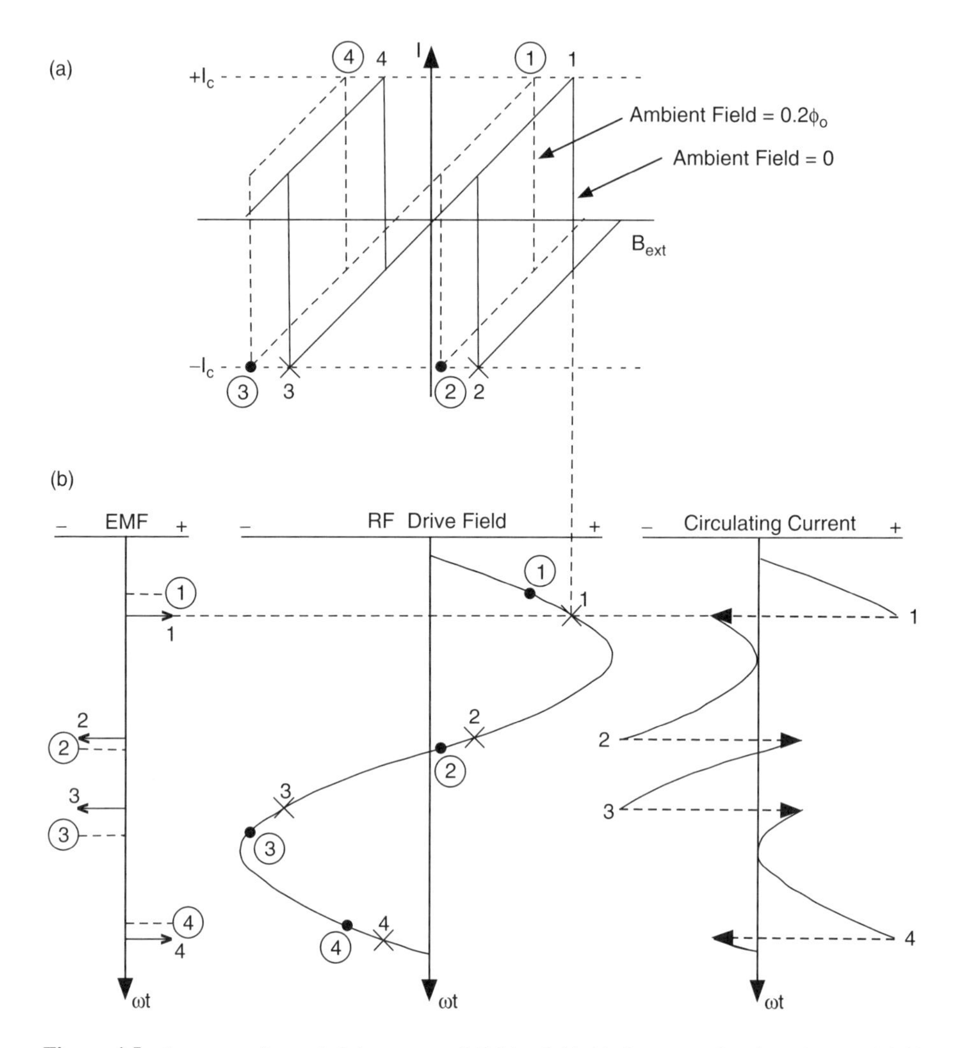

Figure 4.5 Response of a weak link sensor to RF drive field. (a) Current as function of external field without (full lines) and with (dashed lines) applied DC field. (b) Voltage spikes (EMF) caused by polarity switching of the circulating SQUID current in response to the RF drive field. Switching times indicated by numbers 1 to 4. The frequency spectrum of these switching times is modulated; that is, the switching times are displaced (numbers 1 to 4 in circles), as soon as an additional DC field acts upon the sensor. (From Goree and Fuller, 1976.) © American Geophysical Union. Reproduced by permission of American Geophysical Union.

SQUID magnetometer running at liquid nitrogen temperature constructed by Forschungsgesellschaft für Informationstechnik, F.I.T. (Germany; http://www.finoag.com/). The SQUID sensor has a very small effective measuring area (64 μm^2) and is placed at the very bottom of the dewar (Tinchev, 1992). Hence measurement resolution is mainly determined by the dewar wall thickness of about 1 mm. The natural remanent magnetization of individual discrete ferromagnetic grains in a

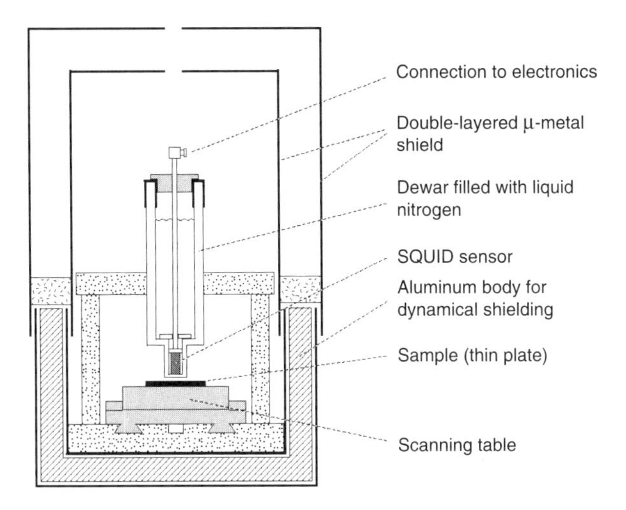

Figure 4.6 Section view of the FIT scanning magnetometer used for remanence measurements on material surfaces such as soils or rocks. (From Egli and Heller, 2000.) © American Geophysical Union. Reproduced by permission of American Geophysical Union.

granite and the fine texture of layered sandstones have been analyzed using this instrument (Egli and Heller, 2000).

3. The flux of a sample may be coupled indirectly via a superconducting transformer coil circuit. Most modern SQUID magnetometers use this arrangement because pickup coils can be constructed that fit the requirements of the particular application. Helmholtz coil geometries (Hermann von Helmholtz, 1821–1894) are especially valuable because they provide homogeneous measurement space (Fig. 4.7a). The entire pickup–transformer circuit may be superconducting and acting as a noiseless *B*-field amplifier for the flux emanating from a sample inserted into the pickup coils. The current induced in the Helmholtz pickup coil may be amplified in the transfer coil by a factor of up to 50. Finally, the transfer coil flux will be measured by the SQUID. Instruments such as the 2G commercial magnetometer (http://www.2genterprises.com/) contain up to three pairs of orthogonal Helmholtz coils that measure instantaneously and simultaneously the three orthogonal vector components. Typical measurement times range between 10 seconds and 1 minute. They depend on the stacking procedure and measurement technique, possibly involving measurements in different sample positions. Typical sensitivity of a SQUID magnetometer is about $(0.1–1) \times 10^{-6}\,\mathrm{Am}^{-1}$ for a 1-inch sample (11.4 cm^3).

The physical construction of SQUID magnetometers may vary greatly. In Fig. 4.7b, a vertical instrument has been sketched. The sample is inserted into the measuring region from the top via the room-temperature access. Its *B* field causes a permanent flux change in the superconducting Helmholtz coils, which are connected to the transfer-output circuit as shown in Fig. 4.7a. The whole signal pickup unit is located in a magnetically shielded volume inside the helium-filled instrument. The shield being also superconducting is essential for the remanence measurements

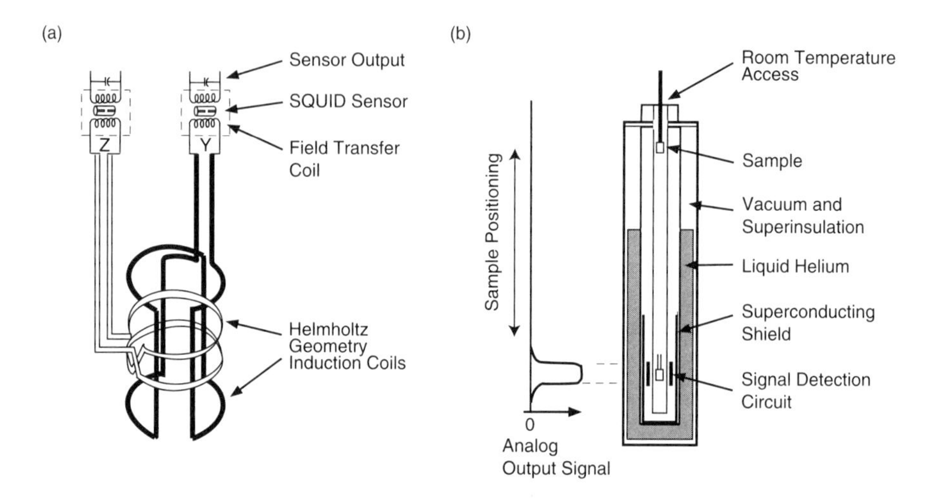

Figure 4.7 (a) Cryogenic magnetometer 2 axes pickup coil configuration including Helmholtz coils where samples are passed through, field transfer coil, SQUID, and output circuit. (b) Design of a vertical cryogenic magnetometer that is housed in a superinsulated vacuum space. (From Goree and Fuller, 1976.) © American Geophysical Union. Reproduced by permission of American Geophysical Union.

because it excludes the external field upon cooling through the critical temperature. It provides field attenuation factors of more than 10^8 independent of frequency from DC to gigahertz. Horizontal design of cryogenic magnetometers has become more popular during recent years because with a helium reservoir open on both ends it is possible to pass sediment cores—often in the shape of so-called u-channels—through the magnetometer. Continuous measurement along the core will provide a remanence record concerning environmental and/or geomagnetic field change for the time span of deposition of the core material. The instruments are also often built inline with an AC demagnetizer. This arrangement allows automatic progressive demagnetization of the remanence whether it is natural and to be tested for long-term stability or is laboratory produced to be analyzed for grain size spectra that may preserve environmental information.

4.2.3 High-Field Techniques

High-field methods are most commonly used to identify magnetic mineral phases. Mineral type and grain size are of fundamental importance when interpreting the environmental significance of materials under investigation.

When experimenting with laboratory-produced remanent magnetization, DC field–generating solenoids, impulse magnetizers, electromagnets, or superconducting magnets are used for studying the field dependence of isothermal remanent (IRM) or anhysteretic remanent (ARM) magnetization in progressively higher fields (see later). Air-cored solenoids may produce fields up to 100 mT; conventional electromagnets approach 2.5 T. Impulse magnetizers produce magnetic fields up to 5 T by rapid discharge of energy from a capacitor bank through a coil surrounding the sample.

Similar fields may be reached using superconducting solenoids. Fields of this order of magnitude are sufficient to saturate most natural ferromagnetic minerals [except certain goethites (Dekkers, 1988)].

The stability or coercivity distribution of laboratory-induced remanences may be studied when samples are exposed to alternating fields that decrease from preset high peak fields to zero. Alternating field demagnetization equipment consists basically of an air coil within which the sample is placed, a capacitor to tune the coil, a power supply whose current can be reduced electronically to zero, and some shielding (coils, μ-metal). Commercial equipment is available that generates AC peak fields of about 100 to 200 mT. These fields are required in order to demagnetize low-coercivity enviromagnetic minerals. Collinson (1983) has given all the technical details of magnetizing and demagnetizing apparatus in his comprehensive textbook. Information on commercial instruments may be found easily in the Internet home pages of paleomagnetic laboratories worldwide.

The temperature dependence of high-field magnetization can be measured using various instruments. The Curie balance has been used since the early susceptibility measurements of Michael Faraday (1791–1867), Louis-Georges Gouy (1854–1926), and Pierre Curie (1859–1906) (Collinson, 1983). Principally it measures the translational force exerted on a magnetized material by a nonuniform field. In a horizontal suspension-type balance (Fig. 4.8)—where gravity may be neglected—the force increment due to a horizontal gradient field in the translation direction x is:

$$dF_x = \kappa dv B_y(\partial B_y/\partial x)$$

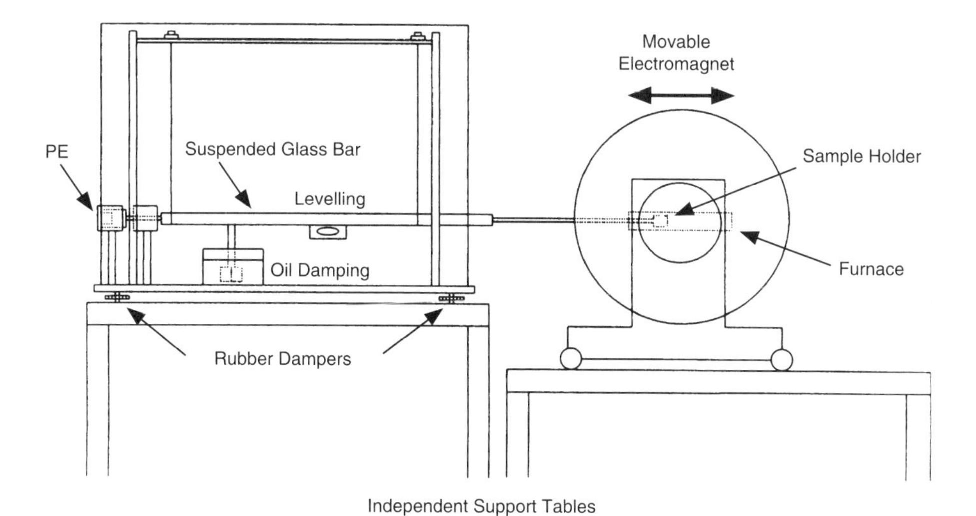

Figure 4.8 Schematic cross section of a horizontal translation Curie balance. PE = position encoder including automatic feedback system (electric connection not shown). The sample is loaded on the right end of the suspended glass bar and is located inside the gradient field of the electromagnet and the furnace.

where κ denotes the susceptibility and *v* the volume of the sample. Thus, the balance basically consists of a coil or electromagnet providing a variable field and field gradient and a sensitive transducer. This position encoder acts exactly like a seismometer. Through an automatic feedback system, it always keeps the sample in the same position in the applied magnetic field and records the force exerted on the sample. The corresponding output voltage is a measure of magnetization during heating and cooling cycles where eventually Curie temperatures are observed.

If the field is strong enough to saturate the sample, the force is linearly related to saturation magnetization. Using normal electromagnets, this can be achieved without difficulty only for low-coercivity minerals. Paramagnetic and antiferromagnetic minerals generally will not be saturated. This behavior has been utilized in modern Curie balances to separate the contributions from saturable (e.g., magnetite) and nonsaturable minerals (e.g., clay minerals or hematite). The samples are cycled periodically beyond the saturation field of the saturable minerals, yielding a temporally stable signal for the saturable minerals and a time-varying signal for the nonsaturable minerals (Mullender *et al.*, 1993; Exnar, 1997).

The magnetic properties of matter are fundamentally and most conveniently characterized by measuring in applied fields of variable magnitude. The field-dependent magnetization curves of ferromagnetic substances generally show nonlinear response and irreversible changes (see Chapter 2). These hysteresis loops provide information about coercive force, saturation magnetization, and saturation remanence. Structure-dependent magnetic properties such as domain state, grain size distribution, and internal stress may be derived from hysteresis observations.

For many years, the most versatile instrument for measuring hysteresis properties has been the vibrating sample magnetometer (VSM), the workhorse of the rock magnetist, as Fuller (1987) calls it. Basically, the VSM works by vibrating a sample in the homogeneous region of a laboratory electromagnet. Any resulting magnetic moment causes a proportional signal to be induced in adjacent pickup coils. Figure 4.9 schematically illustrates a vertical vibration system showing the position of the sample with respect to the magnet pole pieces and four pickup coils. The four-coil array provides an astatic arrangement that is only minimally affected by the homogeneous but variable field. It is, however, very sensitive to sample motion, which may have amplitudes between one tenth and several millimeters depending on the mechanical drive mechanism of the VSM (loudspeaker, motor). The maximum magnetic fields are of the order of 1 T with strict requirements for ripple-free ramping to keep the electronic noise level low.

Under computer control, variable and automatic measurement schemes and programs may be chosen. For example, different intregrating measurement time intervals may be selected in order to improve sensitivity or, vice versa, to reduce measurement time. Looping in different field ranges or backfield curves may be run (see later in this chapter). Low-temperature cryostats, high-temperature ovens, and fully integrated temperature control extend the VSM operational range over temperatures between 4.2 and 1000 K (Fig. 4.10). Modern computer-controlled VSMs approach very high sensitivity. The MicroMag VSM produced by Princeton Measurement Corporation claims a sensitivity limit of 5×10^{-9} Am^2 at a 1-second integration time at room

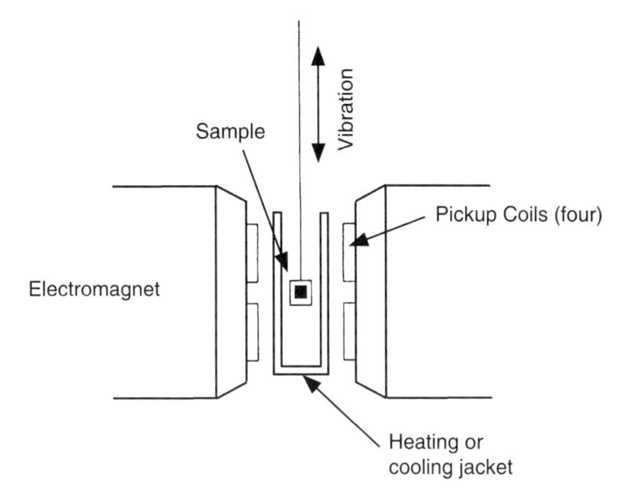

Figure 4.9 Schematic view of a VSM setup. The sample is vibrated in the vertical direction in the uniform magnetic field of an electromagnet. Four pickup coils record flux changes due to the motion of the sample, which may be kept at room temperature or heated or cooled between 4.2 and 1000 K if required.

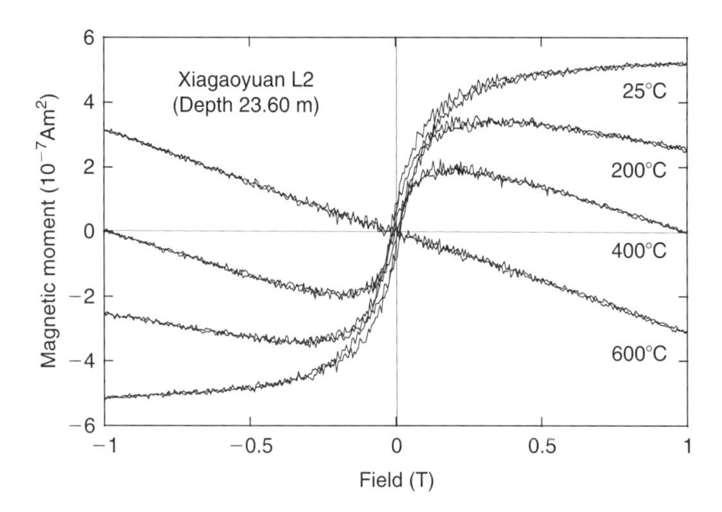

Figure 4.10 Hysteresis curves of a loess sample from Xiagaoyuan (western Chinese loess plateau: 105.0°E, 36.5°N) as a function of temperature. Note that ferromagnetic properties (magnetic moment, coercivity) become weaker with increasing temperature whereas diamagnetism remains steady and eventually dominates the magnetization completely.

temperature. All hysteresis parameters are measured that are necessary for analyzing ferrimagnetic grain size using, for instance, the Day method (Day *et al.*, 1977; see Chapter 2) or, more recently, the FORC technique (e.g., Roberts *et al.*, 2000; see Chapter 2).

The richness of the VSM data set is reduced but experimental simplicity is greatly increased if field application and measurement of the VSM operation procedure are separated. Measurement is then restricted naturally to field-dependent remanent magnetization only. This technique has become very popular because a magnetometer and magnetizing and demagnetizing equipment such as solenoids, pulse magnetizers, or electromagnets are universally available in paleomagnetic laboratories.

Isothermal remanent magnetization (IRM) acquired at room temperature in successively increasing direct fields contains a wealth of information about magnetic hardness as expressed by the distribution of coercivity of remanence. The fields — sometimes applied successively in three orthogonal directions — may be high enough to saturate all ferromagnetic minerals contained in a sample. Analysis of the IRM acquisition curve and subsequent demagnetization by reversed DC fields, by alternating fields, or by heating to temperatures near the maximum Curie temperatures will then reveal spectra of remanent coercivity and of unblocking temperatures. They can be utilized to characterize and identify type and grain size of the ferromagnetic minerals present and even magnetic interactions between grains (Dunlop, 1972; Heller, 1978; Cisowski, 1981; Lowrie, 1990; Robertson and France, 1994; Kruiver *et al.*, 2001; Egli, 2003).

Another instrument has become available that does, in fact, measure the field dependence of both induced and remanent magnetization simultaneously in one experiment (Burov *et al.*, 1986; Jasonov *et al.*, 1998). This so-called coercivity meter is based on simple induction principles when a 1-cm^3 sample housed on a nonmagnetic rotation disk is spun between the pole tips of an electromagnet (Fig. 4.11). The magnetic field is continuously changed and monitored, and the sample is measured both outside and inside the applied field by two induction coils. A large number (up to 10,000 readings) of remanent and induced magnetization values may be recorded in one run up to the maximum field ($B_{max} = 500\,mT$) and to the corresponding backfield, and coercivity and coercivity of remanence can be measured easily, quickly, and with high sensitivity. The remanence channel offers a sensitivity of $2 \times 10^{-3} Am^{-1}$ and induced magnetization is measured down to $2 \times 10^{-2} Am^{-1}$. IRM coercivity spectra are extremely well documented, allowing spectral analysis and modeling (Robertson and France, 1994; Eyre, 1996; Kruiver *et al.*, 2001; Egli, 2003) as well as interpretation according to the Preisach–Néel theory (Fabian and von Dobeneck, 1997).

Another remanence type is the anhysteretic remanent magnetization (ARM, see Chapter 2). It is generated by placing a sample in a small steady field (of the order of the Earth's magnetic field) and superimposing an AC field whose amplitude is steadily reduced from a preset initial value to a zero end value (dashed curve in Fig. 4.12). All particles with remanent coercivity equal to or less than the maximum field will become magnetized along the direction of the biasing DC field. Again as in the IRM case, the procedure may be repeated, the maximum AC field being increased progressively and the ARM measured after each step. The resulting ARM acquisition curve may be analyzed and the final ARM treated in the same manner as the IRM. The ARM, however, is not equivalent to the IRM because its acquisition depends on the response time of the remanence carriers due to the frequency of the alternating field (Lowrie and Fuller, 1971; Dunlop and Özdemir, 1997; Egli and Lowrie, 2002). This will be illustrated later in this chapter.

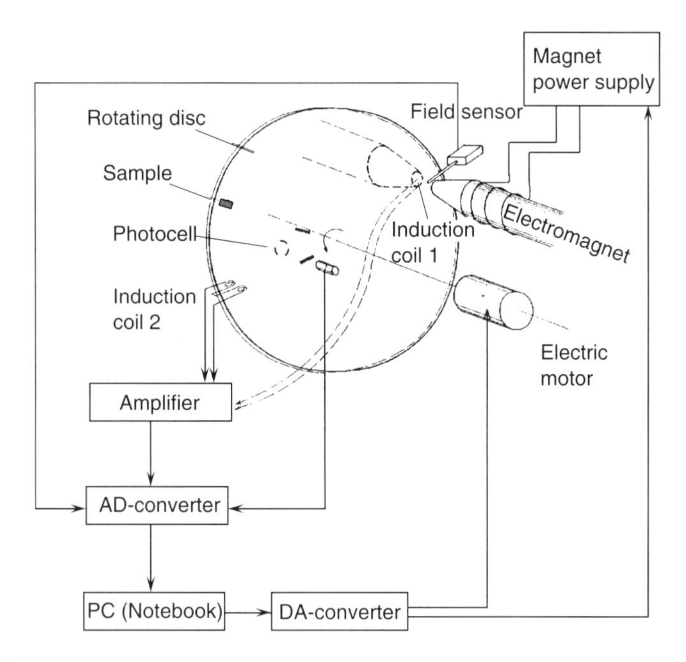

Figure 4.11 Schematic diagram of an induction coercivity meter (Jasonov *et al.*, 1998). The sample (gray rectangle) sits on the circumference of a rotating disk. At each turn it is magnetized by the field of the electromagnet. Induced magnetization is measured inside the pole tips (induction coil 1) and remanent magnetization is measured outside the electromagnet in the μ-metal–shielded induction coil 2. Measurement, motor rotation, and field regulation are computer controlled.

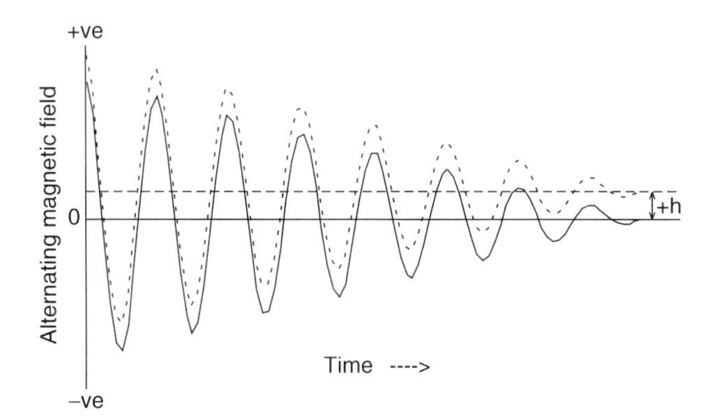

Figure 4.12 Application of decreasing AC fields. Solid curve indicates field behavior during demagnetization experiments in zero DC field. Dashed curve is displaced by applying the DC field h, which causes an asymmetric AC field decrease from a preset initial value to zero. The resulting anhysteretic remanent magnetization (ARM) is proportional to and directed along the biasing field h.

In order to achieve experimentally large alternating fields of the order of 100 mT, the alternating field (AF) coil has to be tuned to resonance by a capacitance bank. The tuned circuit minimizes the impedance of the coil and allows maximum current throughput at a given voltage (for details see Collinson, 1983). The DC fields applied for ARM production are usually of the order of the Earth's magnetic field ($\sim 50\,\mu$T).

The solid curve in Figure 4.12 illustrates the principle of alternating field demagnetization. It follows the same principles as the ARM acquisition process but ideally in the absence of any DC field. Any DC fields are compensated using Helmholtz coils or excluded by μ-metal shields during the reduction of the AC field to zero. In many experimental arrangements, the samples are rotated simultaneously around several axes in order to avoid any influence of stray biasing DC fields. Progressive increase of the initial AC field amplitude, always followed by field reduction to zero, will randomize the existing remanence connected to coercivities less than or equal to this field. This stepwise demagnetization procedure will reveal the coercivity distribution of the remanence.

Demagnetization procedures are based principally on the time stability of remanent magnetization, which varies with relaxation time τ as introduced in Chapter 2. Using the same nomenclature as before, Eq. (2.6) may be rewritten explicitly as

$$\tau = f\,\exp(Kv/kT)$$

that is, a mineral grain with given volume v will have a relaxation time that depends on the ambient absolute temperature. The relaxation time τ may be reduced to very small values of the order of fractions of seconds simply by raising the temperature. At that moment the unblocking temperature of remanence, T_{ub}, is reached. Again this experiment is usually done in steps to progressively higher temperatures, keeping the sample in zero field and appropriate atmosphere in order to avoid possible thermal alteration. The remaining remanence is measured after each step, and the spectrum of unblocking temperatures is revealed, which provides information about the grain size spectrum of the magnetic minerals in the sample. This procedure is what we call thermal demagnetization of remanence. Numerous kinds of furnaces have been constructed for this purpose. They have been designed to hold a reasonable number of samples (10–50), to have noninductive winding of the heating wires in order to reduce stray magnetic fields, and to provide uniform temperature distribution especially near the expected Curie temperatures. Cooling must be done in a field-free environment to avoid remagnetization of the samples. This is mostly achieved by multilayered μ-metal shielding of the heating and cooling chamber of the furnace.

4.3 MAGNETIC PARAMETERS USED IN ENVIRONMENTAL STUDIES

In order to illustrate the range and significance of the magnetic parameters that can be measured and analyzed using the techniques introduced in preceding sections, we present some typical data from the three main environmental settings, the terrestrial, lacustrine, and marine realms. We will not interpret these measurements here with respect to their environmental significance but simply discuss their general physical significance with respect to the underlying magnetic mineralogy. Our purpose is to give the reader a rapid overview of how environmental magnetists go about their business. More complete interpretations of numerous examples occupy much of the remaining chapters.

4.3.1 Loess/Paleosol Sequences

The eastern European section at Roxolany (30.4°E, 45.8°N) has been recognized as one of the most complete Quaternary records in the Black Sea area (Tsatskin *et al.*, 1998). It is made up of the typical alternation of loess and paleosol layers through the upper 30 m of the profile (Fig. 4.13). This interval represents approximately the last 500,000 years of Quaternary history. Samples have been taken at intervals of 5 to 10 cm throughout the profile for laboratory magnetic measurements. Low-field susceptibility, ARM, IRM, and hysteresis properties have been measured.

The magnetization intensity–related parameters χ_{lf}, χ_{ARM}, SIRM, and M_s display very similar variations throughout the profile. Their highs in paleosols and lows in the loess vary by one order of magnitude and simply indicate concentration changes of the magnetic minerals. Thus, magnetic enhancement is recorded in these paleosols. One should keep in mind, however, that only M_s is directly concentration related, whereas the magnitude of the other three parameters depends on grain size as well. Both χ_{lf} and M_s are induced magnetization parameters; that is, they contain contributions from all minerals present whether they be dia-, para-, or ferromagnetic. The M_s presented here originates from the ferrimagnetic saturation magnetization only. The ferrimagnetic hysteresis curve usually reaches saturation between 100 and 200 mT (Fig. 4.14), which is typical for magnetite or maghemite. The high-field magnetization, which still rises beyond the closure of the ferrimagnetic hysteresis, is due to the presence of paramagnetic or antiferromagnetic minerals. The gradient of the high-field magnetization curve above 200 mT can be approximated as a linear segment of the hysteresis curve. It represents the high-field susceptibility χ_{hifi}. It can be subtracted from the hysteresis curve in order to approximate the ferrimagnetic saturation magnetization M_s and coercive force B_c.

The high-field susceptibility (χ_{hifi}) for the Roxolany profile varies much less than the other intensity parameters. Some increase is observed in the soil horizons (especially in pedocomplex PK_4). Hence, not only ferrimagnetic but also paramagnetic or antiferromagnetic minerals are enriched here. Subtracting the high-field susceptibility χ_{hifi} from the low-field susceptibility χ_{lf} yields the magnitude of the ferrimagnetic susceptibility χ_{ferri}. Ferrimagnetic minerals clearly dominate the low-field susceptibility of loess (about 3 times χ_{hifi}) and paleosols (about 10 times χ_{hifi}). This does not imply that the quantity of ferrimagnetic minerals is high compared with paramagnetic and antiferromagnetic minerals because their intrinsic susceptibility is much higher. The volcanic material of the thin tephra layer at 11.3 m depth is also characterized by a sharp and strong increase of the magnetic mineral content.

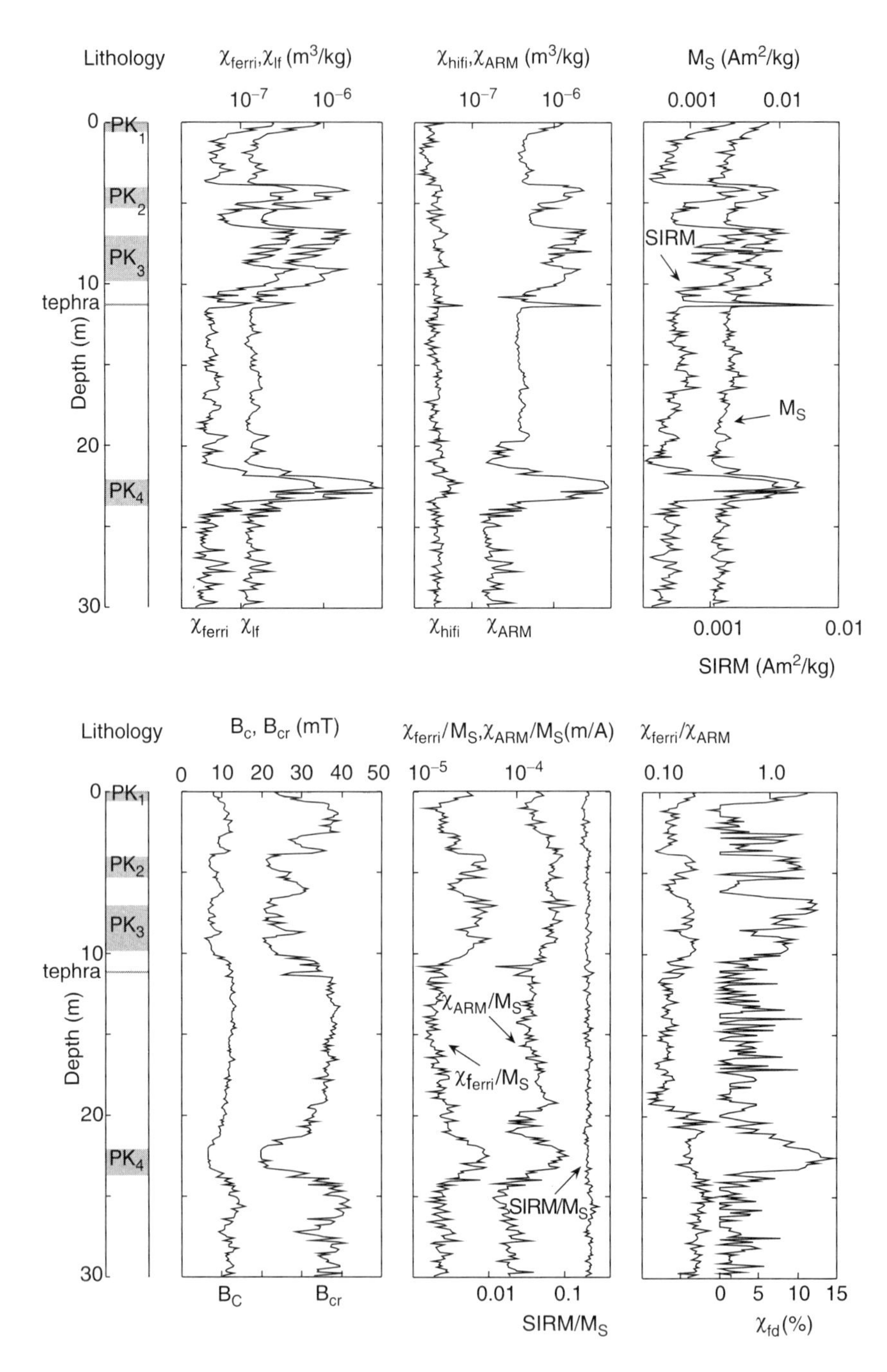

Figure 4.13 Magnetic parameters measured throughout the upper 30 m of the loess section of Roxolany (Black Sea, Ukraine). Lithological variations between loess (white) and paleosols or pedocomplexes (PK, gray) are shown schematically. (Top) Concentration-dependent parameters: low-field susceptibility

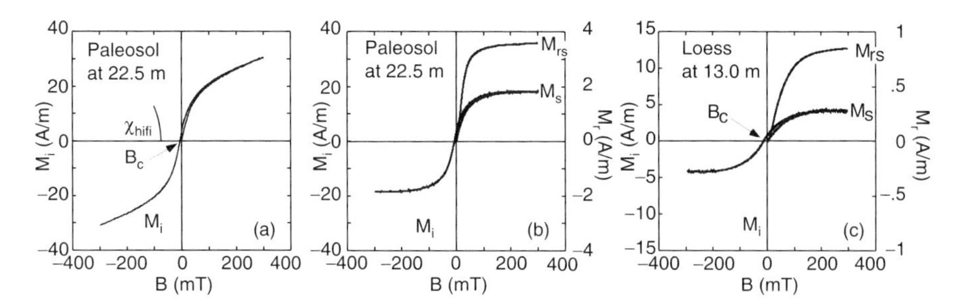

Figure 4.14 Hysteresis loops of a paleosol and a loess sample from Roxolany measured using the coercivity meter constructed by Jasonov *et al.* (1998). Both remanent magnetization and induced magnetization are measured simultaneously at 0.5-mT intervals. Above about 200 mT the curves of induced magnetization M_i are closed and change linearly with applied field. (a) The high-field susceptibility χ_{hifi} can be derived from the gradient by linear regression at fields > 200 mT. (b and c) M_s is calculated after subtracting the product $B\chi_{hifi}$. This procedure also shifts the coercivity value B_c to higher absolute values as seen most clearly in the loess sample. The remanences $M_r(B)$ approach saturation around 100 to 150 mT but do not reach full saturation at 300 mT because high-coercivity minerals are present in addition to the ferrimagnetic fraction.

Coercive force B_c and coercivity of remanence B_{cr} vary by about a factor 2. Compared with the intensity variations, an inverse relation is observed along the profile: high coercivities in loess and lower coercivities in paleosols. Thus, not only is enrichment of magnetic minerals occurring in the pedogenic horizons but also the material accumulating there is magnetically softer than in the unaltered loess horizons and hence is of a different quality. This can also be recognized from the behavior of remanent magnetization in the hysteresis loops (Fig. 4.14). The paleosol sample is very close to saturation at 150 mT, whereas the loess sample remanence increases less quickly with field and still rises distinctly above 150 mT. The relative contribution from high-coercivity minerals (e.g., antiferromagnetic hematite) is larger in the loess.

The concentration-independent parameters (Fig. 4.13, bottom) can be taken as indicators of relative grain size variation. The frequency dependence of low-field susceptibility χ_{fd} is always high in the paleosol layers and indicates enrichment of SP particles near the SD–SP boundary. This parameter is also very "noisy" in the loess layers (e.g., between 10 and 20 m). Apparently the enhancement process is partly active in the loess layers, too. The ratio χ_{ferri}/M_s is also related to variations in the content of SP particles of magnetite/maghemite. Low-field susceptibility in ferrimagnetic minerals is rather constant from MD to SD particles but increases by

(χ_{lf}, measured at 470 Hz using a Bartington MS2 meter), high-field susceptibility (χ_{hifi}, measured using the induction coercivity meter described in Fig. 4.11), ferrimagnetic susceptibility χ_{ferri}, ARM susceptibility χ_{ARM}, saturation magnetization M_s, and saturation isothermal remanent magnetization SIRM (acquired in a field of 0.3 T). (Bottom) Concentration-independent coercive force B_c and coercivity of remanence B_{cr} and normalized magnetization ratios such as ferrimagnetic susceptibility over saturation magnetization χ_{ferri}/M_s, anhysteretic susceptibility over saturation magnetization χ_{ARM}/M_s, saturation isothermal remanent magnetization over saturation magnetization SIRM/M_s, ferrimagnetic susceptibility over ARM susceptibility χ_{ferri}/χ_{ARM}, and frequency-dependent low-field susceptibility $\chi_{fd} = 100(\chi_{470\,Hz}\text{-}\chi_{4.7\,kHz})/\chi_{470\,Hz}$. These ratios indicate variations in type and grain size of the carrier minerals.

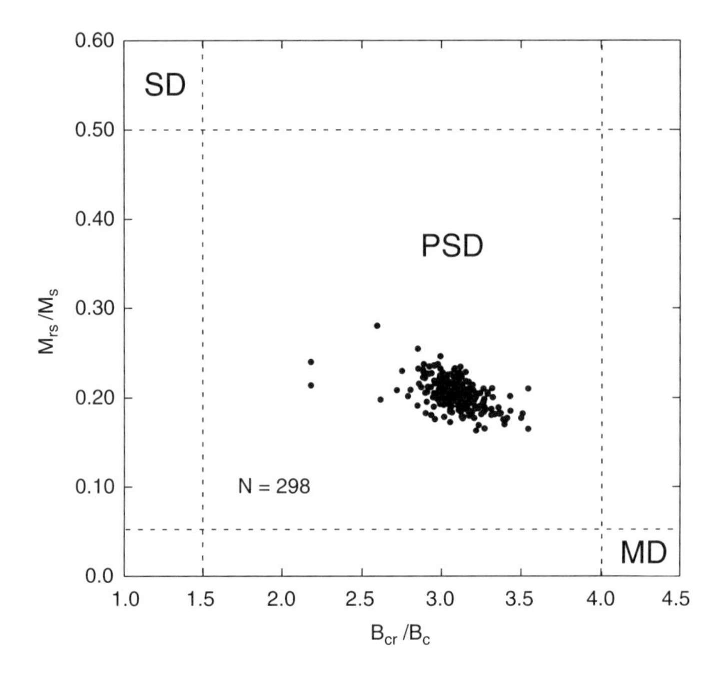

Figure 4.15 Hysteresis properties of the Roxolany loess/paleosol sequence. SD, PSD, and MD fields according to Day *et al.* (1977). All samples regardless of lithology cluster tightly in the PSD field (see discussion in Chapter 2).

about one order of magnitude in the SP range (Maher, 1988; Heider *et al.*, 1996), where high-magnetic-moment vector mobility in alternating fields occurs due to very small relaxation times (see Chapter 2). The larger magnitudes of χ_{ferri}/M_s in the paleosols confirm the evidence provided by χ_{fd}.

The ratio χ_{ARM}/M_s shows trends similar to but less distinct than those of χ_{ferri}/M_s. Because ARM resides preferentially in SD particles, it can be concluded that the enrichment of ferrimagnetic particles also occurs in the SD range. Thus, fine particles of SP and SD size have been produced in the pedogenically altered layers in addition to the existing magnetic loess material, the SP contribution being especially pronounced in the paleosols as also evidenced by the increased values of the ratio χ_{ferri}/χ_{ARM} in the paleosols and in the lower part of the profile below 20 m. The fine particle production in the loess here must be higher than in the upper loess layer between 10 and 20 m.

The ratio SIRM/M_s is rather constant over the whole profile. This indicates that the grain size variations are not very strong in general. Therefore the Day plot of Fig. 4.15 has all samples regardless of lithology in a rather restricted area within the PSD field of magnetite. The bivariate plot of χ_{ARM} versus χ_{ferri} shows most data along the 0.1-μm magnetite size line in accordance with the hysteresis ratio result (Fig. 4.16).

The presence of magnetite is supported by Curie temperature measurements of a paleosol sample from the nearby section in Novaya Etuliya (Moldavia; 28.4°E,

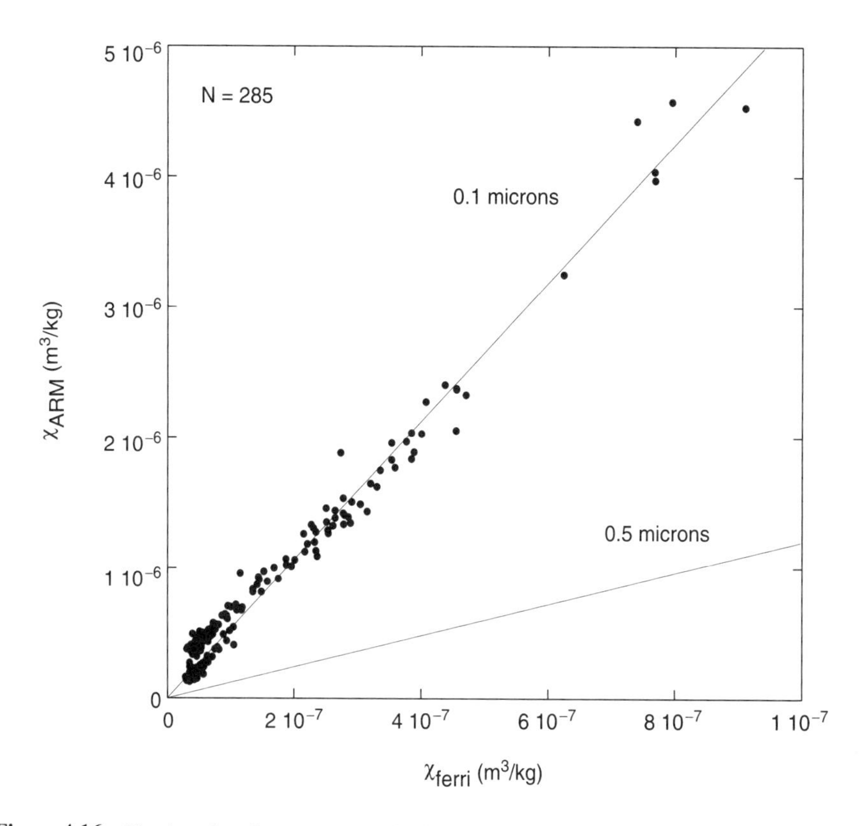

Figure 4.16 Bivariate plot of χ_{ARM} vs χ_{ferri}. The lines corresponding to a magnetite grain size of 0.5 μm and 0.1 μm follow those derived by King *et al.* (1982).

45.5°N) (Fig. 4.17). A Curie temperature of 600°C of the ferrimagnetic mineral phase present indicates magnetite that is slightly oxidized, possibly due to low-temperature oxidation on grain surfaces (van Velzen and Dekkers, 1999). The field-dependent high-field magnetization follows Curie's law for paramagnetics. Therefore the high-field susceptibility observed in Fig. 4.14 is considered to be largely of paramagnetic origin.

At this stage it is concluded that the magnetic mineralogy in the loess/paleosol sequence at Roxolany varies quite distinctly and fine grain sizes of a ferrimagnetic mineral—slightly oxidized magnetite—are found enhanced in the pedogenically altered horizons (Fig. 4.13). This is confirmed by the grain size tests according to the methods proposed by Day *et al.* (1977) and King *et al.* (1982). In the loess, however, these tests give only a general picture of prevailing fine grain sizes but are not able to resolve the subtle variations seen in some of the ratio plots of Fig. 4.13. Paramagnetic and high-coercivity minerals (probably clay minerals and hematite) also play a significant role in the magnetic properties of the loess/paleosol section at Roxolany.

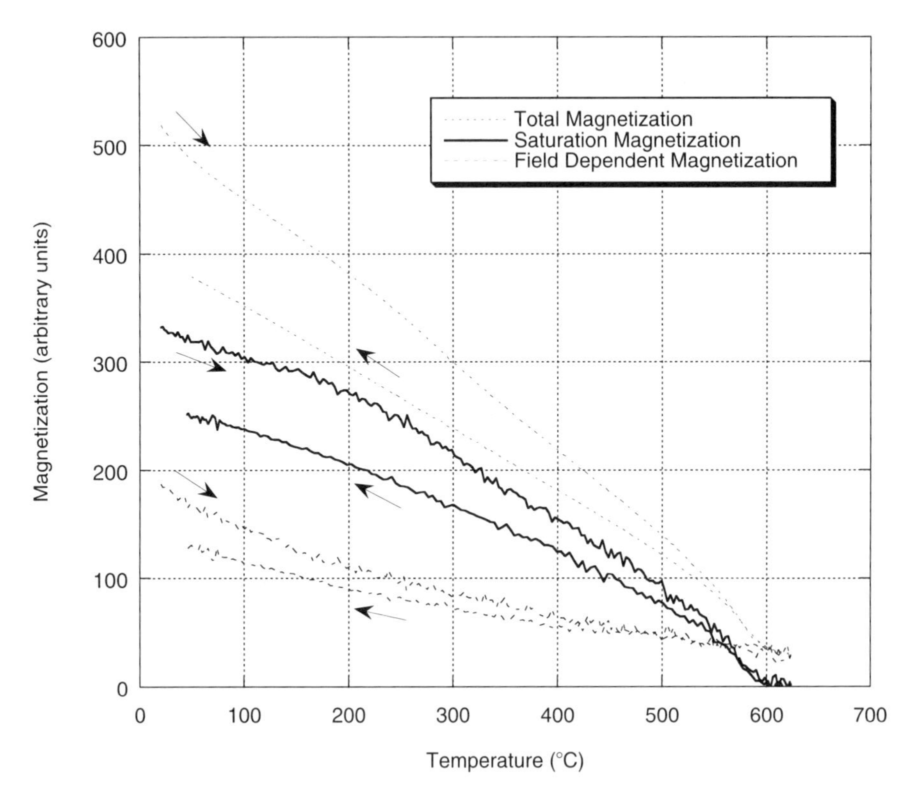

Figure 4.17 High-field magnetization vs temperature of a paleosol sample from the loess/paleosol section at Novaya Etuliya (Moldavia) using the separation technique of Mullender *et al.* (1993) and Exnar (1997). The ferrimagnetic saturation magnetization vanishes at the Curie temperature of about 600°C, slightly above that expected for pure magnetite. The field-dependent high-field magnetization follows the paramagnetic Curie law. Both magnetization components are not fully reversible upon heating and cooling because of thermal alteration during measurement in air. Applied field varies between 0.2 and 0.4 T.

4.3.2 Lacustrine Deposits

The Late Pleistocene and Holocene paleoclimatic history of the Zunggar desert (northwest China) has been deduced from sediment cores drilled in Lake Manas (86°00′E, 45°45′N), which has been dry since the 1960s. The variegated lithology (Fig. 4.18) points to rather abrupt changes between lacustrine (laminated muds) and fluviatile (sand) deposition in oxygenating to stagnant and, in recent times, even saline waters (Rhodes *et al.*, 1996). The magnetic mineralogy of the sediments of Lake Manas is quite variable and reflects these strong paleoenvironmental and paleohydrological changes (Jelinowska *et al.*, 1995).

Analyzing the variations of bulk magnetic parameters such as susceptibility κ_{lf}, κ_{ARM}, and SIRM measured in 10-cm intervals in two cores from Lake Manas, Jelinowska *et al.* (1995) suggested that the sediment column be divided magnetically into three zones (Fig. 4.18). These zones are also recognized in the behavior of

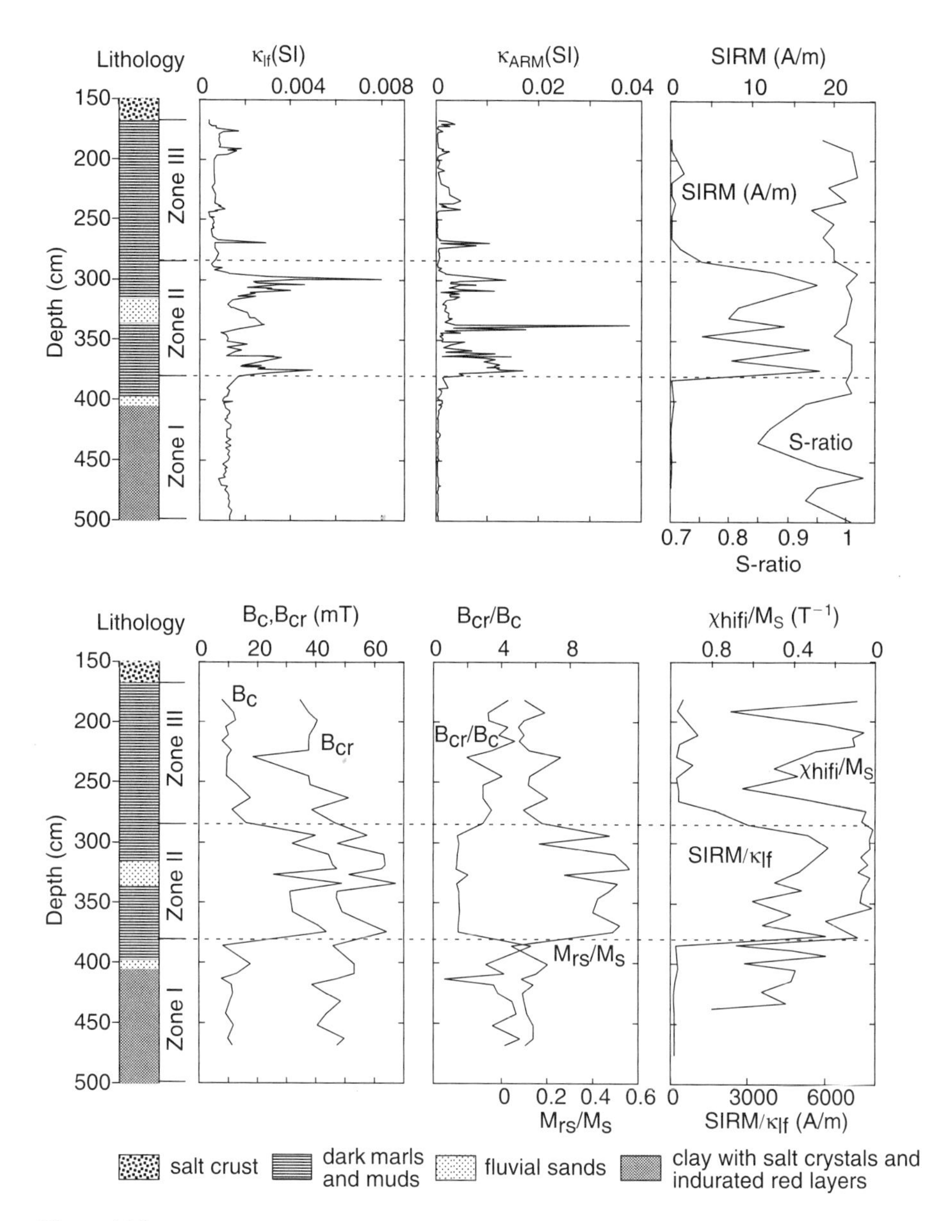

Figure 4.18 Magnetic parameters measured in two cores from Lake Manas (Jelinowska *et al.*, 1995; original data kindly provided by Alina Jelinowska and Piotr Tucholka). (Top) Concentration-dependent parameters such as low-field susceptibility (κ_{lf}, measured at 470 Hz using a Bartington MS2 meter), anhysteretic remanence susceptibility (κ_{ARM}, measured in a 50-μT bias field in the presence of an alternating field decaying from 50 mT to 0 mT), and saturation remanent magnetization (SIRM, in a field of 0.5 T, saturation generally being achieved at 0.3 T). The *S*-ratio ($IRM_{0.3T}/SIRM$) has been plotted in addition to the SIRM data. (Bottom) Concentration-independent coercive force B_c and coercivity of remanence B_{cr}, coercivity ratio B_{cr}/B_c and magnetization ratio M_{rs}/M_s, $SIRM/\kappa_{lf}$, and χ_{hifi}/M_s. No data are available from the salt crust (0–160 cm). © American Geophysical Union. Reproduced by permission of American Geophysical Union and the authors.

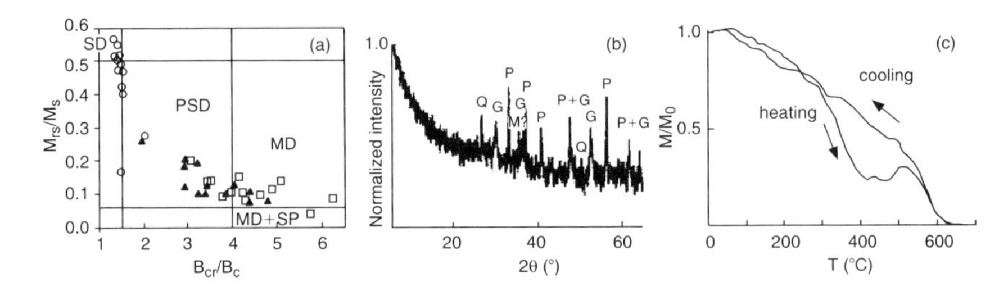

Figure 4.19 Magnetic mineral classification in Lake Manas sediments (Jelinowska *et al.*, 1995). (a) Magnetic grain size characteristics from hysteresis properties: squares, zone I; circles, zone II; triangles, zone III; MD, multidomain; PSD, pseudo–single-domain; SD, single-domain; SP, superparamagnetic grains according to Day *et al.* (1977). (b) X-ray diffractogram for magnetic extract from zone II; P, pyrite; G, greigite; Q, quartz; M?, possibly magnetite. (c) Thermomagnetic curve of the same extract in argon atmosphere, applied field 0.25 T.

hysteresis properties B_c, B_{cr}, M_s, M_{rs}; the high-field susceptibility χ_{hifi}; and some normalized parameters such as B_{cr}/B_c, M_{rs}/M_s, SIRM/κ_{lf}, and χ_{hifi}/M_s, derived from these quantities. The parameter χ_{hifi}/M_s is expected to give information on the relation between paramagnetic and ferrimagnetic mineral fractions.

The bottom of the sequence (zone I from 498 to 380 cm) is weakly magnetic with low κ_{lf}, κ_{ARM}, and SIRM values, which suggest a low concentration of ferromagnetic minerals. Low coercivity B_c around 10 mT indicates the presence of magnetite (Peters and Thompson, 1998a), and hysteresis properties (squares in Fig. 4.19a) assign PSD to MD grain size to this mineral. Coercivity of remanence B_{cr} always exceeds 40 mT, and the *S*-ratio drops sometimes to values distinctly below 1. Both of these properties point to partial oxidation of magnetite or to the presence of minor amounts of high-coercivity minerals such as goethite or hematite (note occasional red sediment color). Higher values of χ_{hifi}/M_s (note the reversed scale) indicate relatively large quantities of paramagnetic minerals.

Zone II (380–285 cm) is characterized by generally higher magnetization intensities (κ_{lf}, κ_{ARM}, and SIRM) and hence increased concentration of ferromagnetic minerals, which drop in a few horizons (note the sand layer in Fig. 4.18). Hysteresis parameters indicate mostly fine grain size in and near the SD field (circles in Fig. 4.19a), and the *S*-ratio is always very close to 1, suggesting the presence of a ferrimagnetic mineral fraction, but coercivity values B_c around 40 mT are too high for SD or PSD magnetite (Dunlop and Özdemir, 1997). X-ray diffractograms and the thermomagnetic behavior where a magnetization drop is observed between 300 and 400°C and subsequent transformation into a ferrimagnetic phase (probably maghemite) takes place give independent evidence of the presence of greigite in this zone (Fig. 4.19b and c). The observed thermal destruction behavior and the hysteresis properties and high SIRM/κ_{lf} ratios are expected for natural greigite (Roberts, 1995; Snowball, 1991). Paramagnetic minerals play a minor role as suggested by the low χ_{hifi}/M_s ratios.

In zone III (285–170 cm), magnetization intensities (κ_{lf}, κ_{ARM}, and SIRM) return mostly to low intensity and low concentration of ferromagnetics. Hysteresis properties (Fig. 4.18 and triangles in Fig. 4.19a) and the *S*-ratios are consistent with the presence of PSD magnetite. The contribution of paramagnetic minerals (χ_{hifi}/M_s ratio) is quite variable in this zone.

The magnetic measurements in the Lake Manas sediments show distinct horizons with different magnetic mineral phases. They range from magnetite of varying grain size to maghemite and even greigite. High-coercivity minerals may also be present, as are paramagnetic minerals, which contribute to a variable extent. The identification of this magnetic mineralogy is extremely useful when diagnosing the drastic environmental changes in the region during the Holocene (Jelinowska *et al.*, 1995; Rhodes *et al.*, 1996).

4.3.3 Marine Sediments

A 1098-cm-long marine core (P-094) extending into oxygen isotope stage 5 was collected from a small deep channel on the Labrador rise (45°41.2′W, 50°12.5′N, water depth 3448 m) in order to reconstruct Late Pleistocene geomagnetic field intensity for regional and global correlation (Stoner *et al.*, 1995a). The hemipelagic mud sedimentation at this locality is frequently interrupted by rapidly deposited detrital carbonate layers, some of which are correlated with North Atlantic Heinrich events (see Chapter 7). The magnetic signature of the sediments reflecting paleoclimatic changes has been studied thoroughly and interpreted in terms of environmental change by Stoner *et al.* (1995a, 1996). We present this marine data set as an excellent example to demonstrate how rock magnetic properties document the presence of magnetite in variable concentration and grain size.

Core P-094 has been sampled continuously in 7-cm^3 plastic boxes, and several bulk magnetic properties have been measured at this high resolution along the whole core length (Stoner *et al.*, 1995a, 1996). These authors report *S*-parameters (see Chapter 2) throughout the whole core that are predominantly very close to -1 and hence point to the presence of a low-coercivity mineral such as magnetite. Figure 4.20a shows the warming of a high-field low-temperature remanent magnetization from 10 to 160 K for four samples of the two main lithologies, hemipelagic mud and detrital carbonate layers. Distinct kinks are observed in all curves at temperatures between 100 and 120 K that are due to vanishing of magnetocrystalline anisotropy near the Verwey transition of magnetite and hence provide convincing evidence that the major low-coercivity mineral is indeed magnetite. The transition is consistently more strongly pronounced in the detrital layers. It results from coarser MD magnetite grain size here. The relatively small gradients above and below the transition are taken as evidence that superparamagnetic magnetite grains are virtually absent. Different grain size populations are also indicated in the bivariate plot κ_{ARM} versus κ in Fig. 4.20b. The hemipelagic material displays a nearly linear relation with a steep gradient over a restricted susceptibility range, which is compatible with magnetite PSD grain size behavior, whereas samples from the detrital layers are much more scattered with no straightforward κ_{ARM}–κ relationship. This is again due to larger grain size in these layers as confirmed in Fig. 4.20c, where the hysteresis parameters of these samples preferentially plot toward the MD field of the Day plot.

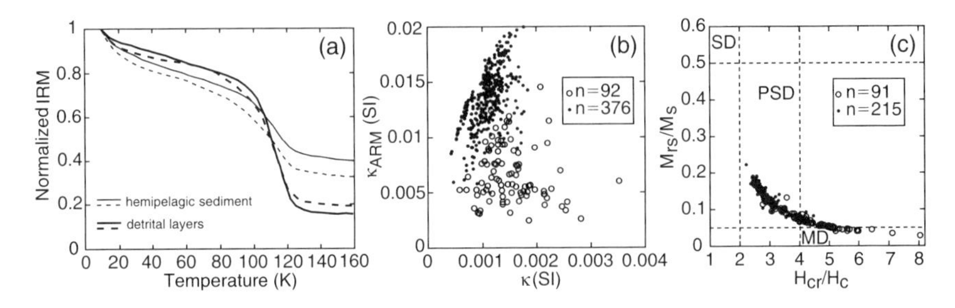

Figure 4.20 Magnetic mineral characterization in Labrador piston core P-094 (Stoner *et al.*, 1995b; 1996). (a) Normalized high-field remanence (500 mT DC field at 10 K) during warming from 10 to 160 K for two samples each of hemipelagic mud and detrital carbonate layers. Note apparent lithology dependence of magnetite low-temperature (Verwey) transition. (b) Bivariate plot of anhysteretic susceptibility κ_{ARM} vs low-field susceptibility κ for hemipelagic (small dots) and detrital carbonate layers (circles). (c) Magnetic hysteresis parameters for the two lithologies [symbols as in (b)] as bivariate plot of M_{rs}/M_s (M_{rs} saturation remanence; M_s saturation magnetization) vs H_{cr}/H_c. Magnetic grain size fields according to Day *et al.* (1977). © American Geophysical Union. Reproduced by permission of American Geophysical Union and the authors.

Low values of SIRM and ARM and also of the coercivities H_c and H_{cr} are observed in the detrital carbonate layers (Fig. 4.21). This correlation is not so clear for the low-field susceptibility, κ, which in general tends to increase slightly in the carbonates. Stoner *et al.* (1996) exclude paramagnetic susceptibility contributions because of the small high-field gradients of the hysteresis curves measured. The reduced coercivities indicate grain size coarsening in these layers, which leads to the conclusion that the detrital layers are apparently often enriched in susceptibility-enhancing coarse MD magnetite, which, however, because of the enlarged grain size, does not carry stable remanence. Hence susceptibility may increase due to higher magnetite concentration when at the same time SIRM and ARM decrease. The M_s data, which do not depend on grain size at all, run downcore very much in parallel with the susceptibility data, arguing for enrichment of magnetite in the detrital carbonate layers.

The parameter ratios (Fig. 4.21, bottom) finally prove that coarser magnetic material is indeed prevailing in the detrital layers because SIRM/κ and ARM/κ are always reduced compared with the hemipelagic mud. This also holds true for the ARM/SIRM ratio, which generally discriminates between SD-PSD and MD magnetite, becoming smaller with increased MD contribution (see Chapter 2). Finally, the larger H_{cr}/H_c ratios in the detrital layers simply reiterate the points made when discussing the grain size evidence of the Day plots (Fig. 4.20c).

Thus, overwhelming and consistent evidence of increased magnetic grain size in the detrital layers has been obtained. An interpretation of the occurrence and significance of the magnetic grain coarsening in these layers in terms of depositional environment has been given by Stoner *et al.* (1996).

4.4 MAGNETIC PARAMETERS UNMIXED

In the foregoing sections we have seen that rock magnetic parameters are controlled by a variety of magnetic minerals of different grain sizes and mineral-specific

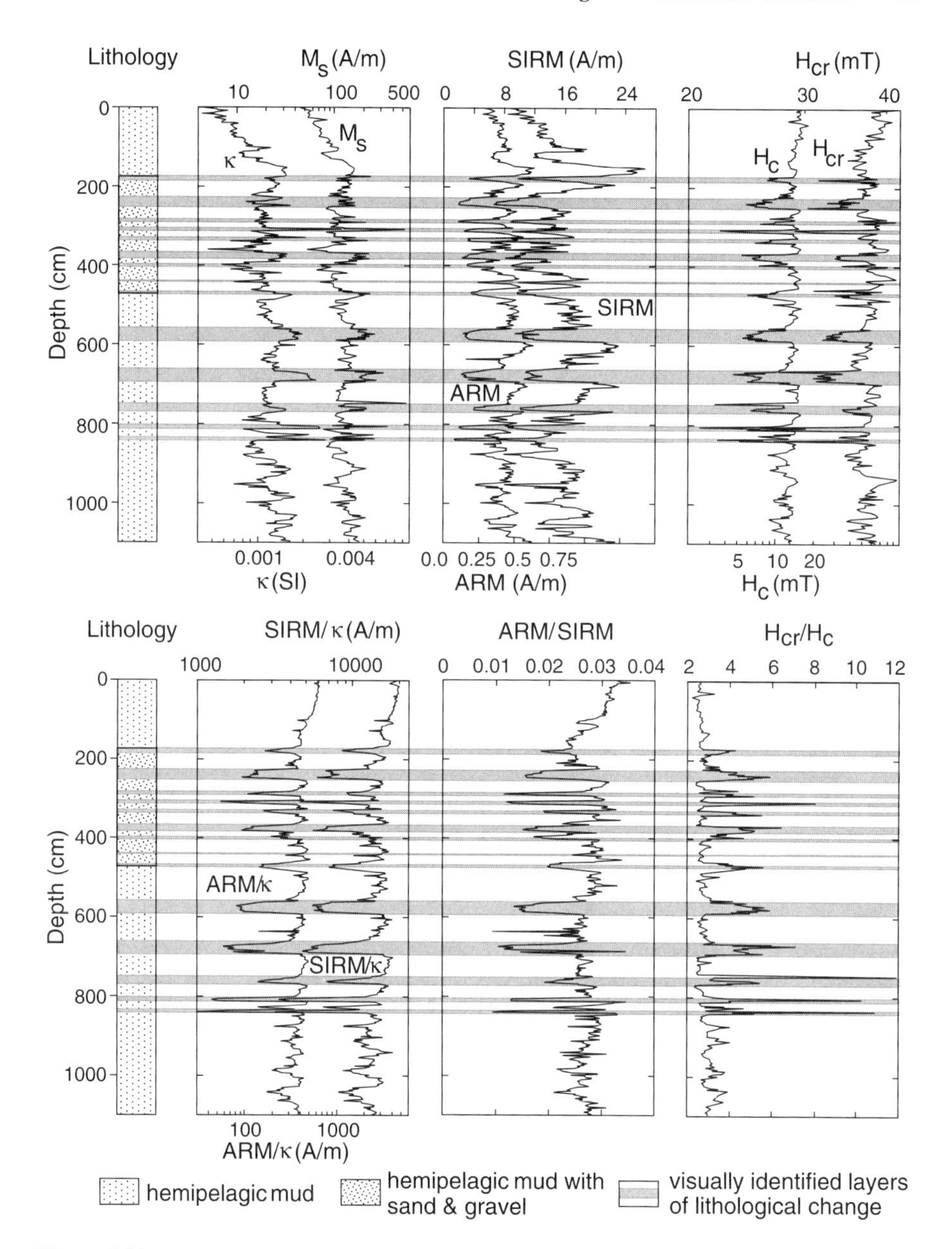

Figure 4.21 Magnetic parameters as a function of depth in Labrador Sea piston core P-094 (Stoner *et al.*, 1995a; original data kindly provided by Joe Stoner). Simplified lithology profile with gray-shaded detrital carbonate layers plotted across the magnetic data set. (Top) Concentration-dependent parameters such as volumetric low-field susceptibility κ, saturation magnetization M_s, saturation remanent magnetization SIRM (DC field: 1 T), and anhysteretic remanence ARM (99 mT AF peak field; 0.04 mT DC bias field) and in addition coercivity H_c and coercivity of remanence H_{cr}. (Bottom) Grain size–dependent parameter ratios such as SIRM/κ, ARM/κ, ARM/SIRM, and H_{cr}/H_c. © Elsevier Science, with permission of the authors and the publishers.

magnetic properties of mineral species occurring in different concentrations. In order to analyze the magnetic composition of these mixtures, some quantitative and semi-quantitative unmixing techniques have been proposed and applied to lacustrine and eolian sediments (Thompson, 1986; Robertson and France, 1994; Peters and Thompson, 1998a; Kruiver *et al.*, 2001; Carter-Stiglitz *et al.*, 2001; Egli, 2003; Spassov *et al.*, 2002). Linear additivity of the various magnetic parameters (e.g., ARM, IRM, susceptibility) is an important prerequisite for these studies because it permits simply structured cumulative magnetization curves. Violation of the linearity condition by magnetic interaction can be checked by the *R*-parameter of Cisowski (1981), which tests the stability of IRM during acquisition and alternating field demagnetization, respectively, or by FORC analysis (e.g., Roberts *et al.*, 2000). Mixing experiments attest linear additivity of magnetization parameters for artificial magnetite mixtures of variable size (Carter-Stiglitz *et al.*, 2001).

Robertson and France (1994) observed that IRM acquisition curves of individual magnetic minerals conform, in general, to a cumulative log-Gaussian (CLG) distribution. Hence, measured IRM acquisition or demagnetization curves can be approximated by a number of CLG curves that are characterized by their SIRM intensity and its coercivity distribution (specified by its mean and associated dispersion). In order to differentiate quantitatively magnetizations connected to different coercivities, Kruiver *et al.* (2001) present an elegant curve-fitting method that analyzes IRM acquisition curves on a logarithmic field scale. The IRM data are then expressed in gradient form and finally transferred onto a probability scale (see Box 4.3). Kruiver and her coauthors statistically evaluate the number of magnetic components required for an optimal fit to the measured IRM acquisition data by comparing the variance of the squares of the residuals for different fits. Their analysis algorithm is available as an *Excel* workbook for public use at: http://www.geo.uu.nl/~forth/.

As an example of the unmixing success, we select a soil sample from the study of Kruiver *et al.* (2001). It originates from a hydromorphous soil layer in a continental red bed sequence, samples of which were exposed to laboratory fields up to 2.5 T (Fig. 4.22). Although saturation has not been achieved in the maximum applied field,

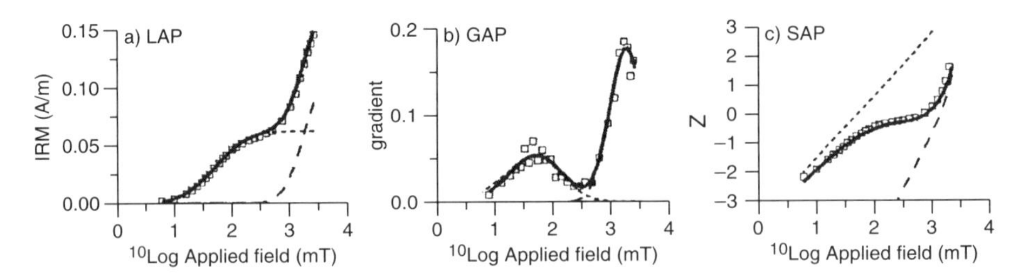

Figure 4.22 Coercivity components of a hydromorphous soil sample. (a) IRM acquisition on a log-linear scale (LAP). Squares and solid line represent measured and modeled data, respectively. Short-dashed line represents saturated oxidized magnetite; long-dashed line is due to nonsaturated goethite. (b) Gradient curve (GAP). (c) Standardized acquisition plot (SAP) with dashed straight lines for two unimodal distributions corresponding to magnetite and goethite (see Box 4.3). (From Kruiver *et al.*, 2001.)

Box 4.3 Principles of Magnetic Unmixing

The grain size distribution of magnetic minerals in rocks usually follows a logarithmic law. Kruiver *et al.* (2001) show that the grain assemblage of a single magnetic mineral can be characterized by an IRM acquisition curve that may be plotted conveniently on a log-linear scale (LAP). Its gradient curve (GAP) is centered at $B_{1/2}$ and has a width that may be described by the dispersion parameter DP representing one standard deviation. If the IRM acquisition curve follows a cumulative log-Gaussian function, the field values may be converted to their logarithmic values and the linear ordinate may be converted to a probability scale. The standardized acquisition plot (SAP) has the probability scale on the right ordinate and a corresponding z-score scale on the left. A unimodal distribution is represented on the SAP by a straight line. If a SAP does not plot on a linear path, the IRM acquisition curve needs to be fitted with more than one magnetic component. As Kruiver *et al.* (2001) point out, the nonequidistant cumulative percentage scale is replaced for convenience by the equivalent equidistant z-scores (Swan and Sandilands, 1995). Fifty percent of the cumulative distribution corresponds to a standardized value of $z = 0$ at field $B_{1/2}$; 84.1% (one standard deviation from the center) corresponds to $z = 1$ at field $B_{1/2}$+DP, etc. Values of $|z| > 3$ represent only $2 \times 0.13\%$ of the distribution.

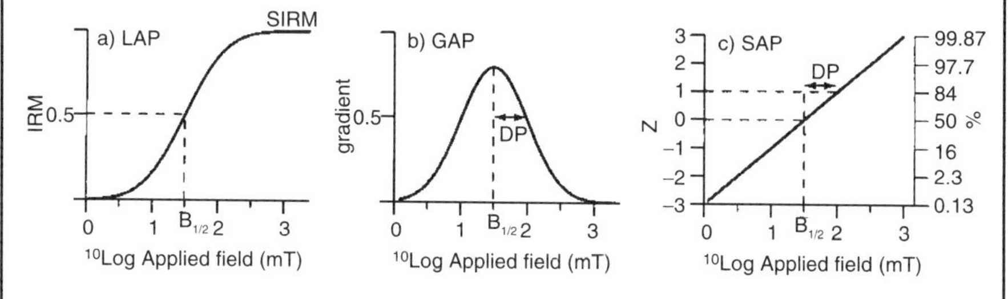

In the actual analysis, curve fitting of the preceding curves may involve more than one mineral component. It is performed by forward modeling using an *Excel* workbook (http://www.geo.uu.nl/~forth/). Estimated initial values for SIRM, $\log(B_{1/2})$, and DP are used for calculating theoretical IRM acquisition curves, which are optimized interactively by minimizing the squared differences between actual data and model curves.

the IRM gradient curve shows two clearly separated coercivity distributions, which the authors assign to the presence of oxidized magnetite [SIRM = 0.062 A/m, $\log(B_{1/2}) = 1.70$ mT, $B_{1/2} = 50$ mT, DP = 0.46] and goethite [SIRM = 0.124 A/m, $\log(B_{1/2}) = 3.25$ mT, $B_{1/2} = 1800$ mT, DP = 0.28]. The presence of these minerals has been independently confirmed by thermal demagnetization experiments. According to their analysis, magnetite and goethite contribute 33 and 67%, respectively, to the SIRM.

Whereas Kruiver *et al.* (2001) unmix the magnetic mineralogy using preassessed coercivity distributions, Egli (2003) proposes a method that makes no assumptions

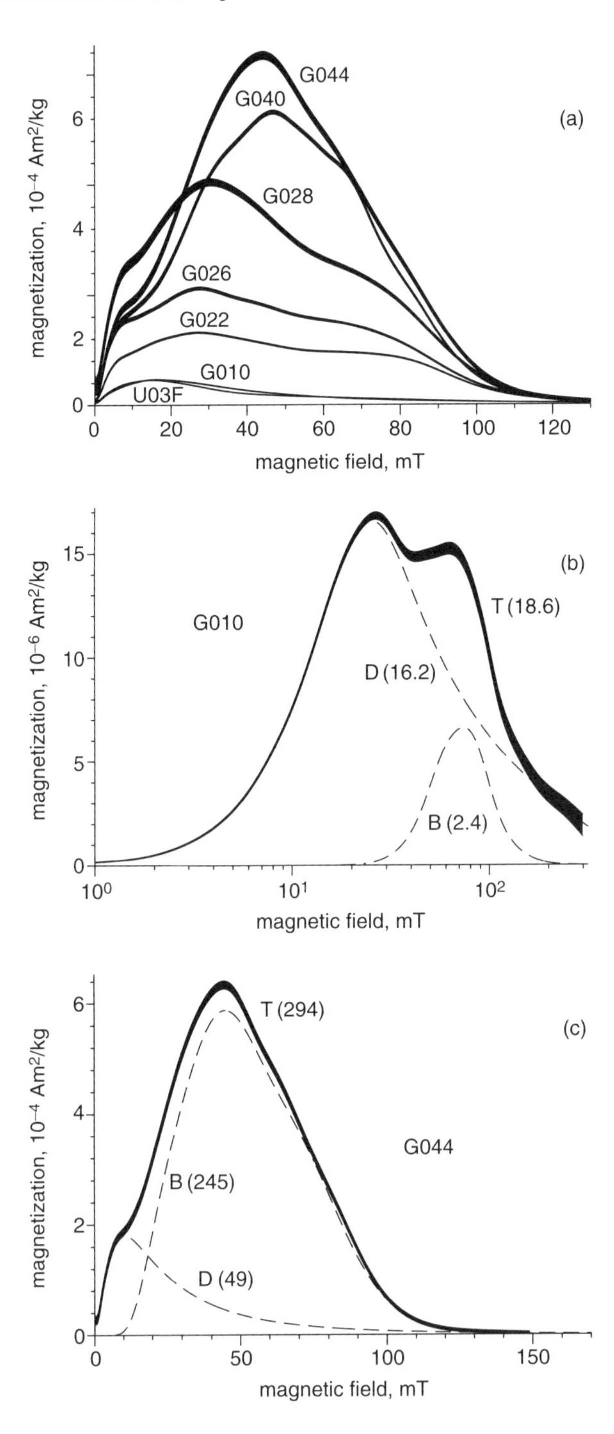
magnetization, 10⁻⁴ Am²/kg
magnetic field, mT
G044
G040
G028
G026
G022
G010
U03F
(a)
magnetization, 10⁻⁶ Am²/kg
G010
T (18.6)
D (16.2)
B (2.4)
(b)
magnetization, 10⁻⁴ Am²/kg
T (294)
B (245)
D (49)
G044
(c)

about the number and distribution characteristics of the coercivity components. His resulting coercivity distributions therefore do not depend on specific model distributions (e.g., logarithmic Gaussian distributions). The acquisition or demagnetization curves of ARM or IRM are linearized by rescaling the field and magnetization axes. Appropriate filtering of the linearized curves removes measurement errors prior to the evaluation of the coercivity distributions. Using Pearson's χ^2 goodness-of-fit test, even components with strongly overlapping coercivity distributions can be separated. They are unmixed and estimated precisely using generalized—skewed Gaussian—distribution functions that are able to fit a large number of different statistical distributions. Despite the careful data treatment and filtering, multiple solutions are still possible.

Egli (2003) was able to unmix quantitatively detrital and biogenic coercivity components in a short 1.2-m-long gravity core taken from Baldeggersee (a small lake in Switzerland) representing the last 200 years of depositional history. The unmixing of the two populations of particles required high-precision measurements and good error estimation because the two coercivity distributions overlap considerably (Fig. 4.23). The biogenic component (B), which formed in an oxic environment, is predominant in the lake before anthropogenic eutrophication (see Box 5.5) produced an anoxic environment (preserved in the youngest sediments) where detrital (D) ferromagnetic minerals survive almost exclusively. The detrital component is identified by comparison with a sample representing the catchment area, taken from the delta of a small river flowing into the lake. Typically, it has a lognormal shape with a wide range of coercivities (< 3 to > 300 mT). Population B accounts for up to 85% of the total ARM in the older part of the core and progressively diminishes with increasing eutrophication (Fig. 4.23a). Egli (2003) postulates an authigenic origin and identifies the B population as magnetite particles of biogenic origin because of the narrow coercivity range (30–80 mT), high values of independently measured ARM/SIRM that are characteristic for magnetosomes (Moskowitz *et al.*, 1988), and TEM observations of intact magnetosome chains.

Figure 4.23 (a) ARM coercivity distributions of six lake sediment samples from Baldeggersee (Switzerland) that formed in various degrees of anoxia over the last 200 years and one sample (U03F) taken from the delta of a small river flowing into the lake (data kindly provided by Ramon Egli). The shape of the curves changes with increasing eutrophication (G044 → G010). The thickness of the lines represents the estimated error of the distributions as calculated from eight repeat measurements for each sample. (b and c) Unmixing of the total spectrum (T) yields two major components D and B. D predominates in the most anoxic sample G010, where component B is subordinate. The B component is very strong in the most oxic material G044. The numbers in parentheses denote the contribution in (μAm^{2}kg^{-1}) to the ARM of each sample. (Modified from Egli, 2003.)

5

PROCESSES AND PATHWAYS

5.1 INTRODUCTION

The environmental archives of interest to us are encoded in soils, eolian deposits, and water-laid sediments as magnetic responses to complex physical, chemical, and biological mechanisms. This chapter outlines the main processes and pathways that give rise to these magnetic signals. Enviromagnetic minerals get into these natural archives in two major ways. Either they already exist elsewhere and are transported by wind, rivers, or ocean currents into the location in question or, because of the extremely high biological and chemical reactivity of iron, they are created or transformed *in situ* by chemical processes, which are often biologically mediated. The transformations of the iron minerals influence the (bio-)chemistry and sedimentary occurrence of other elements, such as phosphorus and sulfur, as well as the magnetic properties of the sediments that record the environmental and geomagnetic history in recent times and in the geological past. Thus, the record bears witness to (1) various detrital processes caused by erosion in different climates and environments; (2) the results of elemental solubility and subsequent precipitation in fluctuating redox conditions, which again depend on local or regional environmental circumstances; and (3) the history of the geomagnetic field—if ferromagnetic minerals in the sediments carry stable natural remanent magnetization.

5.1.1 Depositional Processes

Detrital ferromagnetic minerals originate ultimately from igneous rocks, in which they form a characteristic fraction depending on rock type, rock chemistry, and the conditions of rock formation. Titanomagnetites, titanohematites, and pyrrhotites in a wide range of chemical composition and grain size are common accessory minerals in such rocks. Their occurrence is governed by magma composition, oxygen partial pressure, and cooling rate. Because titanomagnetites are sensitive to changing p–T conditions, they may oxidize at lower temperatures to titanomaghemites, thereby keeping the original crystal structure, or they may oxidize and unmix at high temperatures, giving rise to very characteristic exsolution structures. All these alteration products may survive even long transportation from the original emplacement area to the point where they are finally fixed in a sedimentary environment.

During deposition and eventual lithification, the detrital magnetic minerals may become aligned parallel to the Earth's magnetic field and acquire a detrital remanent magnetization (DRM). Physical alignment of the magnetic particles starts when they sink through the water column to the sediment surface and ends at a late diagenetic stage when the motion of the embedded particles is restrained during dewatering and consolidation of the sediment (see Box 5.1). The DRM process is subject to many physical forces resulting from the magnetic field, gravity, viscous drag, and Brownian motion but also from environmental factors such as current directions, bioturbation, porosity, and pore water circulation. Because of differing grain sizes and shapes and different specific magnetic properties, the interplay of the mechanical forces controls the final lock-in of the particles in a very complex manner.

Notwithstanding this complexity, DRM-bearing sediments are among the most useful recorders of geomagnetic field behavior in the past because often one can demonstrate that they have been deposited continuously over long time intervals, thus providing complete records of the direction and intensity of the geomagnetic field. These may be essential for dating the sediments by comparison with the geomagnetic polarity timescale or with well-known features of paleosecular directional and intensity variations throughout the Quaternary, an important prerequisite when studying the environmental history of a sediment.

The particulate magnetic influx also gives evidence of environmental change that may be related to paleoclimatic variability. Depending on fluvial, glacial, or eolian transport activity, the quantity and composition of the detrital fraction may fluctuate. The magnetic mineral fraction may be diluted or concentrated in the sediments, resulting in variations of susceptibility, remanence, and other magnetic parameters. For instance, erosional processes may be facilitated in cold and arid climate periods when vegetation is reduced, making more material available to removal by rivers, whereas in warmer times, the land surface is protected by vegetation (e.g., Thouveny *et al.*, 1994). Alternatively, the magnetic mineral grain size in eolian sediments may change when, for example, in cold times stronger winds carry not only more material but also generally coarser grain size fractions into the sedimentation basins (Chen *et al.*, 1997). This will also change the magnetic signature of the deposits in a characteristic manner that may be utilized for reconstruction of the environmental change in detail.

Sedimentation processes may also find expression in the development of a magnetic fabric that results from anisotropy of susceptibility or remanent magnetization. A commonly observed fabric is that due to compaction, wherein the triaxial anisotropy ellipsoid is flattened and the minimum axes are aligned vertically (e.g., Tarling and Hrouda, 1993). On the other hand, well-grouped — usually horizontally oriented — maximum axes give evidence of paleocurrent directions in fluviatile and marine environments.

Box 5.1 DRM and pDRM

Detrital remanent magnetization is due to the alignment of particles with magnetic moment, m, during sedimentation and lithification in the presence of the Earth's magnetic field, B. The magnetic particles sink through the water column, reach the sediment/water interface, and come finally to a rest in the pore space of the sediment. Only a few centimeters of water depth are generally required for optimum alignment of a depositional remanent magnetization (DRM). Elongated particles have a tendency to be orientated horizontally due to gravitation and hence produce too shallow inclination (inclination error). Water currents may also distort the long grain axes with respect to the field B (azimuthal error).

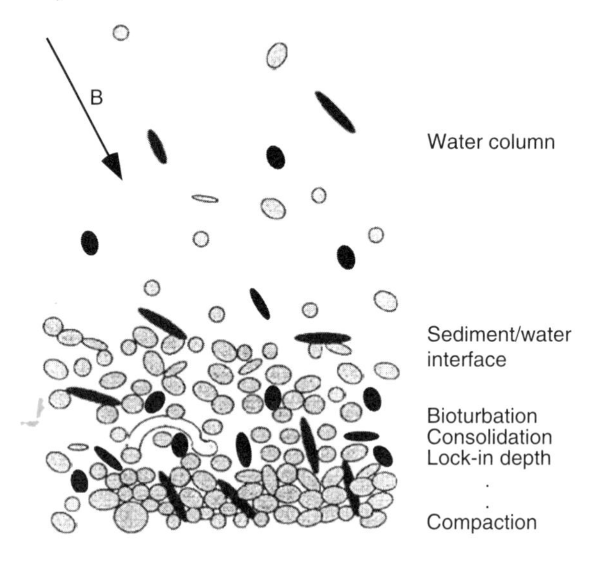

The magnetic particles may still be mobile after deposition in the water-filled substrate due to Brownian motion of the water molecules. A postdepositional remanent magnetization (pDRM) will eventually be locked in when the water content falls below a critical value so that physical contact of the particles hinders further motion.

The lock-in depth position is controlled by factors such as grain size distribution of the magnetic minerals and the sediment matrix, sedimentation rate, or bioturbation (Guinasso and Schink, 1975; Hyodo, 1983; Bleil and von Dobeneck, 1999). In fine-grained terrigenous clays the pDRM may be locked in nearly simultaneously during ongoing sedimentation (Clement *et al.*, 1996); in coarser grained silts such as loess sediments the pDRM may be delayed by up to several meters (Spassov *et al.*, 2001, 2002).

Sediment compaction will probably cause inclination errors, too (Jackson *et al.*, 1991).

5.1.2 (Bio-)Chemical Processes

Chemical production of magnetic minerals is the other major source of magnetism in sediments. The ferromagnetic minerals either form from preexisting iron-bearing minerals by alteration or oxidation or precipitate directly from iron solutions. In the latter case, biogenic mediation through bacterial activity is often important, but authigenic inorganic mineral formation must also be considered. The growth of new magnetic mineral phases will obviously influence the sediment magnetic properties: for instance, paramagnetic substances may be converted into strongly magnetic ferrimagnetic minerals; alternatively, strongly magnetic ferrimagnetic minerals may be transformed into weakly magnetic antiferromagnetic or even diamagnetic minerals. The growth of magnetite from green rust under reducing conditions (Pick and Tauxe, 1991); the precipitation of goethite, hematite, and lepidocrocite from ferrous bicarbonate solutions (Hedley, 1968); the oxidation of magnetite to maghemite by oxygen addition or iron removal; the dehydration of iron oxyhydroxides; and the oxidation of Fe^{2+}-bearing oxides to minerals containing only Fe^{3+} are some natural examples that have also been simulated in the laboratory at room temperature (Froelich *et al.*, 1979). Biogenic magnetite production has been demonstrated to occur in marine and lacustrine sediments (e.g., Petersen *et al.*, 1986; Hawthorne and McKenzie, 1993).

The iron transformation processes can indeed be explained by (1) inorganic chemical processes and (2) biologically mediated chemical processes (e.g., Bingham Müller, 1996), which may finally fix the ferromagnetic minerals firmly in the sediments.

5.1.2.1 *Inorganic chemical processes* At pH > 5, ferrous iron oxidizes in air or water to the highly insoluble ferric form, which may be reduced in turn to the ferrous state under anoxic conditions. This oxidation/reduction cycle at the oxic/anoxic boundary is important in sedimentary environments and especially in lacustrine and marine water columns with seasonal or permanent anoxic bottom-water conditions. The reaction may be described by:

$$Fe^{2+} + 3H_2O \rightarrow Fe(OH)_3(s) + 3H^+ + e^-$$

Ferric iron may also dissolve without a change in oxidation state under acidic (pH < 4) conditions or form soluble Fe^{3+} ligand complexes in highly productive systems that are characterized by extensive microbial activity. In the photic zone of lakes and oceans, ferric iron may be reduced under oxic conditions by organic solvents in the presence of light (Sigg *et al.*, 1991). The rate of ferric iron dissolution is controlled either by mineral surface reaction rates or by transport processes, depending upon the dissolving or reducing agent.

5.1.2.2 *Biologically mediated chemical processes* Because of its high reactivity, iron is involved in many biological processes. The mobile ferrous form is easily transported across cell membranes and is used directly for nutritional needs. As Bingham Müller (1996) explains, cells use iron in the enzymatic transfer of electrons to catalyze reactions, and some bacteria use iron proteins in the reduction of nitrogen. Many bacteria couple the degradation of organic matter with the reduction of ferric iron. Some bacteria are direct iron reducers, magnetotactic bacteria utilize chelators, and others reduce ferric iron indirectly (e.g., sulfate reducers). The transformation of iron generally depends on the environment. Direct microbial iron reduction may occur preferentially in suboxic marine or in methanogenic lacustrine environments where previously produced sulfides have been removed and hence sulfate reduction does not take place. Sulfate reduction, on the other hand, regulates the reductive ferric iron dissolution in most anoxic marine systems and within acid lakes, saline and brackish wetlands in which organic matter and sulfates are amply available.

If the size of chemically or biologically grown ferromagnetic minerals exceeds the threshold volume necessary for SD behavior, a crystallization remanent magnetization (CRM) will be acquired (see Box 5.2). In natural sediments, however, the time of origin of such a CRM is always a matter of debate. Because diagenetic alteration or authigenetic formation of magnetic minerals may take place long after deposition, the information about environmental conditions and the geomagnetic field during deposition may be masked or even completely lost (e.g., Tarduno, 1994). Such a *secondary* CRM may overprint the remanence acquired at the time of deposition. Its value for sediment dating is much diminished compared with DRM, which can often be assumed to lock in at shallow depth in the sediment column shortly after deposition. Nevertheless, the CRM may give significant clues about the timing and diagenetic conditions of the alteration processes (as discussed in Chapter 8).

5.2 SOILS AND PALEOSOLS

Soils originate from the interaction of physical, chemical, and biological processes of weathering. When sediments and rocks are exposed at the land surface, their outermost structures begin to loosen due to skin-deep penetration of gas and water. Physical weathering is due to expansion of rocks and sediments starting at joints and cracks that facilitate fluid flow in the changing pore space. Soils also undergo similar physical changes: clay minerals swell when wetted and strong temperature variations caused by fires or resulting from freezing water also affect the physical soil structure.

Soil-forming chemical reactions occur in four major processes: hydrolysis, oxidation, hydration, and dissolution (Table 5.1). The corresponding backward reactions are alkalization, reduction, dehydration, and precipitation.

- *Hydrolysis* usually describes reaction of carbonic acid with cation-rich minerals to form clay and free cations. It is the main way in which silicate minerals are chemically destroyed.
- *Oxidation* can be visualized as converting ferrous ions (Fe^{2+}) to ferric ions (Fe^{3+}) whereby the lost electron becomes available to other compounds. Minerals containing ferrous iron, such as olivine or pyroxene, usually display green colors, whereas ferric iron as found in goethite or hematite gives rise to yellowish, brown, and even red soil coloration. Because ferric cations are relatively insoluble in aqueous solutions, their oxides and hydroxides may remain stable in soils.

Box 5.2 CRM

Changing environments may cause neoformation or alteration of ferromagnetic minerals. Weathering can cause mineralic Fe^{2+} ions to be dissolved and oxidized by abiotic or biogenically mediated processes. They may end up in purely trivalent iron–bearing minerals such as maghemite or hematite.

Dehydration of goethite is a simple example of these processes:

$$2\alpha\text{-FeOOH} \rightarrow \alpha\text{-Fe}_2\text{O}_3 + \text{H}_2\text{O}$$

which may result in a typical color change from yellow (limonite) to red (hematite) being observed macroscopically.

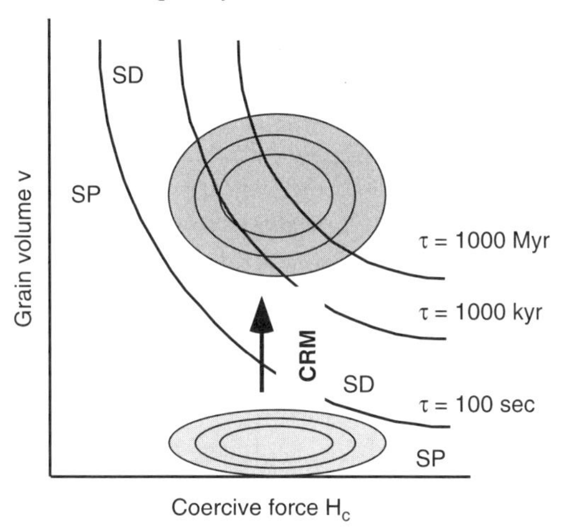

The new minerals grow from a microscopic germinal state into larger grain sizes, thereby crossing the boundary region from superparamagnetic (SP) to stable single-domain (SD) magnetic state. When this growth occurs in the presence of the Earth's magnetic field well below the Curie temperature (e.g., at room temperature), a chemical or crystallization remanent magnetization (CRM) arises when the new particles exceed a critical blocking volume. The persistence of the CRM has been described by Néel (1955) using a relaxation time formalism:

$$\tau = \frac{1}{C} \exp\left(\frac{M_s H_c v}{2kT}\right)$$

where τ denotes the relaxation time of the magnetization, C is a frequency factor of the order of 10^{10} s^{-1}, M_s the saturation magnetization, H_c the coercivity, v the grain volume, k Boltzmann's constant, and T the absolute temperature. Crystallizing grain populations wander from the bottom to the top of the diagram crossing the isolines of τ which are proportional to $v \cdot H_c$, thereby acquiring a CRM that may be stable over millions of years. (Figure adapted from Butler, 1992.)

Table 5.1 Examples of Common Chemical Reactions during Weathering

I. Hydrolysis

$$2NaAlSi_3O_8 + 2CO_2 + 11H_2O \rightarrow Al_2Si_2O_5(OH_4) + 2Na^+ + 2HCO_3^- + 4H_4SiO_4$$

albite, carbon dioxide, water → kaolinite, sodium ions, bicarbonate ions, silicic acid

II. Oxidation

$$Fe^{2+} \rightarrow Fe^{3+} + e^- \text{ (partial reaction)}$$

ferrous ion → ferric ion, electron to other element

$$2Fe^{2+} + 4HCO_3^- + 1/2O_2 + 4H_2O \rightarrow Fe_2O_3 + 4CO_2 + 6H_2O$$

ferrous ions, bicarbonate ions, oxygen, water → hematite, carbon dioxide, water

III. Hydration and dehydration

$$2FeOOH \Leftrightarrow Fe_2O_3 + H_2O$$

goethite ⇔ hematite, water

$$CaSO_4 \cdot 2H_2O \Leftrightarrow CaSO_4 + 2H_2O$$

gypsum ⇔ anhydrite, water

IV. Dissolution

$$CaCO_3 + CO_2 + H_2O \Leftrightarrow Ca^{2+} + 2HCO_3^-$$

calcite, carbon dioxide, water ⇔ calcium ion, bicarbonate ions

From Garrels and Mackenzie, 1971; compiled by Retallack, 1990 (© Kluwer Academic Publishers, with kind permission of Kluwer Academic Publishers and the author).

• *Hydration* involves addition of water bound in the crystal lattice, as exemplified by the goethite/hematite reaction.

• *Dissolution* most commonly involves the removal of calcite by a weak solution of carbonic acid derived from atmospheric carbon dioxide and water. Bicarbonate and Ca^{2+} ions may easily precipitate again under alkaline conditions to form, for instance, calcium carbonate nodules at the bottom of a paleosol (or to block your kitchen faucet when a temperature change reduces the solubility of $CaCO_3$).

As Retallack (1990) points out, the influence of organisms (plants, burrowing organisms, bacteria) is so overwhelming that biological processes can hardly be differentiated from other soil-forming processes. Three of the most important biological factors are humification, nutrient availability, and bioturbation.

• *Humification* in modern soils can be characterized by the extent to which identifiable plant fragments are destroyed and by the amount of humic and fulvic acids or amorphous organic carbon present.

• *Nutrient availability* determines the biological productivity of soils because plants and animals need a number of nutrient elements that are provided from cations in solution.

• *Bioturbation* describes the degree of physical reworking of a soil by organisms. It depends on the amount of biomass available and on the biological activity (e.g., depth and frequency of root penetration or animal burrowing).

Given the wide variety of starting bedrock materials, pedogenic processes, and environmental factors (rainfall, temperature, topography, drainage), there is great variability in the soil finally produced. Nevertheless, there are some broad features displayed by all soils, in particular, the ubiquitous development of more or less horizontal layers, or *horizons* (see Box 5.3). A specific sequence of horizons constitutes

Box 5.3 Soil Horizons

It is often observed that soils—and paleosols—present a pattern that basically consists of three horizons. Starting at the top, these are conventionally labeled A, B, and C according to the material they consist of. They are illustrated for the schematic Chernozem profile here. Other horizons are sometimes differentiated, particularly the organic (O) layer found at the top of modern soils.

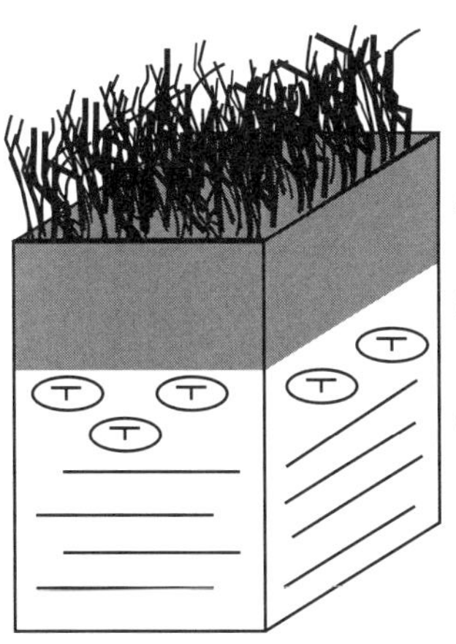

Chernozem profile

In addition to these main horizons (always indicated by uppercase letters), it is common practice to specify certain attributes by means of one or more lowercase letters. For example, t indicates an accumulation of clay, c the presence of concretions or nodules, k the enrichment of carbonates, and so on. The whole system provides a powerful shorthand which—with the addition of the appropriate thicknesses—provides the reader with a clear mental picture of any given situation. Full details can be found in any soil science textbook (for paleosols, see Retallack, 1990). Figure © Kluwer Academic Publishers with permission of the publishers.

a *soil profile*, and different profiles are indicative of different soil-forming regimes. This leads to the recognition of certain major soil types found on the land surface of the earth (e.g., *Chernozem* = dark-colored soil of the Russian steppe and parts of the North American prairie grassland; *Ferralsol* = red, brown, and yellow soil of the humid tropics, notably in Brazil and west–central Africa). For an excellent introduction to this complex subject, see Fitzpatrick (1986). These major soil types have been classified by different authorities into different systems of taxonomic nomenclature, four of the most important being as follows:

1. *The Handbook of Australian Soils* (Stace *et al.*, 1968)
2. The soil map of the Food and Agriculture Organization 1:5,000,000 (FAO, 1971–1981)
3. The soil taxonomy of the U.S. Soil Conservation Service (Soil Survey Staff, 1975)
4. The soil classification of the former USSR (Egorov *et al.*, 1977)

Both field and laboratory data are necessary to arrive at the correct identification within a specific classification scheme. This is not always a simple matter, even for an expert. For example, the U.S. system (which is hierarchical, like zoological and botanical classification schemes) consists of 10 *orders* divided into 47 *suborders*. These, in turn, are divided into 230 *great groups*, which are themselves further divided into *subgroups* and then *families*.

Ferromagnetic mineral formation in soils and paleosols often leads to enhanced soil magnetism, a good example of which is illustrated in Box 5.4. This enhancement is always connected to an increasing content of ferrimagnetic minerals such as magnetite, maghemite, or greigite. However, it should be kept in mind that these strongly magnetic minerals carry only a fraction of the iron available in soils and paleosols (e.g., Evans and Heller, 1994; Virina *et al.*, 2000). With respect to magnetic enhancement, several pathways and connections with iron supply from soil substrates and climate have been discussed for many years, but an exhaustive general theory cannot yet be given because of the diversity, complexity, and interaction of the processes involved. Five major processes are discussed at present (see Dearing *et al.*, 1996):

1. Detrital input is provided from the atmospheric fallout of fossil fuel–burning power plants, metallurgical industries, and cement factories (e.g., Hulett *et al.*, 1980; Flanders, 1994; Strzyszcz *et al.*, 1996). Relatively coarse-grained (1–20 μm) spherules of magnetite (and partly hematite) are generated by these various industrial sources and are transported as dusts and aerosols over variable distances to settle eventually on the soil surface. After landing there, the particles penetrate into the soil profile and accumulate mostly in the uppermost fermentation and humic subhorizons. In this case, magnetic low-field susceptibility will show very little frequency dependence because of the large particle size.

2. Natural fires or crop burning may cause thermal transformation of weakly magnetic iron oxides, hydroxides, and carbonates to ferrimagnetic magnetite or maghemite in the presence of organic matter (Schwertmann and Heinemann, 1959;

Box 5.4 Magnetic Enhancement

The accompanying figure illustrates the increase in magnetic susceptibility caused by the formation of a soil from a preexisting substrate. In this particular case, the starting material is windblown silt (loess, white fields in lithology column) in China. The entire package of sediments is constituted by an alternation of loess and soil layers; the soils are covered by younger loess and hence are called paleosols (hatched areas in lithology column). Also plotted is the stratigraphic variation in sediment grain size, which is also diagnostic, loess being systematically coarser than paleosol. The grain size gives information about the wind regime during deposition, whereas susceptibility reflects detrital magnetic input *and* influence of alteration processes after deposition of the eolian material.

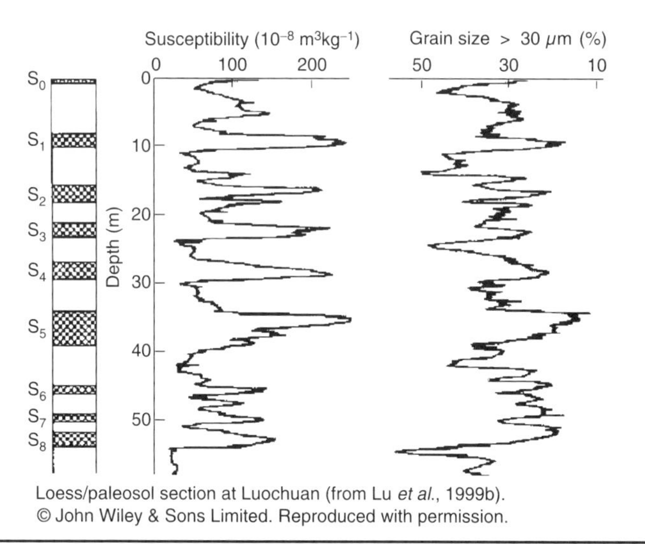

Loess/paleosol section at Luochuan (from Lu *et al.*, 1999b).
© John Wiley & Sons Limited. Reproduced with permission.

Le Borgne, 1960; Kletetschka and Banerjee, 1995). Iron hydroxides dehydrate to, or fine-grained hematite may be reduced to, magnetite, which can subsequently oxidize to maghemite. Fires affect the topmost soil layers, and the degree of magnetic enhancement is highly variable depending on organic matter content, temperature of burn, availability of preexisting iron minerals (especially iron hydroxides), and soil porosity.

3. Inorganic *in situ* formation of ultrafine-grained magnetite has been postulated by Maher and Taylor (1988) for some English soils. Laboratory experiments (Taylor *et al.*, 1987) show that magnetite can indeed be synthesized through controlled oxidation of Fe^{2+} solutions at room temperature and near-neutral pH. The synthetic material was similar in chemical composition, morphology, and grain size to the soil analogues that are thought to be related to soil erosional inputs. Brennan and Lindsay (1998) confirm that transfer of electrons from decomposing organic matter to Fe^{3+} in soils increases the activity of Fe^{2+}. They find experimentally that exposure

of soils to atmospheric oxygen causes them to become sufficiently oxidized that amorphous ferrihydrite forms, which appears to control Fe solubility. Crystalline Fe^{3+} oxides including hematite and goethite can control Fe solubility under highly stable oxidizing conditions. However, with prolonged soil submergence and adequate organic substrate, the formation of magnetite may control Fe solubility in reducing environments when oxygen is greatly restricted. Such cycles of highly variable redox conditions may be met by changing soil moisture and aeration in variable climate sequences.

4. Bacterial microorganisms influence the precipitation of Fe^{3+} oxides, cause Fe^{2+} oxidation, and utilize organic ligands of Fe^{2+} and Fe^{3+} compounds or dissolve iron compounds by reducing Fe^{3+}, by forming iron complexes, by releasing complex organics, or by utilizing iron for metabolism (Fischer, 1988). This activity may lead to another source of *in situ* enhancement of magnetic soil signals. If anaerobic conditions prevail, dissimilatory bacteria may produce extracellular ultrafine SP and SD magnetite grains (Lovley *et al.*, 1987) or greigite may form due to microbial reduction (Stanjek *et al.*, 1994). Such conditions may be met in periodically waterlogged gleys and strongly podzolized humic-iron podzols (Dearing *et al.*, 1996). Those soils may not contain high concentrations of magnetite, but the ultrafine SP grains are expected to be uncontrolled in size because of the locally varying anoxic microenvironment (Maher and Taylor, 1988). Magnetotactic assimilatory bacteria that produce intracellular chains of SD magnetite magnetosomes can exist under more oxygenated conditions in topsoils (Fassbinder *et al.*, 1990). However, their low population densities in modern topsoils do not seem to provide a substantial contribution to the magnetic enhancement of soils.

5. The most prominent process of magnetic mineral formation in soils has been formulated by Schwertmann (1988a,b) and discussed in detail by Dearing *et al.* (1996). Weathering of iron-bearing minerals during soil wetting and drying cycles, that is, change in pedoenvironmental factors such as temperature, soil water activity, pH, organic matter content, and release rate of iron, may produce Fe^{2+} solutions that are oxidized and—favored by the presence of organic matter—form ferrihydrite ($5Fe_2O_3{\cdot}9H_2O$) when critical concentrations are achieved. Ferrihydrite is the most easily reduced iron oxide and occurs at relatively high Eh values under short-lived periods of anaerobicity in well-drained soils (Fischer, 1988). Iron-reducing bacteria utilizing ferrihydrite and other iron oxides/hydroxides as terminal electron acceptors will liberate the Fe^{2+} ions (see Chapter 9). This microbial iron reduction during metabolic respiration is considered to be the essential process to produce soluble Fe^{2+} ions in soils (Schwertmann, 1988a):

$$4FeOOH + CH_2O + H_2O \rightarrow 4Fe^{2+} + CO_2 + 8OH^-$$

It represents the so-called *fermentation mechanism*, which was postulated many years ago by Le Borgne (1955) and Mullins (1977). Partial dehydration and reduction of ferrihydrite to magnetite will finally take place in the presence of excess Fe^{2+}, giving rise to enhancement of soil magnetic properties as summarized in the model by Dearing *et al.* (1996). Figure 5.1 illustrates a schematic sequence in which ferrihydrite

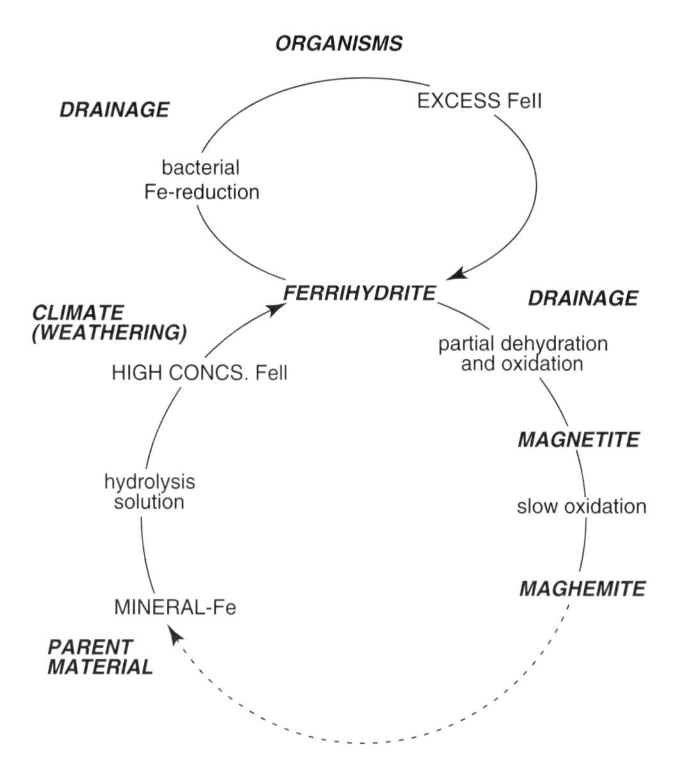

Figure 5.1 Sequence for secondary ferrimagnetic mineral formation in temperate soils as proposed by Dearing *et al.* (1996). Main iron phases (uppercase lettering), important processes (lowercase lettering), and factors (italic lettering) at different stages are indicated. The possible weathering of metastable secondary ferromagnetic minerals is shown as a dashed line.

forms under weathering conditions after iron becomes available from the parent material by hydrolysis and dissolution and through bacterially mediated iron reduction and then reacts with excess Fe^{2+} to form magnetite, which may be partially oxidized to maghemite.

Hydrolysis and dissolution in soils are often connected to precipitation. Hence, in soils or paleosols developing on substrates with uniform iron mineral content and rock texture such as many loess sequences, the iron supply and the production of new magnetite/maghemite during pedogenesis may semiquantitatively reflect precipitation in variable (paleo-)climates (Dearing *et al.*, 1996). Magnetic sediment properties may be used successfully as proxies for paleoclimate reconstruction if the paleosols reach comparable maturation. Appropriate magnetic climofunctions have been established (Singer *et al.*, 1992; Heller *et al.*, 1993; Maher and Thompson, 1995; Han *et al.*, 1996; Evans *et al.*, 2002).

Not only enhancement but also magnetic depletion is observed in soils. Leaching is often encountered in waterlogged soils and natural or artificial (e.g., in rice fields) gleization leads to reduced magnetic signals (Babanin *et al.*, 1995). Oxidation may provoke the formation of weakly magnetic minerals from strongly ferrimagnetic minerals. These weakly magnetic phases, together with paramagnetic minerals, mask clear correlation between magnetic susceptibility and the total iron content of the soil. There is a weak correlation between total iron and magnetic susceptibility in Chinese loess/paleosol samples, but the correlation increases significantly if one considers only dithionite-soluble iron or oxalate-soluble iron (Fig. 5.2). The latter quantities are determined by means of standard soil science procedures involving certain reagents able to remove iron bound in specific target minerals. In the case of dithionite, the main target is maghemite. For oxalate, the targets are ferrihydrite, lepidocrocite, and (according to Canfield, 1989) magnetite. This complex behavior underlines the great need for caution when using sediment magnetic properties as quantitative climate proxies.

5.3 MARINE SEDIMENTS

Magnetic minerals in marine sediments are of terrigenous, chemical, and biogenic origin (Henshaw and Merrill, 1980). Terrigenous material consists largely of granular (clastic) material eroded off the continents by water, ice, and wind. Much of this continental debris is introduced by rivers and is eventually distributed widely by the general ocean circulation. Near the mouths of rivers, sedimentation is relatively high (e.g., for the Mississippi it exceeds 1 cm/year). On the other hand, in the deep oceans, pelagic clay accumulates very slowly, typically at a rate of less than 1 mm/year. Volcanic ashes and eolian dusts may also contribute to sedimentation rates of the order of several $mg/cm^2/year$, as noted in the Yellow Sea, for example (Hovan *et al.*, 1989).

In addition to the imported terrigenous detritus, magnetic minerals are created and transformed in the marine realm during diagenetic alteration of iron-bearing minerals and authigenesis. In hemipelagic regions, organic matter dominates the geochemical system. Oxidation and fermentation of organic constituents control Eh and pH and become the driving mechanism for chemical reactions. Elemental supply may come from seawater or from volcanic and hydrothermal vents at seamounts and midoceanic ridges where marked increases of heavy metals such as iron, copper, nickel, and manganese are found. Thus, hydrogenous or authigenic iron–manganese oxides, oxyhydroxides, and iron-rich clays may form that are very sensitive to changes in the geochemical environment.

Precipitation and dissolution under variable redox conditions influence the chemistry and physical properties of new magnetic mineral phases. Change of grain volume and magnetic domain structure will give rise to stable, unstable, or completely viscous remanent magnetization; to variable susceptibility behavior; or to hysteresis properties that are indicative of environmental change (e.g., Tarduno, 1994).

In order to understand the sediment geochemistry and to reconstruct the paleoenvironment, the chemical stability of the authigenically formed minerals must be known. The stability diagram of Fig. 5.3 shows that a variety of magnetic minerals

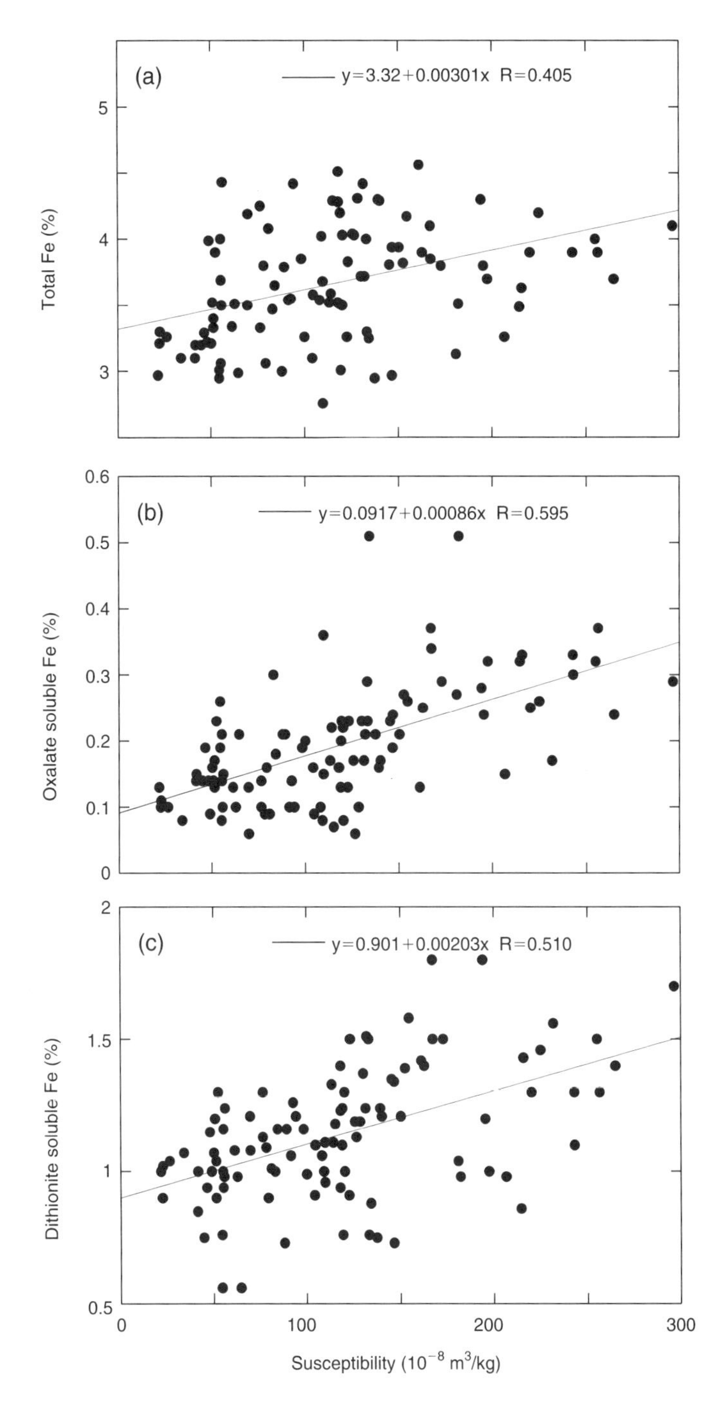

Figure 5.2 Iron content (wt %) versus susceptibility in loess/paleosols from Luochuan (Chinese Loess Plateau). (a) Total, (b) oxalate-soluble, and (c) dithionite-soluble iron.

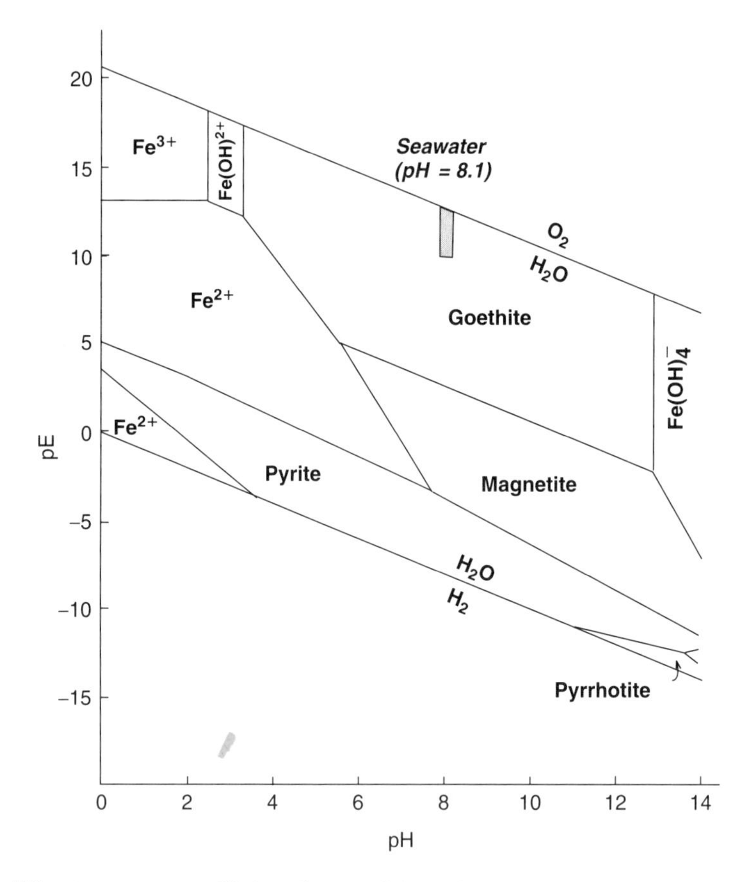

Figure 5.3 pE versus pH equilibrium diagram of the Fe-S-H_2O system, using the hydroxide–oxide mineral change of goethite to magnetite. Assumed concentrations of dissolved Fe and S in seawater 3.5×10^{-8} mole/l and 2.8×10^{-2} mole/l, respectively. $pO_2 = 0.20$ atm, $T = 25°C$. pH < 7 indicates acidic, pH > 7 basic, pE > 0 oxidizing, and pE < 0 reducing conditions; stability fields for precipitation of goethite, magnetite, pyrite, and pyrrhotite are shown. (From Henshaw and Merrill, 1980.) © American Geophysical Union. Reproduced by permission of American Geophysical Union.

can form authigenically in seawater (pH ~ 8.1). Because the oxidation state in marine sediments can vary considerably, it is possible that magnetite and/or goethite (the iron oxy-hydroxide polymorph to which other iron oxy-hydroxides convert upon aging) form in this environment. Hence ferromanganese oxides and oxyhydroxides are the most important minerals in the marine environment except in anoxic regions, where, among other iron sulfides, magnetic pyrrhotite may prevail. The oxidation potential may also be strong enough to cause maghemitization of titanomagnetites derived from oceanic basalt.

The organic matter content and type play a central role in determining the oxidation state of the marine sediments. On a large scale this is expressed in the Pacific Ocean by two major types of sedimentary environment (Henshaw and Merrill, 1980). In the hemipelagic zone near the continents, a two-layer sediment column exists where the uppermost sediments are oxidized due to contact with oxygenated bottom waters. The sediments below are rich in organic matter and become anoxic because of rapid burial due to high sedimentation rates that prevent their oxidation. Seaward into the ocean basin, the oxidized layer thickens until in the deep sea the anoxic zone virtually disappears because of low organic input and very slow sedimentation rates.

A relatively simple and idealized model of the geochemistry and the ionic mobility and remobilization reactions in the sediment column due to organic matter decomposition has been introduced by Froelich *et al.* (1979). It takes as starting organic matter a standard reference compound representative of the average oceanic plankton (Redfield, 1958; Fleming, 1940). This so-called *Redfield composition* has atomic ratios carbon/nitrogen/phosphorus = 106:16:1 [i.e., $(CH_2O)_{106}(NH_3)_{16}(H_3PO_4)$]. During diagenesis, this is first oxidized by the oxidant yielding the greatest free energy change per mole of organic carbon oxidized. When this oxidant is depleted, oxidation proceeds utilizing the next most efficient (i.e., energy-producing) oxidant until all oxidants are consumed or oxidizable matter is depleted. The reactions, which are bacterially mediated, use available electron acceptors in the following order: O_2, NO_3^-, MnO_2, Fe_2O_3, FeOOH, and SO_4^{2-}. The sequence of these chemical reactions, which proceed from aerobic respiration to sulfate reduction and methane production, is outlined in Table 5.2. The chemistry of a schematic pore water profile overlain by oxygenated waters results in a zonation as shown in Fig. 5.4.

Table 5.2 Decomposition Reactions of Sedimentary Organic Matter in Marine Sediments

Process	Reaction
Aerobic respiration (zone 1)	$(CH_2O)_{106}(NH_3)_{16}(H_3PO_4) + 138O_2 \Rightarrow$ $106CO_2 + 16HNO_3 + H_3PO_4 + 122H_2O$
Nitrate reduction (zone 2)	$(CH_2O)_{106}(NH_3)_{16}(H_3PO_4) + 94.4HNO_3 \Rightarrow$ $106CO_2 + 55.2N_2 + H_3PO_4 + 177.2H_2O$
or	$(CH_2O)_{106}(NH_3)_{16}(H_3PO_4) + 84.8HNO_3 \Rightarrow$ $106CO_2 + 42.4N_2 + 16NH_3 + H_3PO_4 + 148.4H_2O$
Manganese reduction (zones 3 and 4)	$(CH_2O)_{106}(NH_3)_{16}(H_3PO_4) + 236MnO_2 + 472H^+ \Rightarrow$ $236Mn^{2+} + 106CO_2 + 8N_2 + H_3PO_4 + 366H_2O$
Iron reduction (zones 6 and 7)	$(CH_2O)_{106}(NH_3)_{16}(H_3PO_4) + 212Fe_2O_3$ (or 424FeOOH) $+ 848H^+ \Rightarrow 424Fe^{2+} + 106CO_2 + 16NH_3 +$ $H_3PO_4 + 530H_2O$ (or $742H_2O$)
Sulfate reduction	$(CH_2O)_{106}(NH_3)_{16}(H_3PO_4) + 53SO_4^{2-} \Rightarrow$ $106CO_2 + 16NH_3 + 8S^{2-} + H_3PO_4 + 106H_2O$
Methane production	$(CH_2O)_{106}(NH_3)_{16}(H_3PO_4) \Rightarrow$ $53CO_2 + 53CH_4 + 16NH_3 + 8S^{2-} + H_3PO_4$

From Leslie *et al.*, 1990, following the zonation of Froelich *et al.*, 1979.

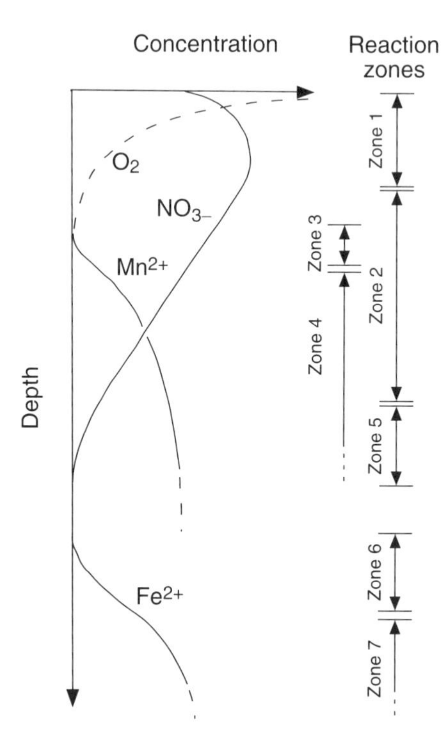

Figure 5.4 Schematic representation of chemical composition trends in pore water profiles of pelagic sediments according to Froelich *et al.* (1979). Depth and concentration axes in arbitrary units. The reaction zones and the characteristic curvature of the reaction gradients are discussed in the text. Typical reactions are listed in Table 5.2. © Elsevier Science, with permission of the publishers.

The sequence starts with oxidation of organic carbon in the topmost oxic environment, followed by consumption of nitrate and labile MnO_2. Then iron reduction takes place in suboxic environments, and further reactions in deeper zones of anoxic environment may follow. As we are interested primarily in the iron cycle, we note the following iron reduction reaction according to Froelich *et al.* (1979):

$$(CH_2O)_{106}(NH_3)_{16}(H_3PO_4) + 212Fe_2O_3(\text{or } 424FeOOH) + 848H^+ \Rightarrow$$
$$424Fe^{2+} + 106CO_2 + 16NH_3 + H_3PO_4 + 530H_2O\ (\text{or } 742H_2O)$$

with

$$\Delta G^{o'} = -1410\text{kJ/mole} \quad (\text{hematite, } Fe_2O_3)$$
$$-1330\text{kJ/mole} \quad (\text{limonitic goethite, } FeOOH)$$

where $\Delta G^{o'}$ denotes Gibbs free energy changes.

In more detail (see Fig. 5.4): In zone 1, oxygen is utilized for the oxidation of organic matter and is used up toward the bottom of this zone. In zone 2, nitrate reduction commences and oxic diagenesis ceases. In zones 3 and 4, deposited MnO_2, which has been buried down to zone 4, is reduced to Mn^{2+} and diffuses up to zone 3, again to be oxidized and redeposited. This process is repeated and hence acts as a sedimentary manganese trap that eventually creates a highly enriched manganese layer. By analogy with the MnO_2 reduction (which may be missing if manganese ions are absent), zone 7 represents the production of dissolved Fe^{2+} by reduction of ferric oxides during carbonate oxidation (see preceding reaction). Dissolved iron diffuses upward to be consumed near the top of zone 7. Because only a small portion of total iron is mobile (e.g., surface coatings of iron oxyhydroxides), iron trapping is small. Deposition of dissolved iron in zone 6 may result from oxidation by still-existing O_2 or more likely by nitrate moving down from zone 5. Hence, an electron acceptor must be involved. Alternatively, Fe^{2+} consumption may occur by incorporation of reduced iron into solid phases such as mixed carbonates (siderite), iron-rich smectites, and/or glauconites. In the absence of strongly anoxic conditions, slight SO_4^{2-} reduction may produce S^{2-}, which binds with excess Fe^{2+} to form FeS. Alternatively, if Mn is available with excess Mn^{2+}, formation of MnS takes place.

Thus, as we go down this idealized sediment column, we expect (1) an increase in the NO_3^- concentration to a maximum, (2) an increase in dissolved Mn^{2+} and a decrease in NO_3^- to zero, (3) an increase in dissolved Fe^{2+}, and a fairly monotonic increase in total CO_2 and PO_4^{3-} (Froelich *et al.*, 1979).

The characteristic times for these processes vary between 1 and 700 years, and the extent of these microbially mediated reactions is determined by the organic supply, sedimentation rate, and availability of reactants.

The effects of iron–sulfur diagenesis throughout the sediment column largely control the magnetic mineralogy. In the oxic zone, authigenic Fe-Mn-oxyhydroxides may precipitate inorganically from ferrihydrite precursors. As the reaction sequence progresses into the suboxic zone through Mn and Fe reduction, authigenic magnetite forms on top of the iron reduction zone at the Fe^{3+}/Fe^{2+} redox boundary (Karlin *et al.*, 1987). Macroscopically, these reduction changes are reflected in corresponding color changes from brown to tan and then from tan to green (Lyle, 1983; Sahota *et al.*, 1995). The position of the Mn and Fe redox boundaries depends largely on the organic carbon flux. If this flux decreases, reductants are depleted less rapidly and redox boundaries are shifted to greater depths. Organic carbon flux variations have been observed on glacial–interglacial timescales in the deep sea when the nature and depth of redox boundaries have changed, leading to non–steady-state magnetic mineral reduction (Tarduno and Wilkison, 1996). Authigenic magnetite formation, however, is considered to be an integral part of the organic matter decomposition as it is produced between the nitrate and iron reduction zone (Karlin, 1990). This milieu is optimal for the activity of iron-digesting bacteria to produce magnetite biogenically as observed by Petersen *et al.* (1986).

In the suboxic zone, microaerophilic magnetotactic bacteria form intracellular chains of magnetite by iron assimilation while using oxygen or nitrate as the terminal electron acceptor for metabolism (Blakemore *et al.*, 1985). According to Karlin *et al.* (1987), a sharp increase in authigenic magnetite formation can be observed just below the zone of nitrate reduction and continuing into the zone of iron reduction. In the anoxic zone, both anaerobic, NO_2-using, magnetotactic bacteria and dissimilatory bacteria that reduce amorphic ferric oxide to extracellular magnetite during metabolic organic matter oxidation have been proposed as primary producers of authigenic magnetite (Lovley *et al.*, 1987; Stolz *et al.*, 1990).

Magnetite production by aerophilic magnetotactic bacteria in the oxic and suboxic zones is extremely sensitive to environmental change. Different species of magnetotactic bacteria have different tolerances to oxygen concentration and prefer different habitats, thereby producing different magnetosome morphologies. Hesse (1994) demonstrated that during the Brunhes period the concentration and shape of bacterial magnetosomes in Tasman Sea sediments changed following Pleistocene climate variations. Cold periods were marked by lower pore water oxygen concentration so that elongate magnetosomes were produced in a less oxygenated environment whereas equant magnetosomes were found in more oxygenated conditions during warmer intervals (Fig. 5.5). Figure 5.5 also shows the habitats preferred

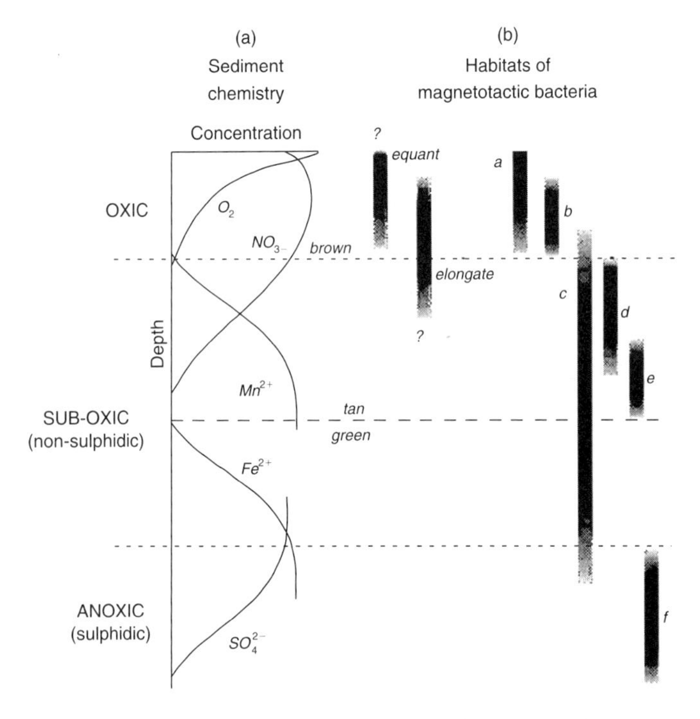

Figure 5.5 Schematic diagram of the chemical environments (a) and bacterial habitats (b) in the upper marine sediment column after Hesse (1994). Sediment chemistry from Froelich *et al.* (1979) extended into the anoxic sector. Equant and elongate magnetosome-producing bacteria apparently prefer high- and low-oxygen conditions, respectively (Hesse, 1994); a, laboratory culture of freshwater strain AMB-1 (Matsunaga *et al.*, 1991); b, laboratory culture of freshwater *Aquaspirillum magnetotacticum* (Blakemore *et al.*, 1985); c, laboratory culture of MV-1 strain from estuarine environment (Bazylinski, 1991); d, living magnetotactic bacteria in South Atlantic deep-sea sediments (Petermann and Bleil, 1993); e, inferred bacterial concentration in marine sediment (Karlin *et al.*, 1987); f, laboratory culture of sulfate-reducing freshwater strain RS-1 (Sakaguchi *et al.*, 1993).

with respect to redox zones of a number of aerophilic and anaerobic as well as dissimilatory bacteria as observed in the laboratory and in marine sediments when oxygen availability is reduced down the sediment pile in the anoxic zone.

Sulfate is thermodynamically the least preferred acceptor but is widely available, and hydrogen sulfide produced by bacterial sulfate reduction is common in organic-rich marine sediments (Leslie *et al.*, 1990). During active sulfate reduction, increased amounts of reduced sulfur are created so that H_2S in the pore water increases with depth. As soon as the boundary from an iron-rich ($Fe^{2+} > H_2S$) to a sulfur-rich ($H_2S > Fe^{2+}$) system in the anoxic zone is crossed, dissolution of magnetite begins.

The rate of magnetite dissolution is proportional to dissolved pore water sulfide concentration, which is related to the sulfide production rate, sedimentation rate, and intensity of bioturbation and depends on pH conditions and the surface area of magnetite grains so that the "half-life" of magnetite in anoxic sediments ranges between 50 and 1000 years (Canfield and Berner, 1987). Hydrogen sulfide easily reacts with Fe^{2+} to form mackinawite and greigite (Leslie *et al.*, 1990). Persistently high dissolved sulfide concentration will accomplish dissolution, sulfidization, and finally pyritization of magnetite. On the other hand, magnetite may be preserved if H_2S production is absent or is at a maximum when sediment burial is rapid. In the latter case, magnetite may move quickly through the dissolution zone when dissolved H_2S has not enough time to build up but is constantly taken up by detrital ferric iron minerals that are more reactive than magnetite (Canfield and Berner, 1987).

5.4 RIVERS AND LAKES

Integrated study of the magnetic aspects of substrates, soils, and sediments of lakes and their catchment areas may provide useful insight into the processes affecting the origin, transport, transformation, and accumulation of lake sediments as well as tools for elucidating their paleoecological significance (Oldfield, 1977). The framework of studies of this type is provided by the concept of ecosystems. An ecosystem is a regional unit of nature in which the biotic community of all organisms living in the area and their nonliving environment function together in order to maintain and develop this ecological system (Odum, 1997). Here *biogeocoenosis*, the coupled functioning of life and earth, takes place. Bormann and Likens (1969) describe the ecosystem "as a series of components, such as species populations, organic debris, available nutrients, primary and secondary minerals, and atmospheric gases, linked together by food webs, nutrient flow, and energy flow." Following these authors, the catchment area of a lake may be assigned a number of individual watershed ecosystems that interact with the lake ecosystem itself. This group of ecosystems results in an interacting catchment–lake ecosystem that may be called a lake–watershed ecosystem as proposed by Oldfield (1977).

The input–output processes of a single ecosystem have to be recognized if we are to understand its energy and nutrient relationships; the dynamic effect of geological processes such as erosion and deposition, mass wasting and weathering; the effects of meteorological changes on its behavior; and the interrelation with other ecosystems.

For the lake–watershed ecosystem — a system consisting already of several subecosystems but still being much simpler than the extremely complex and open oceanic ecosystems — input may be of meteorological (precipitation and dry fallout), biological (moved by animals or humans), and geological (dissolved or particulate stream load, alluvial and colluvial matter) origins. Biological output in an undisturbed natural system would generally balance the biological input. Geological output would consist of dissolved and particulate matter and could be estimated from hydrological and chemical measurements. Of course, the mass transfer from the higher energy catchment subecosystems into the lower energy lake subecosystem is of prominent importance for the nutrient budget of the lake. Figure 5.6 illustrates some of the input–output relations between the lake sedimentary record and the contributory processes and materials.

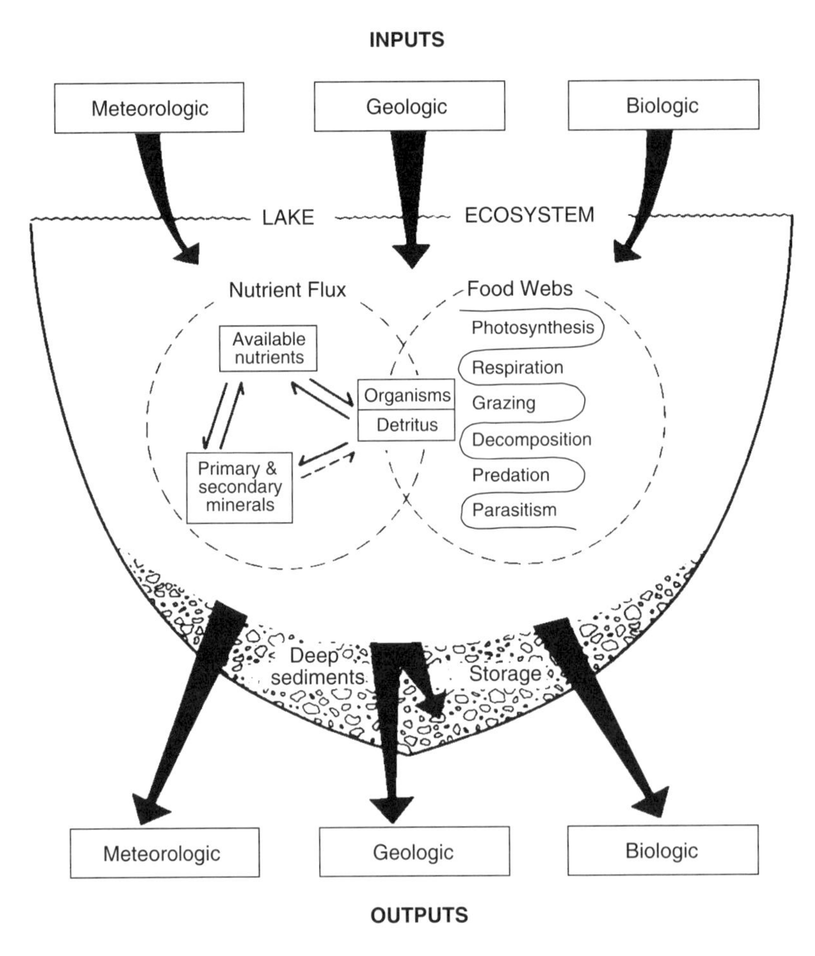

Figure 5.6 Simplified model for nutrient and energy output/input in a watershed–lake ecosystem. Major sites of accumulation and exchange pathways within the ecosystem are shown. [Following Likens and Bormann (1975) and Oldfield (1977).] © Springer-Verlag and Arnold Publishers, with permission of the publishers.

The origin of the ferromagnetic minerals in lake sediments depends largely on three major input sources:

1. Geological material produced by erosion of the catchment substrate or soils provides minerogenic material input to the lake ecosystem. A maar lake such as Lac du Bouchet situated in the southeast of the volcanic province of the Auvergne region of France receives large amounts of titanomagnetite derived from the surrounding strongly magnetic basaltic rocks, which are heavily eroded during cold climate periods when the surface vegetation cover is reduced (Thouveny *et al.*, 1994). Indeed, the Hubbard Brook (White Mountains of New Hampshire) watershed ecosystem experiment of Bormann and Likens (1969) showed that nearly all the iron — being a tiny fraction of the drained dissolved matter — was brought in as inorganic particulate matter.
2. Air-transported particulates form a substantial part of direct precipitation input. Atmospheric fallout of industrial man-made particles may contribute an additional ferromagnetic mineral fraction in lake sediments deposited during the last 150 years (Oldfield and Richardson, 1990). In low-sedimentation areas such as lake ridges, eolian dust influx may become important. This factor has been positively identified from increased concentration of high-coercivity minerals (goethite, hematite) on the Academician Ridge of Lake Baikal during cold climate episodes (Peck *et al.*, 1994).
3. Some iron may also reach the lake sediments in dissolved form when anaerobic conditions develop in the soils of the catchment area, for instance, as a result of podsolization or peat formation. Particulate iron, however, may also dissolve and precipitate, depending on pH and redox conditions and biogenic activity before or after deposition and burial in the sediment column (Karlin, 1990). Thus, ferromagnetic minerals may grow authigenically at or near the mud–water interface in a manner similar to that already discussed for the marine environment.

The first two sources are controlled directly by external forcing by climate, weathering, and wind activity. They will be mirrored by increasing or decreasing concentration or dilution of detrital ferromagnetic minerals of characteristic type and grain size as seen, for instance, in Lough Neagh, Northern Ireland, where the magnitude of susceptibility correlates with yearly rainfall (Dearing and Flower, 1982) or Lake Baikal (see earlier).

The third magnetic mineral source — the authigenic mineral formation in lake sediments — depends on the aquatic productivity, which is directly related to nutrient availability. Primary and secondary minerals may chemically decompose to form available nutrients, or secondary minerals may be formed from available nutrients with or without mediation through the activity of organisms (Likens and Bormann, 1975). Anaerobic and acid conditions keep ferrous iron and phosphate compounds soluble, but with oxygenation under more alkaline conditions the solubility product of ferrous phosphate (vivianite) may be reached (Anderson and Rippey, 1988). This process may cycle reversibly with seasonal changes in oxygen content of the hypolimnion of certain lakes (see Box 5.5). Hilton (1990) argues that greigite forms in

Box 5.5 Anatomy of a Lake

Many of the biological and chemical processes that take place in lakes are controlled by the depth to which light penetrates into the water and hence the depth to which photosynthesis occurs. Thus, the *epilimnion* is dominated by phototrophic algae that fix inorganic carbon to manufacture organic compounds. Deeper down, the *hypolimnion* is where organic compounds are broken down rather than synthesized: respiration and decomposition dominate. As a result, an oxygen profile is established as shown in the diagram. The epilimnion/hypolimnion boundary migrates up and down according to the season. The situation depicted here refers to typical summer stratification. As winter approaches, surface waters are cooled and sink through the epilimnion, eventually eliminating the hypolimnion by progressive erosion from above. Oxic conditions then reach to the bottom of the lake and may even penetrate into the sediments.

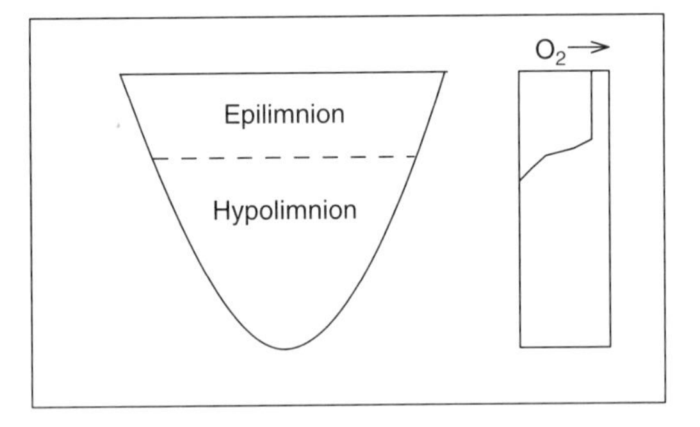

Lakes can be classified in terms of their biological activity or trophic state. Typically, a lake with a chlorophyll content of $\sim 20\,\mu g$ per liter is said to be *eutrophic*, whereas values about 10 times smaller indicate *oligotrophy*. For comprehensive discussions, see Wetzel (2001) and Dodds (2002).

eutrophic lakes with enough labile carbon to lower the redox and enable SO_4^{2-} reduction to occur and with labile iron and sulfur and an oscillating oxycline within the sediments. Alternating oxidizing and reducing conditions in connection with a steep diffusion gradient to get high H_2S seem to be necessary to produce first acid-volatile sulfide in reducing conditions and then elemental sulfur from the sulfide under oxidizing conditions, especially when iron oxide or manganese oxide is present, before forming greigite under reducing conditions again. This oscillation occurs in the millimeter- to centimeter-thick surface layers of productive sediments when the oxycline moves seasonally into and out of the sediment. Greigite may be produced inorganically or may be mediated biologically by sulfate-reducing microorganisms (Mann *et al.*, 1990a), most likely in eutrophic lakes.

Ultrafine magnetite grains extracted from lake sediments deposited under oligotrophic and oxic water conditions (marl lake) and near the sediment–water interface are biogenically precipitated magnetite produced by microaerophilic bacteria such as *Aquaspirillum magnetotacticum* (Bingham Müller, 1996). This type of magnetite will be preserved, thereby being able to record paleomagnetic signals. Eutrophication of a marly lake, however, will lead to dysaerobic conditions with increased productivity and deposition and preservation of organic matter (Hollander *et al.*, 1992). Anaerobic respiration becomes the dominant metabolic pathway for organic matter oxidation within the uppermost sediments (Froelich *et al.*, 1979), leading to reducing sulfidic diagenetic conditions that force coarsening of ferromagnetic grain size [e.g., in Lough Augher as discussed by Anderson and Rippey (1988)] and favor destruction of biogenic but at a later stage also of detrital magnetite. Sulfur may be released from the breakdown of organosulfur complexes during organic matter mineralization and may become available for the formation of mackinawite (Fig. 5.7).

Bingham Müller (1996) discusses four phases of iron transformation that may take place seasonally or in the course of progressive eutrophication of a lake, which may possibly be induced by human activity. Her evidence is developed from studies of modern Lake Greifen (Switzerland) and applied to the Holocene Lake Greifen and ancient marine systems such as the Mississippi delta. These phases model the response of the lake waters and the diagenesis of the lake sediments to processes that occur during

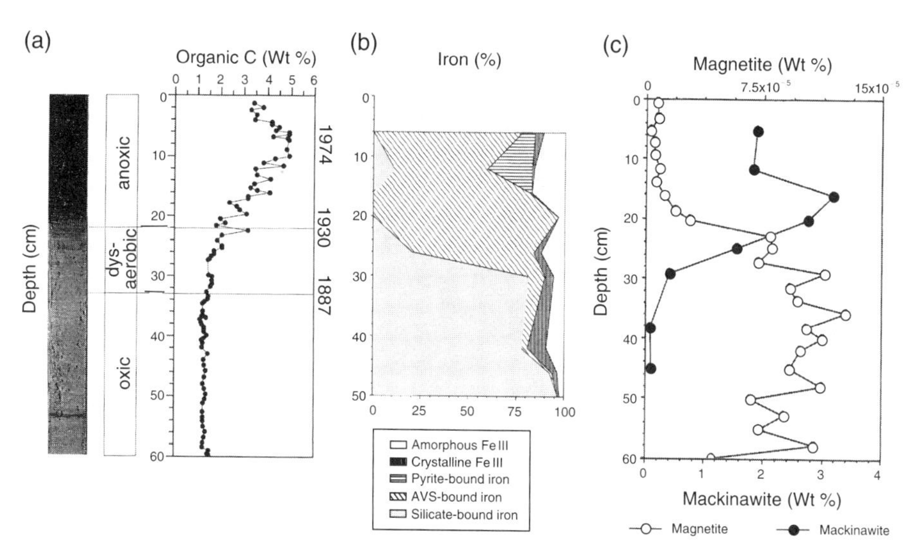

Figure 5.7 (a) Photograph of the split section of a short gravity core from Lake Greifen (Switzerland) showing the lithologic changes associated with the changing redox state of the water column from oxic (bottom) to dysaerobic (middle) and anoxic (top) sediments. The increase in organic carbon is related to increasing productivity due to progressive eutrophication since 1887. (b) Solid iron species plotted as percentage of total iron. Iron bound to acid-volatile sulfide (AVS) and pyrite increases and iron bound to silicates decreases at 30 cm depth or near the boundary between bioturbated marls and transition sediments of the dysaerobic zone. (c) Magnetite and mackinawite content versus depth. The up-core decrease in magnetite concentration from laminated marls (below 22 cm) to organic carbon–rich varved sediments correlates to an increase in acid-volatile sulfide. (All data from Bingham Müller, 1996.)

1. Periods of low to moderate rates of productivity and permanently oxic waters
2. Transitional periods characterized by moderate rates of productivity and seasonally anoxic waters
3. Periods of high productivity and degree of anoxia
4. Periods of extreme water-column anoxia, due either to extremely productive conditions or to stagnant waters in which sulfate becomes limiting to iron sulfide formation

In phase 1 (low productivity), the iron and phosphorus cycles are coupled, and ferric (oxyhydr)oxide phosphate and organic colloid complexes are formed. The iron cycle is not coupled with the sulfur cycle, and an iron–phosphate barrier is built. As far as ferromagnetic minerals are concerned, biogenic magnetite is formed by iron-chelating bacteria in the topmost aerophilic sediments and under suboxic conditions and remains preserved together with detrital (titano)magnetite. Geomagnetic field variations (secular variations, polarity changes) during sediment formation may be recorded with high fidelity. Reactive ferric iron species, such as ferrihydrite or lepidocrocite, are common and are associated with phosphorus. Metabolism of organic matter is oxic, and hence the pH in the sediments is neutral. Sulfide is produced minimally yet as a result of sulfate reduction and is oxidized and lost to the water column.

The following microbial and chemical processes are some of the pathways by which ferromagnetic minerals are formed or destroyed in phase 1 (and in oligotrophic lake ecosystems):

Sedimentation of (titano)magnetite
Authigenesis of magnetite via iron chelation by *A. magnetotacticum*

$$FeOOH \Rightarrow Fe_3O_4$$

Bacterial iron reduction

$$FeOOH + 2H^+ + 1e^- \Rightarrow Fe(OH)^+ + H_2O$$

In phase 2 (moderate productivity), the iron and sulfur cycles are coupled rapidly so that phosphorus is released from the sediments. As far as ferromagnetic minerals are concerned, biogenic magnetite, formed in the topmost sediments, is rapidly sulfidized and destroyed, and detrital magnetite undergoes dissolution and sulfidization even in systems with low sulfate availability, low sulfate reduction rates, and high sedimentation rates as a result of internal sulfur cycling. Ferric or ferrous iron competes with phosphate for ferric iron adsorption sites, and sulfur binds with ferrous iron to form either acid-volatile sulfide or sulfide minerals such as mackinawite or pyrite. The geomagnetic record in the sediments deteriorates, depending on the availability of reactive iron. Organic matter is enriched in sulfur, pH conditions are basic, the anaerobic metabolism of organic matter is controlled mainly by bacterial sulfate reducers, and sulfur is internally cycled at moderate rates and retained.

In phase 3 (high productivity), the iron and sulfur cycles are coupled and the sediments become a source of nutrients for further water column productivity. As far as ferromagnetic minerals are concerned, magnetite destruction takes place due to reducing diagenetic conditions as long as sulfate reduction rates are faster than ferric iron reduction; detrital magnetite is completely dissolved by reduction. High concentrations of reduced inorganic iron sulfides, such as mackinawite and pyrite, and extremely low concentrations of reactive iron, associated with sulfate, are present. Reliable geomagnetic signals have disappeared in the sediment magnetization. Organic matter is enriched in sulfur and mineralized anaerobically by microbial sulfate reducers, pH is basic, and sedimentary sulfur is cycled rapidly, resulting in an annually renewable source of sulfur for sulphate reduction.

The following microbial and chemical processes are some of the pathways by which ferromagnetic minerals are formed or destroyed in phases 2 and 3:

Sedimentation of (titano)magnetite
Precipitation and deposition of iron oxyhydroxides (ferrihydrite and lepidocrocite)
Mackinawite formation via sulfidization of reactive iron phases (e.g., ferrihydrite)
Authigenesis of magnetite via iron chelation by *A. magnetotacticum*

$$FeOOH \Rightarrow Fe_3O_4$$

Magnetite sulfidization and subsequent formation of mackinawite

$$\text{(i)}\ Fe_3O_4 + 8H^+ + 2e^- \Rightarrow 3Fe^{2+}_{aq} + 4H_2O$$
$$\text{(ii)}\ Fe^{2+} + HS^- \Rightarrow FeS + H^+$$

Bacterial iron reduction

$$FeOOH + 2H^+ + e^- \Rightarrow Fe(OH)^+ + H_2O$$

Abiotic oxidation of mackinawite or other reduced sulfide species by reactive ferric iron phases:

$$16H^+ + 8FeOOH + FeS + FeO \Rightarrow 9Fe^{2+} + SO_4^{2-} + 12H_2O$$

In phase 4 (extreme lake stagnation), the iron cycle is coupled with both the sulfur and phosphorus cycles. In extreme water anoxia, hydrogen sulfide production may become limited. If reactive iron is available, iron and nitrate reducers produce ferrous iron in concentrations higher than hydrogen sulfide, thereby enabling iron sulfide and even excess ferrous iron formation. The availability of both ferrous iron and phosphorus results in the formation of ferrous complex species such as vivianite ($Fe_3P_2O_8 \cdot 8H_2O$), in addition to iron sulfides. As far as ferromagnetic minerals are concerned, magnetite is sulfidized and destroyed. Alternatively, detrital magnetite may remain intact if it moves quickly across the reduction zone (compare with the marine situation as discussed by Canfield and Berner, 1987), but the geomagnetic sediment record is distorted in any case. Inorganic reduced sulfides and ferrous iron phosphate minerals are present; pH is basic; organic matter is sulfur enriched and mineralized anaerobically by sulfate reducers, iron- and nitrate-reducing bacteria, and methanogenic bacteria. Internal sulfur cycling is restricted due to the stagnation of the waters.

Thus, the Bingham Müller (1996) four-phase model is able to set up close constraints between nutrient availability and ferromagnetic mineral formation and destruction. Hence sediment magnetic properties not only report geomagnetic field variations but also record unforeseen details of environmental change in terrestrial and water-laid sediments.

6

TIME

6.1 INTRODUCTION

In geological problems, one of the most universal difficulties is that of attaching a reliable chronology to the sequence of events under scrutiny. This problem is not new. In the 19th century, the eminent British physicist Lord Kelvin (1824–1907) calculated—on the very soundest scientific principles—that the Earth could not be older than about 20 million years, similar to his own estimate of the sun's age. With the advent of radiometric dating following the discovery of radioactivity in 1896 by Henri Becquerel (1852–1908), it eventually emerged that the Earth is at least 200 times older than Kelvin had estimated. As Stacey (2000) points out, however, the new developments did not invalidate Kelvin's arguments, they merely created a paradox. The source of the sun's energy was not properly understood until the discovery of thermonuclear fusion in the 1930s. The paradox immediately evaporated and a vastly greater age became possible. This whole episode serves as a sobering reminder that many aspects of the earth sciences are historical and that history without a suitable timescale is doomed. Great efforts have therefore been made to develop suitable chronometric techniques.

Many procedures are now well established, and excellent sources of detailed information are available (e.g., Geyh and Schleicher, 1990; Noller *et al.*, 2000). Three areas are of particular importance to us, not only as techniques offering some degree of chronological control but also as subjects whose contents are intimately interconnected and now form an integral part of environmental magnetism—especially the aspects related to paleoclimatology. These are the geomagnetic polarity timescale, oxygen isotope stratigraphy, and Milankovitch cycles. Each of these is dealt with in detail in the sections that follow. Before considering them, however, it is instructive to summarize briefly some of the other methods that provide chronological control in enviromagnetic studies. Of these, the most important are those directly exploiting radioactive decay, particularly ^{14}C and K/Ar but also including uranium-series techniques. In addition, the luminescence methods (TL, OSL, and IRSL = thermoluminescence and optical and infrared stimulated luminescence, respectively) rely indirectly on radioactivity by measuring the integrated dose of ionizing radiation (from the surrounding sediments) to which a sample has been exposed. To these may be added dendrochronology (tree rings), varve chronology (annually laminated lake sediments), and — for very recent events — the historical record itself.

Each technique has its own range of effectiveness. The geomagnetic, oxygen isotopic, and orbital schemes are applicable over the entire range of interest to us — the last 5 to 10 million years. This is also the case for the K/Ar technique (Dalrymple and Lanphere, 1969). Uranium-series dating (using the $^{230}Th/^{234}U$ ratio) has a useful range up to ~350,000 years (Broecker and Bender, 1972), on the same order as luminescence dating (Wintle, 1990). Berger *et al.* (1992) claim reliable TL ages up to 800,000 years, but this has been challenged by Wintle *et al.* (1993). Radiocarbon dating is limited by the relatively short half-life of ^{14}C (5730 ± 40 yr) — it becomes very difficult to determine the decay with sufficient accuracy in samples that are 10 or more half-lives old, although special experimental procedures have succeeded in pushing the limit to ~75,000 years (13 half-lives) (Stuiver *et al.*, 1978). Finally, dendrochronology and varve sediments are generally limited to the Holocene [0–10,000 yr before present (BP)] except in rare cases, such as the varves of the Cariaco Basin in Venezuela, which apparently reach back some 15,000 years (Hughen *et al.*, 1998).

In summary, it is useful to list the age range(s) over which the various techniques are applicable (although such a list is only a rough guide, there being considerable overlap between the various techniques):

- 10^7–10^5 years: geomagnetic, oxygen, astronomical, K/Ar
- 10^6–10^4 years: U series, TL
- 10^5–10^2 years: ^{14}C, palynology
- 10^4–10^0 years: ^{14}C, varves, tree rings, archeology, historical records

This is only part of the story, however. As well as the range of applicability of a given method, it is essential to consider its resolving power. For some climatic events, the historical record may provide vital information on an annual, or even a seasonal, basis. The most comprehensive summary is given by Lamb (1995), but Bradley (1999) and Cronin (1999) also provide excellent discussions. As one would expect, the sources are very diverse and are poorly distributed in space and time. The longest record known refers to the flood level of the Nile, for which stone inscriptions reaching back some 5000 years indicate that the East African summer monsoons produced higher rainfall then than now. Early records are available from the Shang dynasty in China (~3700–3100 yr BP; Chu, 1973). For Europe, Lamb has produced "winter severity" and "summer wetness" indices for the last 1000 years. Two of the best known events described from such sources are the so-called *Maunder minimum* and the *Little Ice Age*. The former was an interval from about 1650 to 1715 when the sun had virtually no sunspots and during which modern calculations suggest that the solar irradiance was about a quarter of a percent lower than the current value of 1367 W/m^2 (Lean *et al.*, 1995). For the same period, there is a plethora of records indicating a prolonged cold interval. For example, people in London skated on the frozen river Thames, and in the decade 1685–1695, Zürich had snow cover for about

70 days per year compared with about half that nowadays (Pfister, 1978). It is precisely natural variations of this kind that make it tricky to establish the validity of supposed anthropogenic forcing of global warming but, at the same time, provide a strong incentive for further study (see, for example, http://www.pages.unibe.ch and http://ngdc.noaa.gov/paleo/paleo).

Annual resolution can, in principle, be obtained from tree ring and varve counting. Luckman (1994), for example, shows how trees that were overrun during an important advance of the Athabasca Glacier in the Canadian Rockies can be used to date the event to AD 1714. On a wider geographic scale, the many investigations of the so-called *Younger Dryas* are illuminating. *Dryas octopetala* is a flower that grows in the Arctic but appears in the pollen record at lower latitudes in Europe and North America during certain cold periods. In paleoclimatic research, two particular cold intervals have been especially important—the more recent of them being universally referred to as the Younger Dryas. Annually layered sediments in the Soppensee (a small lake located in the central Swiss plateau at 8.3°E, 47.1°N) imply that the Younger Dryas cold episode started 12,125 years ago and lasted 1139 years (Hajdas *et al.*, 1993). This astonishing precision is, however, slightly misleading—similar studies on other lakes in Europe do not give exactly the same results. For example, the data from Lake Gosciaz, Poland, imply that the Younger Dryas started 12,520 years BP and lasted 1080 years (Goslar *et al.*, 1995). One should always remember that precision and accuracy are not the same thing! On the other hand, we should not lose sight of the fact that these particular discrepancies amount to only a few percent.

Radiocarbon dating depends on the production of ^{14}C in the upper atmosphere by neutron bombardment of nitrogen ($^{14}N_7 + {}^1n_0 \rightarrow {}^{14}C_6 + {}^1H_1$). The carbon so produced rapidly combines with atmospheric oxygen to form CO_2, which mixes with all the other (nonradiogenic) carbon dioxide and enters the biosphere by many different pathways (Mangerud, 1972). The ^{14}C decays back to nitrogen by the emission of an electron ($^{14}C_6 \rightarrow {}^{14}N_7 + \beta + \text{neutrino}$). Over geological time, an equilibrium has been achieved between creation and decay of ^{14}C. When an organism dies, its ^{14}C content (which, during life, had been in equilibrium with the atmosphere) starts to decay. The carbon clock has been started. In the so-called *conventional* methods, the age of a sample is determined by directly measuring the rate of β decay (i.e., the number of electrons emitted per unit time). To offset low count rates, large samples are required (enough to yield 100 g of carbon for analysis). With the development of the accelerator mass spectrometer (AMS) (Stuiver, 1978), it became possible to date much smaller samples on the order of 1 mg.

As mentioned before, the half-life of ^{14}C limits the method to 75,000 years at most, but it is rare that much confidence can be placed in ages exceeding ~45 ka. In fact, with the advent of large numbers of radiocarbon dates, a much more serious difficulty has emerged. We now know that the cosmogenic production of ^{14}C is not constant. Among other things, it depends on the activity of the sun and the strength of the geomagnetic field (Laj *et al.*, 1996; see also Chapter 12). The net result of the many complications that arise is that *radiocarbon years* must now be independently calibrated into *calendar years*. Back to ~15,000 years BP, such a calibration curve has been deduced by comparison with tree rings, corals, and varves (Stuiver and Braziunas, 1993). For the last 2500 years, differences are not too severe, but for early Holocene times (~10,000 years BP) the discrepancies amount to ~1000 years. Clearly, a calibration curve is an essential tool. Even with it, life is not always simple. This is because it is not monotonic—it has maxima, minima, and plateaux, which mean that a sample with a given ^{14}C age may correspond to several calendar ages, all equally probable.

Uranium-series dating relies on the fact that the natural equilibrium arrived at in an undisturbed system can be upset by differences in the properties of the various decay products. The most significant of these are the virtual insolubility in water of ^{230}Th and ^{231}Pa (the first is produced after two successive α-particle emissions, $^{238}U \rightarrow {}^{234}U \rightarrow {}^{230}Th$; the second by a single α-particle emission, $^{235}U \rightarrow {}^{231}Pa$). These insoluble intermediate decay products are quickly precipitated in sediments. Two particularly important applications of U-series dating are in corals and caves. Trace amounts of uranium are coprecipitated with calcite but with a *daughter deficiency* because the ^{230}Th has already been removed. After calcite formation, the daughter starts to build up again, and because the rate of buildup is known, a chronometer is provided. The development of thermal ionization mass spectrometry (TIMS) in the mid-1980s permitted the use of small samples and led to great improvements in precision (Edwards *et al.*, 1987).

U-series dating of corals has, among other things, helped to establish sea level fluctuation curves. This is a very important topic for paleoclimatology because of the reciprocal link between the total amount of water in the oceans and the total amount of ice locked up in ice sheets and glaciers. For a late Glacial/early Holocene series of raised coral terraces on the Huon Peninsula (New Guinea), Edwards *et al.* (1993) obtained 2σ errors of only 30 to 80 years. Deposits in caves (stalagmites, stalactites, and flowstones, collectively known as *speleothems*) have also been extremely important. From a cave at Mo I Rana (Norway), Lauritzen and Lundberg (1998) describe an excellent example in which dating errors average 10 to 50 years for a record spanning the entire Holocene (0–10,000 years BP).

Luminescence simply refers to the light emitted by certain crystals (mostly quartz and feldspars) when subjected to heat or exposed to light. The emitted light comes from the release of electrons trapped at crystal defects. Dating by this method derives from the fact that the population of trapped electrons is a function of time because they are raised into the traps by ionizing radiation from radioactive materials in the immediate environment. The longer the sample has been receiving the radiation, the greater will be the luminescence observed when the electrons are released by the laboratory treatment. It is like putting money in your savings account: if you put away a fixed sum every month, then by checking the balance you can figure out when you first opened the account (but, remember, no interest is given). Thermoluminescence dating has found wide application in archeology, where it is used to date fired ceramics (Wintle and Aitken, 1977; Garrison, 2001). In geological settings, its main use has been in the dating of loess and other eolian sediments (Wintle, 1990; Singhvi *et al.*, 2001). There are many complicating factors that lead to generally accepted error estimates approaching ±10%. Furthermore, there is a tendency for older materials to appear systematically too young (by up to 15%), a problem which arises from the fact that they have suffered ionizing radiation for so long that all the available electron traps have been filled — the bank has ceased to accept deposits.

6.1.1 An Example

In their enviromagnetic study of Lough Catherine, Northern Ireland (7.5°W, 54.7°N), Snowball and Thompson (1990) investigated a suite of cores spanning the length and breadth of the lake (~1200 × 250 m). The mineral magnetic records of these (susceptibility and IRM) were correlated by sequence slotting (Thompson and Clark, 1989; discussed further in Chapter 7), and a combined master core was established for the purposes of interpretation. Broad chronological control was provided by the knowledge that the top and bottom of the sediments studied represent the present day and the beginning of the Postglacial, respectively (i.e., 0 and 10,100 years BP). For the core they illustrate in detail, this interval spans 508 cm, implying an average accumulation rate of 0.50 mm/yr. However, the chronological control provided by radiocarbon, pollen, and historical records (Thompson and Edwards, 1982; Pilcher and Larmour, 1982; McClintock, 1973) indicates that deposition was far from constant (Fig. 6.1). In the first 4900 years, only 66 cm of sediment

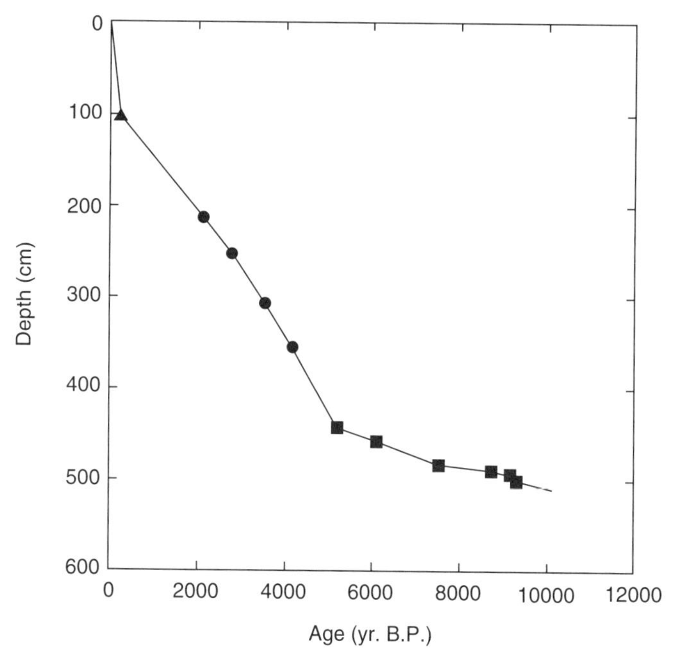

Figure 6.1 Chronometry of Lough Catherine sediments based on written historical records (triangle), radiocarbon dates (circles), and pollen events (squares). In order of increasing age these are the documented afforestation of the Marquis of Abercorn's estate, four ^{14}C dates, and six palynological events based on the appearance (= rise) and disappearance (= fall) of the corresponding pollen [the *Ulmus* (elm) fall, the *Alnus* (alder) rise, the *Quercus* (oak) rise, the *Corylus* (hazel) rise, the *Betula/Salix* (birch/willow) rise, and the *Juniperus* (juniper) rise]. (Data from Figure 2 of Snowball and Thompson, 1990.) © Arnold Publishers, with permission of the publishers.

accumulated (0.14 mm/yr), whereas the next 5000 years provided 341 cm of sediment (0.68 mm/yr). Finally, the uppermost 101 cm took only 200 years to accumulate (5 mm/yr). Age assignments based solely on linear interpolation between the end points could be in error by more than 3500 years—a very misleading situation for an environmental record spanning only 10,000 years.

6.1.2 Another Example

Holzmaar is a maar lake (i.e., a body of water occupying the crater of an old volcano) in the Eifel district of Germany (6.9°E, 50.1°N). Several aspects of the sediments it contains have been studied for a number of years. Zolitschka *et al.* (2000) have reported a very thorough study of the chronology of these sediments using varves, accelerator mass spectrometer (AMS) ^{14}C dating, volcanic ash (tephra) "fingerprinting," luminescence dating (both TL and OSL), and paleomagnetic measurements. Back to 13,000 years BP, the sedimentation is well constrained by 41 AMS ^{14}C dates (Fig. 6.2), but deeper sediments could not be dated due to the paucity of organic remains. This upper part of the age–depth curve is consistent with two OSL dates and two tephras identified by geochemical and mineralogical matching to be the Ulmener

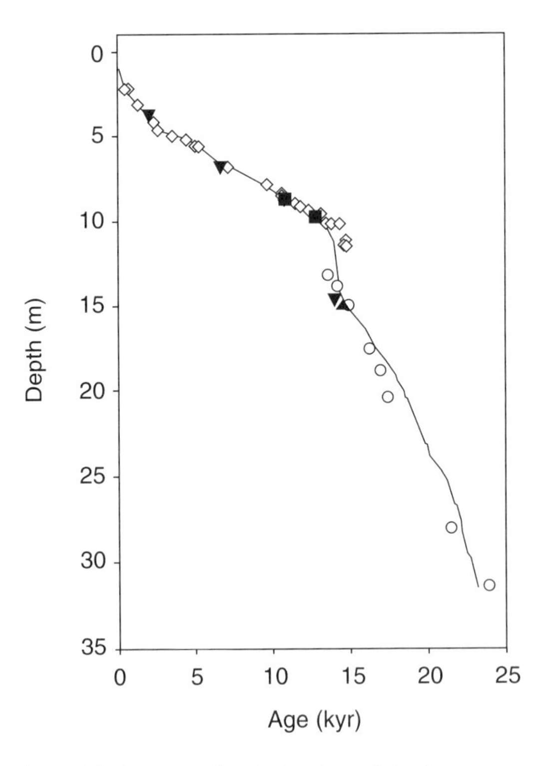

Figure 6.2 Chronology of Holzmaar sediments based on dates from varve counting (solid line), AMS ^{14}C (diamonds), OSL (inverted triangles), TL (triangle), tephras (squares), and paleomagnetism (circles). (Modified from Zolitschka *et al.*, 2000.)

Maar Tephra and the Laacher See Tephra (the ages of which are known from elsewhere). Between the two tephras 1560 varves were counted, but in nearby Meerfelder Maar, the same interval contains 1880 varves. It was concluded that a 320-year hiatus exists in Holzmaar. For the older part of the record, a further OSL date and a single TL date are available, but the majority of the material below ~12 m is calibrated by the observed paleosecular variation pattern (see later), which can be matched to corresponding data from Lac du Bouchet in France (Thouveny *et al.*, 1990; see also Chapter 7). As with the Irish example previously, the control provided by the several methods demonstrates that the rate of deposition varies significantly, ranging from less than 1 to almost 10 mm/yr. Zolitschka *et al.* (2000) go on to show how important these variations in sedimentation rate are for interpreting the Holzmaar climate record based on their measurements of carbon content (both organic and inorganic), organic pigments, and magnetic susceptibility.

6.2 TEMPORAL CHARACTERISTICS OF THE GEOMAGNETIC FIELD

To a reasonable approximation, the geomagnetic field is dipolar, with the magnetic axis aligned slightly more than 11° off the spin axis. There are other complications, but they are not important for our present purposes (a fuller discussion is given in Chapter 12). What are important here are the ways in which the field changes with time because it is these that provide chronological control. Figure 6.3 illustrates the main points as far as changes in the direction of the field are concerned. The local

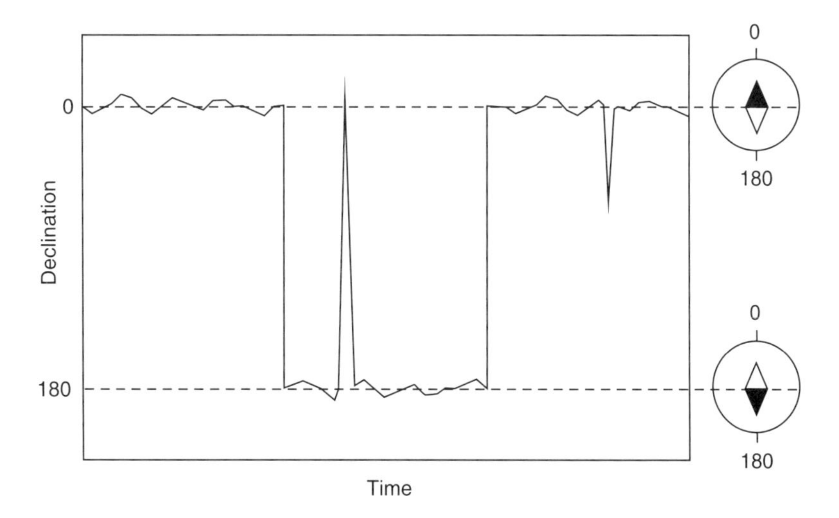

Figure 6.3 Schematic representation of temporal changes in geomagnetic declination due to secular variation, excursions, and polarity reversals. Polarity reversals involve a jump of 180° in no more than a few thousand years. Excursions are probably of even shorter duration and may or may not lead to a complete sign inversion. Secular variation consists of smaller fluctuations (typically of amplitude ±10–20°, but sometimes larger) generally associated with a timescale of a few centuries. For compilations of real data, see www.ndgc.noaa.gov/seg/potfld/paleo.shtml.

field vector also varies in magnitude, but the paleomagnetic determination of such changes requires a somewhat more complicated (*paleointensity*) procedure than that needed for the simple determination of paleodirection. (This is also true, of course, for direct measurements; magnetic north can be obtained immediately from a compass, but the strength of the field is not quite so easy to measure.) For the moment, therefore, we restrict our attention to directional changes, returning later to a consideration of magnitude fluctuations.

Three main directional features are involved (Fig. 6.3). The most prominent of these consist of occasional 180° flips involving a complete change of sign of the planetary dipole. These are the well-known *geomagnetic polarity reversals*. Superimposed on the abrupt changes at times of reversal is a relatively steady background of lower amplitude (10–20°) fluctuations collectively called the *secular variation*. A third phenomenon, resulting in what are usually called *geomagnetic excursions*, is poorly understood [see Jacobs (1994) for a useful discussion]. Regardless of the underlying physics, all three types of behavior offer chronometric possibilities.

6.2.1 Geomagnetic Polarity Reversals

When the geomagnetic field is oriented like that of the present day, it is said to be *normal*, while an oppositely directed field is called *reversed*. In some ways this nomenclature is misleading because one can easily fall into the trap of regarding the opposite of "normal" as being "abnormal," whereas, in fact, the Earth does not prefer one polarity over the other. All the available evidence indicates that the field spends exactly 50% of the time in either polarity. Nevertheless, the nomenclature is now firmly entrenched in the literature and there is no prospect that it will be changed in the foreseeable future.

A more important question than nomenclature concerns the actual length of time the field requires to execute a transition from one polarity to the other. Various lines of evidence indicate that this is geologically short, probably no more than about 5000 years (McElhinny and McFadden, 2000; Coe *et al.*, 2000). Once established, however, a polarity interval may typically last 40 times longer than this, so that, broadly speaking, one can picture a sequence of polarity intervals as a square wave.

For chronological purposes, reversals are by far the most important feature of the Earth's magnetic field. They have been discussed since the 19th century, but it was only with the advent of accurate—and sufficiently widespread—radiometric dating that real progress was made. This took place largely in the 1960s. Since then, there have been occasional adjustments to the temporal pattern of reversals; but broadly speaking, there has been for several decades a workable global geological clock based on what is usually referred to as the *geomagnetic polarity timescale*, or GPTS for short (Cox *et al.*, 1963; Cande and Kent, 1995).

In the early days the GPTS was based mostly on continental basalt lava flows (which generally carry a strong and stable remanent magnetization and which were found to yield reliable K-Ar ages). However, for rocks older than a few million years the experimental error associated with the ages becomes uncomfortably commensurate with the length of the polarity intervals themselves and it is impossible to identify specific intervals. For example, Merrill *et al.* (1996) show that over the last few million years polarity intervals have typically lasted about 200,000 years each, which would require a dating accuracy better than $\pm 2\%$ for the correct identification of a specific polarity interval recorded in rocks 10 million years old. With conventional K-Ar dating, this can rarely be achieved. As a result, the extension of the GPTS relied on marine magnetic anomalies (the *ocean stripes*, as they are often called). These also depend on magnetized basalt, in this case formed at oceanic ridges as a result of sea floor spreading. Direct radiometric dates are not generally available, but chronologies can be worked out in terms of sea floor spreading.

The very nature of plate tectonics ensures that oceanic crust has a limited career — once created at a spreading center it is inexorably drawn toward a subduction zone, where it is subsumed back into the Earth's mantle. The outcome is that the oceanic stripes do not provide a complete record of the Earth's polarity changes throughout the whole of geological time. Indeed, the age of the oldest oceanic crust is less than 5% of the age of the Earth ($< 2 \times 10^8$ years compared with about 4.5×10^9 years). Nevertheless, the oceanic data do permit the GPTS to be extended back to 160 million years, which represents an extremely significant 30-fold increase over what is possible solely on the basis of lava flows extruded on the continents. The complete sequence of reversals constituting the GPTS as currently understood is summarized by Cande and Kent (1995). An example illustrating the most recent 6 million years of the sequence is illustrated in Box 6.1.

Originally, during the initial establishment of what was to become the GPTS, polarity intervals seemed to fall into two groups, one about 10^6 years long, the other about 10 times shorter; the former were called *epochs*, the latter *events*. Epochs were named after important geomagnetists (Bernard Brunhes, 1867–1910; Motonori Matuyama, 1884–1958; Carl Friedrich Gauss, 1777–1855; William Gilbert, 1544–1603); events were named after the geographic location of their discovery. Officially, this naming system has been superseded by the terms *chron* (10^6–10^7 years long) and *subchron* (10^5–10^6 years long), but many of the old names are still current (see Box 6.1).

The whole subject of the accurate dating of the GPTS is currently undergoing an upgrade deriving from developments in radiometric dating. Using modern ultrasensitive mass spectrometers and incremental heating, the new technique of $^{40}Ar/^{39}Ar$ has dramatically reduced the experimental error to the point where it is now possible to obtain 1σ precisions of 0.5%. In this way, Singer *et al.* (1999) have made significant progress in dating lava flows associated with the GPTS, a vivid example being their investigation of a sequence in the Punaruu Valley in Tahiti. Four flows spanning the transition from normal to reversed polarity at the termination of the Jaramillo subchron yield $^{40}Ar/^{39}Ar$ ages of 0.989 ± 0.005, 0.991 ± 0.006, 0.977 ± 0.010, and 0.980 ± 0.011 million years (Ma) (in stratigraphic order starting at the bottom). These are all indistinguishable from one another at the 95% confidence level, leading Singer and his coauthors to conclude that the weighted mean (0.986 ± 0.005 Ma) accurately pinpoints this particular polarity switch. One can therefore anticipate that the GPTS will be gradually refined to a new level of precision (e.g., see Sarna-Wojcicki *et al.*, 2000) and also that new, short-lived features may be discovered — especially for the youngest part of the GPTS, which is most relevant to enviromagnetic research.

Box 6.1 GPTS — Names and Numbers

Historically, two systems of nomenclature arose for labeling specific features of the GPTS, largely as a result of two different applications. Those working on continental volcanic sequences developed a scheme of longer *epochs* and shorter *events*, using names of appropriate scientists for the former and of relevant geographic locations for the latter. Only four epochs were ever named — Brunhes, Matuyama, Gauss, and Gilbert — but about twice as many events were identified [Jaramillo (J), Olduvai (O), Reunion (R), Kaena (K), Mammoth (M), Cochiti (C), Nunivak (N), Sidufjall (S), and Thvera (T)]. This rather cumbersome scheme has stood the test of time and is here to stay.

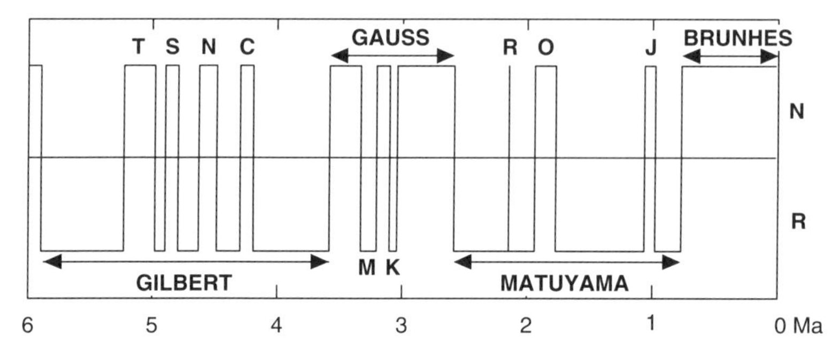

The other camp, working on sea floor spreading, had to contend with a much longer period of time containing many more polarity intervals. They opted for a numbering system based on the scheme used for labeling the marine magnetic anomalies. This starts with anomaly 1 at the active spreading ridge (=Brunhes epoch). It has become standard practice to use the terms *chron* and *subchron*, which refer to actual intervals of time. As more and more detail has been established, this scheme has been modified. For example, the normal interval associated with anomaly 12 is called C12n, with C12r being the reversed interval preceding it. On the other hand, C11n actually consists of two normally magnetized parts (C11n.1n and C11n.2n) separated by a short reversed interval (C11n.1r).

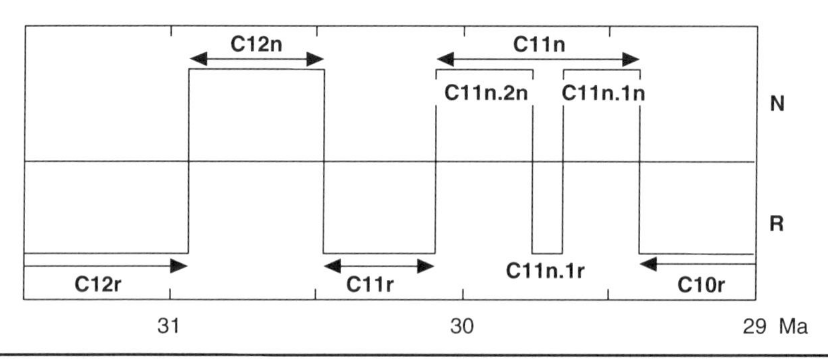

Indeed, as we will see in the following discussion concerning geomagnetic excursions, Singer *et al.* (1999) have already started this process in their interpretation of data from other parts of the sections they investigated.

6.2.2 Secular Variation

In 1634, Henry Gellibrand (1597–1636) observed in London that the compass pointed only 4.1°E of true north, whereas a measurement by William Borough (1513–1599) in 1580 had given 11.3°E. Such changes on a scale of decades to centuries came to be known as the secular variation. They are of great importance to geomagnetists because they reflect the workings of the dynamo that produces the Earth's magnetic field. If they could be suitably calibrated, such angular changes could obviously furnish a geophysical clock. There has, indeed, been some progress in this direction (e.g., the German lake sediments discussed earlier), but it must be said at the outset that chronological control by this means is far less common than in the case of polarity reversals.

Since Gellibrand's day, the deviation of the compass from true north has been carefully monitored, initially by individual pioneers of geomagnetism such as Edmund Halley (1656–1742) in England and Carl Friedrich Gauss (1777–1855) in Germany, but subsequently at national observatories set up for the purpose (http://nssdc.gsfc.nasa.gov/space/model/models/igrf.html). In earlier times, the angular deviation in question was called the *variation*, but this terminology has now been replaced by the term *declination*. In 1544, the German cleric Georg Hartmann (1489–1564) discovered that a magnetized needle arranged to swing about a horizontal axis (rather than a vertical axis, as in a compass) takes up a nonhorizontal orientation. The angular deviation from the horizontal is now called the *inclination*. At any point on the Earth's surface, the secular variation causes these two angles to vary smoothly as

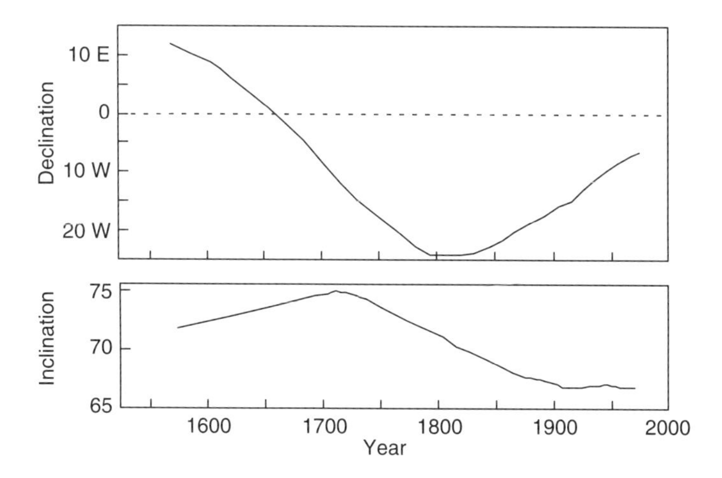

Figure 6.4 Historically recorded secular variation in declination and inclination at London, England. (Redrawn from Malin and Bullard, 1981.)

a function of time, as illustrated in Fig. 6.4. However, unlike reversals—which are global in extent—patterns of secular variation are coherent only on scales on the order of a thousand kilometers. The reason for this is not important here; a fuller discussion is given in Chapter 12.

The direct historical observations (Fig. 6.4) imply that the geomagnetic field at any location varies quasi-periodically on a timescale of a few centuries, so there is a strong incentive to try extending this to earlier times. This can, in fact, be done. Two types of archive have been particularly useful—archeological materials and lake sediments. Lava flows, which were so crucial in setting up the GPTS, have been much less effective in the study of the secular variation because they are extruded at highly irregular rates, depending on the vagaries of volcanic activity: they would make a good tape recorder if only the motor were not so erratic! Even so, there are notable exceptions such as the combined record from three Italian volcanoes (Arso, Vesuvius, and Etna) that spans the last millennium (for a summary, see Incoronato *et al.*, 2002).

An example of an individual archeological result having exceptionally good time control is provided by a pottery kiln in Pompeii (Fig. 6.5) buried by the celebrated eruption of Vesuvius which, according to Pliny the Younger, started during the afternoon of August 25, AD 79. The declination and inclination recorded by the kiln are 358.0° and 59.1°, respectively (Evans and Mareschal, 1989). [The error associated with such a result is usually given as the semi-angle (α) of a circular cone around the mean remanence vector following Fisherian statistics (the three-dimensional analogue of traditional one-dimensional Gaussian statistics). Commonly, the so-called *circle of 95% confidence* (α_{95}) is given: in the case of the Pompeii kiln $\alpha_{95} = 1.7°$.] By comparing the strength of the magnetization carried by the kiln samples with that which was gained in a known field in the laboratory, Evans (1991)

Figure 6.5 The kiln at Pompeii studied archeomagnetically by Evans and Mareschal (1989) and Evans (1991). See color plate.

obtained a value of 61 ± 1 μT for the magnitude of the ancient field at the time of the eruption [for a description of how such *paleointensities* are worked out, see, e.g., Butler (1992) or Merrill *et al.* (1996)]. Thus the AD 79 total geomagnetic vector in the Naples area is known. As it happens, its direction diverged by less than 4° from the present-day field, but its magnitude was 34% greater. Evidently, the geomagnetic field varies in magnitude as well as in direction. As with the directional changes, such temporal fluctuations in the strength of the geomagnetic field are of great interest to those engaged in studying the Earth's core and the geodynamo, but they also provide a further means of chronological control. Some examples are described in a separate section later—for the moment we continue discussion of directional changes.

Hundreds of archeomagnetic results have now been determined from sites of known age, so that the secular variation pattern is quite well determined for the last two millennia for many parts of the world, two of which are illustrated in Fig. 6.6 (Daly and Le Goff, 1996). For earlier times, significant amounts of data are available but spatial and temporal coverage is more sporadic. The data summarized in Fig. 6.6 indicate

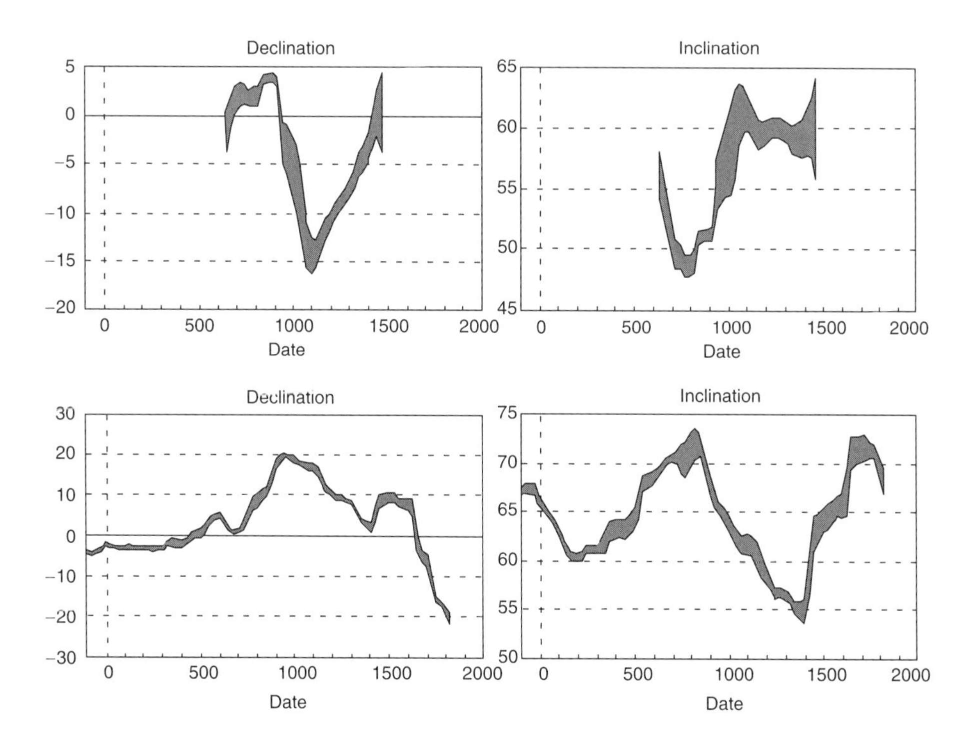

Figure 6.6 Archeomagnetically determined secular variation in declination (East = positive) and inclination for the United States (upper panels) and for France (lower panels). The American data ($N = 155$) come mostly from hearths in Arizona, New Mexico, and Colorado, reduced to a reference point in Arizona (110.0°E, 35.0°N). The French data ($N = 120$) come mostly from pottery kilns, with Paris being the reference point (2.3°E, 48.9°N). The shaded bands represent the 95% confidence limits. Dates are in years AD. Notice how the historical observations at London (Fig. 6.4) follow on smoothly from the French data shown here. (Compiled from Daly and Le Goff, 1996.)

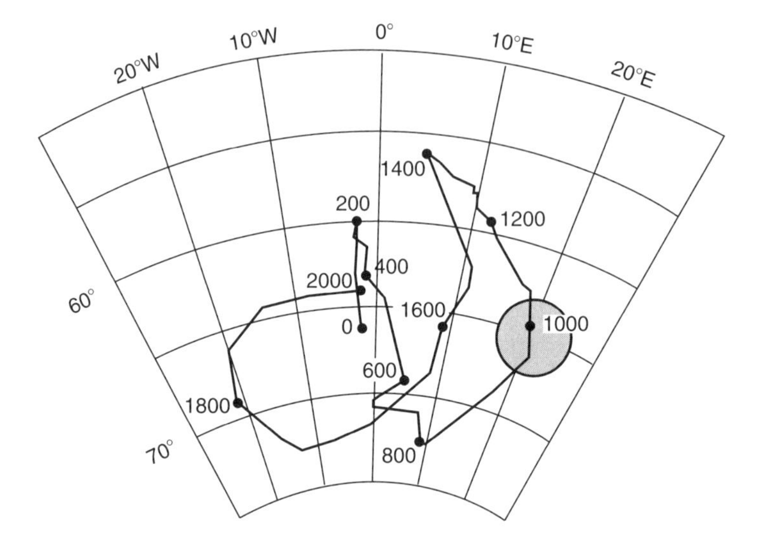

Figure 6.7 Archeomagnetic dating of an ancient quicklime kiln in Belgium (Hus and Geeraerts, 1998). The continuous curve represents the master reference data from Paris (Daly and Le Goff, 1996): it is the same as Fig. 6.6, but with declination and inclination plotted together on a single diagram. Notice, for example, the two inclination minima at AD 200 and 1400. The archeomagnetic direction obtained from the kiln (shown by the 95% confidence circle, shaded) indicates that it was in use about AD 1000.

that, as with the historical observations, the pattern of declination and inclination variations observed at any given site is not representative of the entire Earth. This means that any potential chronological control requires a calibrated reference secular variation curve for the area of interest. An example is given in Fig. 6.7. This concerns a study by Hus and Geeraerts (1998) of a quicklime kiln in a rural Gallo-Roman settlement near Nivelles in Belgium (4.30°E, 50.63°N). The direction of the remanence imparted at the time the kiln was last used (the effects of all prior heatings are eliminated upon reheating) implies an age close to AD 1000, whereas all the archeological evidence indicates that the site was occupied seven to nine centuries earlier. It is a clear case of systematic vandalism—the kiln was used to produce lime from "recuperated limestone building materials" as Hus and Geeraerts delicately put it.

Lake sediments have been exploited as recorders of the secular variation at least since the 1940s (Ising, 1943; Johnson *et al.*, 1948) and an enormous corpus of data has now been amassed. Some representative examples from around the world are mentioned in a different context in Chapter 7 (see Table 7.1). One of the best-constrained results was obtained from lakes in Britain and is illustrated in Fig. 6.8 (Turner and Thompson, 1982). Points of particular note on the declination curve are the westerly (negative) extremum near the top of the record and the easterly extremum about a millennium earlier. These correspond to the historically observed peak in about AD 1820 and to the archeomagnetically determined swing at about AD 1000 (see Figs. 6.4, 6.6, and 6.7). This sort of agreement lends support to the validity of the rest of the lacustrine record, which thus extends knowledge of the secular

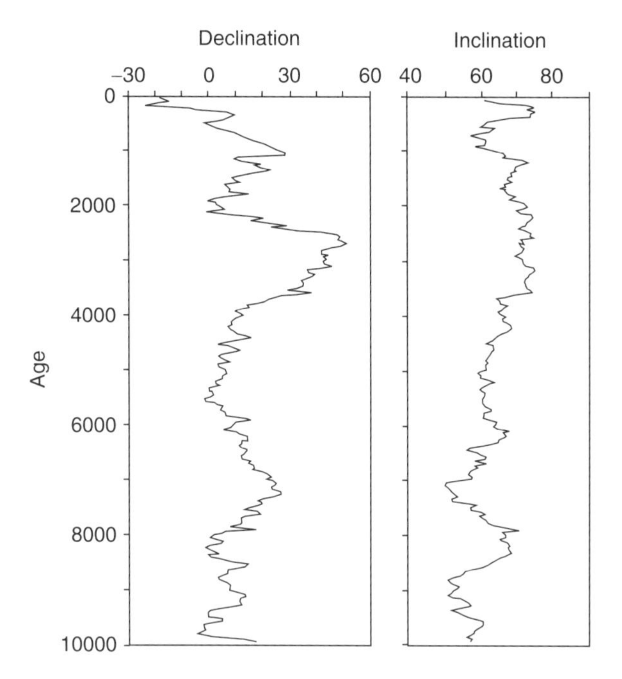

Figure 6.8 Secular variation in declination and inclination determined from sediments in three British lakes (Loch Lomond in Scotland, Llyn Geirionydd in Wales, and Lake Windermere in England). Age is in years before present (BP). Positive declination is east. (Modified from Turner and Thompson, 1982.)

variation through the entire Holocene (0–10,000 years BP). Once again, data of this kind are of considerable value to geomagnetists, but their potential for chronological control has been very little used. Two important exceptions are described by Williamson *et al.* (1991) and Rolph *et al.* (1996).

In the first case, Williamson *et al.* (1991) studied two piston cores from Lake Tanganyika and successfully exploited oscillations in declination and inclination, particularly the latter, to correlate between the two cores and, more important, to tie their record to that from Lake Barombi Mbo (Cameroon) and also to the European record. This enabled them to show that the sediments in question spanned the interval 5000–25,000 years BP. In the second case, Rolph *et al.* (1996) used the secular variation patterns determined from British and French lakes to establish a chronology reaching back some 30,000 years for their otherwise poorly constrained records from Lake Albano (Italy). In both cases, the magnetically derived chronologies were instrumental in allowing paleoclimatic and paleoenvironmental events to be placed in their proper time frame.

6.2.3 Geomagnetic Excursions

As pointed out earlier, the exact nature of geomagnetic excursions is not presently understood; they may represent short intervals of anomalously high secular variation (possibly related to times of anomalously low field strength), they may be "failed" reversals, or they may be an entirely different type of phenomenon. Nevertheless, if they were convincingly found throughout the world — or even significant portions of it — they could serve as useful time markers. Dating techniques are usually not precise enough to offer much control over the actual duration of each excursion. There is universal agreement, however, that they are "short" — which is generally taken to mean less than 10,000 years. Champion *et al.* (1988) argued for the existence of 10 such features within slightly more than the last million years, but subsequent investigations have cast doubt on some of these. Champion and his coauthors originally claimed that the features they identified were genuine reversals and classified them as subchrons, but the current consensus is that this is misleading. Accordingly, most workers prefer to call them excursions. Like the "events" of an earlier era, these excursions have been given geographic names related to the discovery locations. A critical assessment of global data for the Brunhes chron is given by Langereis *et al.* (1997). The excursions that are most observationally secure are indicated in Fig. 6.9. Several other excursions have been suggested (and named), some — or all — of which may eventually be permanently added to the list. This is very much a developing field.

Progress in this area is again exemplified by the $^{40}Ar/^{39}Ar$ ages reported by Singer *et al.* (1999) for basaltic lavas in Tahiti and Hawaii. They argue that two new excursions can now be resolved in the late Matuyama chron, the Punaruu excursion at 1.105 ± 0.005 Ma (between the Cobb Mountain excursion and the onset of the Jaramillo subchron) and the Santa Rosa excursion at 0.922 ± 0.012 Ma (between the termination of the Jaramillo subchron and the Kamikatsura excursion). It should be noted that Singer *et al.* (1999) do not use the term "excursion," preferring instead to use the term "event" in the neutral sense intended by Jacobs (1994, p. 87). In view of its earlier usage (see Box 6.1), this is potentially confusing.

The identification of a particular excursion in a sequence of basaltic lava flows might be possible because thermoremanence is generally very stable. A good example is the Laschamp excursion, originally discovered in the volcanic rocks of the Chaîne des Puys in France (Bonhommet and Babkine, 1967). However, in sediments (which, for enviromagnetic studies, are much more useful than lava flows) there are several perturbing factors. Compaction, bioturbation, and — in cold regions — cryoturbation can all degrade the original signal, but the overriding universal difficulty arises from the so-called *lock-in mechanism* by which the remanence is fixed in the sediment (see Chapter 5). The point is that locking does not happen immediately upon deposition; it requires a certain amount of overburden to accumulate for the sediment to become sufficiently compacted that grain rotation ceases. This may lead to short-lived features being attenuated, or entirely suppressed, as a result of the smoothing accompanying finite lock-in. This is the explanation favored by Vlag *et al.* (1996) for the fact that the Laschamp excursion itself is found in Lac St. Front (Massif Central, France) but not in Lac du Bouchet, which is only 40 km away. They argue that the effect of pDRM smoothing is stronger at Lac du Bouchet than at Lac St. Front because of differences in sedimentation rates ($\sim$25 cm/kyr for Lac du Bouchet compared with $\sim$40 cm/kyr for Lac St. Front).

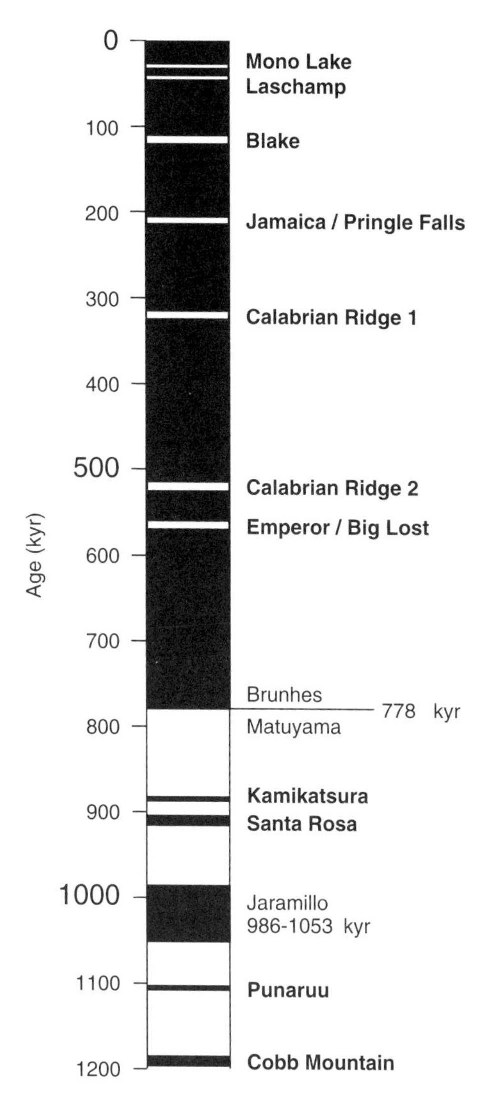

Figure 6.9 Geomagnetic excursions. Above the boundary between the Matuyama and Brunhes chrons, the pattern shown is based on that proposed by Langereis *et al.* (1997), whereas that below follows Singer *et al.* (1999), including their dates for the onset and termination of the Jaramillo subchron. The evidence for each proposed excursion is not always compelling, and it is possible that some of them will not stand the test of time. Several other excursions have been proposed by various authors but have not gained widespread acceptance. Note also that, even for the more robust cases, there are sometimes alternative names.

Even for slower changes, lock-in has a systematic effect. It causes a delay in the magnetic recording leading to a downward displacement of the feature in question, be it an excursion, a secular variation signal, or a polarity reversal. If the lithology is constant, this problem would not be too serious, but where the lithology varies very complex patterns can result (Bleil and von Dobeneck, 1999). This seems to have happened in the important magnetoclimatological sections of China, for example. The Blake excursion (still called by many workers the Blake event) is identified at some sites but not at others; and where it is seen, it sometimes gives a signature involving four polarity jumps, sometimes six (Evans and Heller, 2001). Such complexity may provide sedimentological information but is not a reliable means of investigating genuine geomagnetic behavior. At present, it is best to adopt a cautious attitude toward the use of chronologies based solely on geomagnetic excursions.

6.2.4 Geomagnetic Intensity Fluctuations

An observer at any point on the Earth's surface would, over a period of several centuries, record smooth changes in all three elements of the local geomagnetic field. That is to say, not only would the declination and inclination change but so also would the strength of the field. Together, the changes in these three elements constitute the secular variation. An example was given earlier from a pottery kiln in Pompeii which indicated that the field in Italy had significantly decreased over the last two millennia. Many results (both archeological and geological) from around the world indicate that such fluctuations are both real and widespread. As with the Pompeii kiln, these invariably involve material that carry a thermoremanent magnetization (TRM). This is important because it allows the experimenter to mimic in the laboratory the mechanism by which the material under investigation originally acquired its magnetization. Several problems arise, but to a first approximation it merely requires the comparison of the sample's natural intensity of magnetization with that which it acquires in a known laboratory field. However, for the environmental applications dealt with in this book, TRM-carrying material is of little use. Kilns and lava flows give only instantaneous "snapshots," whereas what is needed are long, continuous sedimentary sequences carrying depositional remanence (DRM, see Box 5.1). These, of course, offer the possibility of good time coverage. But, unlike the situation with TRM, the experimentalist is plagued by the virtual impossibility of mimicking the precise conditions under which the material in question was originally magnetized (e.g., What was the water depth? Was the water turbulent? Did bottom-dwelling organisms stir up the sediment?).

Tauxe (1993) summarizes the theoretical and experimental work germane to paleointensity determination from sediments, starting from the classic work of Johnson *et al.* (1948). Despite the many complicating factors, she concludes that "under certain conditions likely to occur in nature, there is a linear relationship between field and magnetization in sediments." The overriding difficulty arises from compositional variability of the sediments themselves. If, for example, a sedimentary layer contains twice as much magnetic material as the layer below, it will be twice as strongly magnetized (all other things being equal). To have any hope of recovering the actual strength of the geomagnetic field in which these sediments formed, it is necessary to allow for the variations in magnetic content. The problem, therefore, is one of normalization. Even if the search for one, or more, normalizers is successful, it is hardly ever possible to obtain an absolute value for the field. Most workers have therefore settled for determining *relative paleointensities*. The idea, then, is to divide the natural remanent magnetization (NRM) by such normalizers as susceptibility (χ), anhysteretic remanent magnetization (ARM), and isothermal remanent magnetization (IRM). As Tauxe (1993) remarks, "where agreeement is found, investigators can take heart."

Agreement was, indeed, found by Meynadier *et al.* (1994), who applied these three normalizers to sediments from the equatorial Indian Ocean spanning the last 4 million years (a broad temporal framework established, incidentally, on the basis of polarity reversals). The correlation between the various normalized records is always strong ($r > 0.87$). Stoner *et al.* (1995a) obtained similar results from cores in the Labrador Sea spanning the last 110,000 years. Encouraged by their results, Meynadier *et al.* (1994) proposed what they referred to as "a first paleointensity timescale for future stratigraphic studies." This marks an important milestone — if we do not forget to qualify the paleointensities as being "relative." The task of bringing together a global comparison of many data sets has been started by Guyodo and Valet (1996), who compile 17 relative paleointensity records spanning the last 200,000 years. From these, they construct a synthesized pattern of geomagnetic fluctuations which they label Sint-200 (Fig. 6.10). As part of the argument for the validity of their curve, they point out that the low relative field values at 40, 100–110, and 190 ka correspond to the Laschamp, Blake, and Jamaica excursions (see Fig. 6.9). As far as dating control over the last 200,000 years is concerned, their overall conclusion is that Sint-200 provides a reliable correlation/dating tool for geomagnetic features longer than 10,000 years. Once again, however, caution is in order — look, for example, at the difference in the apparent timing of the Jamaica/Pringle Falls excursion in Figures 6.9 (~210 kyr) and 6.10 (~190 kyr).

More recently, the so-called NAPIS-75 record (North Atlantic paleointensity stack for the last 75 ka) leads Laj *et al.* (2000) to argue that "millennial-scale variability of paleointensity records can be used as a global-scale correlation tool," although they warn that this will be "restricted to cores with high sedimentation rate and uniform magnetic mineralogy." On the latter point it is perhaps advisable to err on the side of caution, but it is nevertheless worth pointing out that Haag (2000) argues that "reliable relative paleointensity trends" can sometimes be obtained from sediments that are strongly inhomogeneous. Her data come from a deep-sea

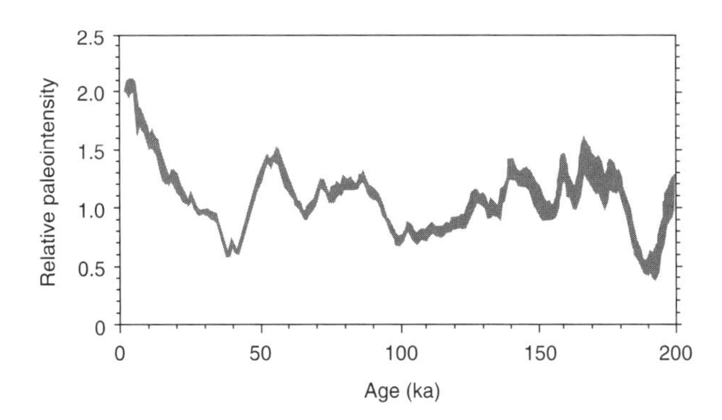

Figure 6.10 Relative paleointensity changes of the geomagnetic field for the last 200,000 years. This curve is the result of stacking 17 individual records from several ocean basins. Guyodo and Valet (1999) have now extended their analysis to cover 33 records spanning the last 800,000 years. (Redrawn from Guyodo and Valet, 1996.) © Elsevier Science, with permission of the publishers.

core collected in the central Atlantic off the coast of Mauritania (17°06.45′W, 25°16.8′N) as part of the SEDORQUA (SEDimentation ORganique marine et changements globeaux au cours du QUAternaire) project. The sediments studied span the interval 0–220 kyr and exhibit variations of a factor of 10 and 25 in natural remanent magnetization (NRM) and anhysteretic remanent magnetization (ARM), respectively.

In summary, it is clear that relative paleointensity investigations (both completed and ongoing) are contributing significantly to our knowledge of the temporal behavior of the geomagnetic field, but their status as a geological clock is still uncertain. The situation, then, is rather like that which existed for the GPTS itself in the 1960s.

6.3 OXYGEN ISOTOPE STRATIGRAPHY

Subtle variations in the measured isotopic composition of certain organic remains in marine sediments have become extremely important in the earth sciences. They do not directly yield a geological age in the same way that radioactive decay schemes do because the isotopes of oxygen are stable. However, coherent patterns of isotopic variations observed in sediments throughout the world's oceans are now so well established that they effectively constitute an independent dating method. The story begins with a suggestion made in 1947 by Harold Urey (1893–1981, Nobel Prize in Chemistry, 1934), that the isotopic ratio of oxygen ($^{18}O/^{16}O$) in the calcium carbonate ($CaCO_3$) of fossil shells could be used as a kind of paleothermometer. This fascinating idea derives from the fact that the two types of atoms are deposited from aqueous solution at different rates because one weighs more than the other. Furthermore, the entire process depends on the temperature of the surroundings. Urey's proposed technique thus relies on temperature-dependent isotopic fractionation. Obviously, the ability to determine temperatures in the remote geological past promised to be a great boon to paleoclimatologists. Within a few years, the basic validity of the technique was verified and the most useful marine organisms identified (Epstein *et al.*, 1953; Emiliani, 1955). These turn out not to be the typical seashells reminiscent of a vacation at the beach but rather the skeletons (or tests) of simple protozoans of the order *Foraminifera*, informally referred to as forams (Fig. 6.11). The disadvantage of the small size of these organisms (typically a few hundred μm) is offset by their great abundance in many parts of the world's oceans, where they constitute a large part of the so-called biogenic ooze. Indeed, as Haynes (1981) points out in his comprehensive textbook, "no other group of fossils is more important to the geologist, and the majority of paleontologists employed in industry and oil exploration are specialists in Foraminifera."

As is so often the case, an elegant initial concept is found to be complicated in practice by many factors (for example, the salinity and depth of the water in which the organisms lived) that threaten to sink the whole enterprise. The quest to overcome these difficulties ultimately delivered an even bigger payback — one that Bradley (1999, p.199) refers to as "the single most important record of past climatic variations for the entire Cenozoic." It was discovered that the $^{18}O/^{16}O$ ratio reflects

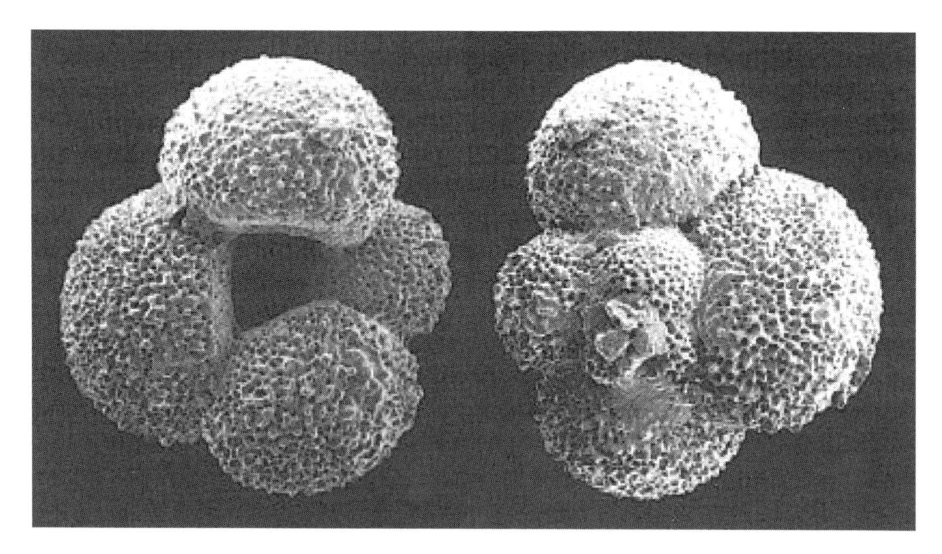

Figure 6.11 A typical foram (*Globigerina bulloides*); its diameter is about a third of a millimeter. A modern database is available at http://www.nmnh.si.edu/paleo/foram/. Nowadays, such images are usually obtained with a scanning electron microscope (SEM). Historically, however, a great deal was achieved with hand-drawn images obtained from ordinary light microscopes. In this regard, it is still worth consulting the report of the foraminifera collected during the H.M.S. *Challenger* expedition (1873–1876) by the English zoologist Henry Bowman Brady (1835–1891) (Brady, 1884).

the total amount of water locked up in ice sheets and thus provides an *ice volume* and *sea level* indicator. During glacial periods, the polar ice sheets expand and sea level consequently falls — by as much as 130 m at the time of the last glacial maximum some 18,000 years ago. The "extra" water necessary for the ice sheets to grow is provided by a net transport of atmospheric water vapor from lower to higher latitudes. Any precipitation produced en route is preferentially enriched in ^{18}O (which condenses more readily because of its lower vapor pressure). As the air masses involved move poleward, they therefore become isotopically lighter and lighter. The overall result is that the ice sheets are enriched in ^{16}O and the ocean water is enriched in ^{18}O. Every meter of sea level drop leads to a relative ^{18}O enrichment of the remaining water by approximately 1 part in 10^5. Such values are usually expressed in parts per mil and written in the form $\delta^{18}O$ (e.g., if during a glacial period sea level falls ~ 100 m, the oceans become $\sim 1‰$ heavier). Subsequent melting in the ensuing interglacial period restores the ocean to its original state. The composition of the carbonate in the marine fauna is determined by the composition of the surrounding seawater and therefore provides a history of the waxing and waning of the ice ages. Unlike temperature — which is subject to local variations — the isotopic ratio of the oceans is a truly global yardstick.

This much was already understood in the late 1960s. What was urgently needed was a reliable timescale to which the accumulated data could be referred. This is where the magnetism comes in. In addition to measuring the isotopic values of foram tests in deep-sea sediments, Shackleton and Opdyke (1973) determined the corresponding magnetic stratigraphy. For the first time in sediments for which both parameters had been studied, they identified the Matuyama/Brunhes boundary (MBB). In other words, the youngest part of the GPTS had been found in the sediments at the bottom of the ocean, giving further proof of the validity of the whole scheme, which had been independently established on the basis of basaltic igneous products in the underlying oceanic crust and on the continents. Because it was known when the Matuyama/Brunhes transition took place, the ice ages could finally be placed in a reliable chronology. Once deciphered, this "Rosetta stone" provided a sound basis for assessing past climatic fluctuations in terms of well-dated maxima and minima in the isotopic composition of seawater. These so-called *marine oxygen isotope (MOI) stages* are now the canonical framework with which all other paleoclimatic proxy results are routinely compared. Even-numbered stages correspond to glacial intervals. The interglacial in which we now live is labeled stage 1, with the previous interglacial corresponding to stage 5 (Fig. 6.12). Stage 3 represents warmer (but not fully interglacial) conditions sandwiched between two colder intervals: it ranks as an *interstadial.* Originally, the sequence of MOI stages numbered no more than a dozen or so, but Shackleton and Opdyke (1973) extended this to 22. Currently, the sequence can be traced back more than 2.5 million years to beyond stage 100. This represents the collective work of a very large number of individuals and research groups, but vital steps along the way were played by two particular projects: CLIMAP (Climate: Long-range Investigation, Mapping And Prediction, see, e.g., McIntyre *et al.*, 1976; CLIMAP Project Members, 1984) and SPECMAP (SPECtral MApping Project, see, e.g., Imbrie *et al.*, 1984; Martinson *et al.*, 1987).

The establishment of the MOI stages revolutionized the whole subject of paleoclimatology. The success of the magnetostratigraphic clock brought two quite separate disciplines together. In the first instance, the magnetists restricted their investigations to *remanent magnetization* (see Chapter 2) because this is what carries the magnetostratigraphic signal. Having made the plunge, however, they soon began to look at other properties. In particular, it often appeared that *magnetic susceptibility* (see Chapter 2 again) was related to climatic variations. Since the 1980s, a variety of

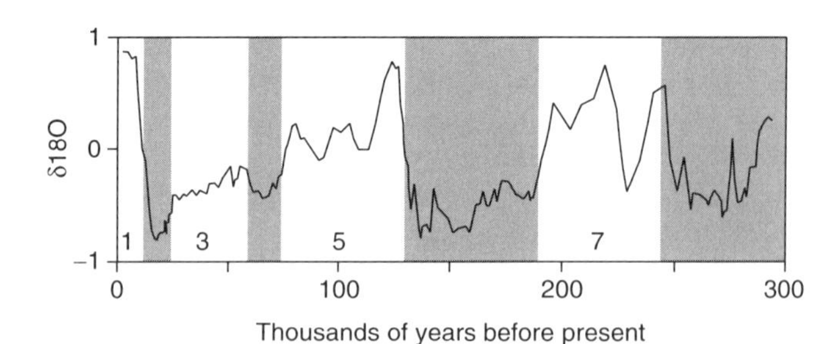

Figure 6.12 Marine oxygen isotope (MOI) stages for the last 300 ka simplified from Martinson *et al.* (1987) (see also http://www.ngdc.noaa.gov/mgg/geology/specmap.html). Odd-numbered (warm) stages are labeled, even-numbered (cold) stages are shaded. The (normalized) curve shown results from the stacking of several records from different ocean basins around the world. © Elsevier Science, with permission of the publishers.

natural geological archives — not only marine sediments — have been investigated by magnetoclimatological techniques. Several examples are dealt with in Chapter 7.

6.4 MILANKOVITCH CYCLES

The discussion concerning past ice ages has a long and checkered history in the scientific literature. Indeed, it was a topic that already interested the Scottish geologist James Hutton (1726–1797) — often regarded as the "father of geology" — in the latter part of the 18th century. The outstanding proponent was the Swiss geologist Louis Agassiz (1807–1873). Starting in 1837, he was for many years indefatigable in his efforts to persuade the leading earth scientists of the reality of an extensive Pleistocene ice sheet covering much of Europe and North America. And yet it has only been during the last few decades that any single theory of what actually causes ice ages has been widely accepted. In its modern form, this theory is due to the Serbian mathematician Milutin Milankovitch (1879–1958). It derives from the idea that slight changes in the Earth's orbital motion cause subtle variations in the distribution of solar energy *(insolation)* arriving at the top of the atmosphere. The orbital changes themselves are caused by the gravitational effects on the Earth of all the objects in the solar system, in particular, the sun and Jupiter. The whole theory — from gravitational forcing to climatic response — is referred to in several ways: *Milankovitch theory, astronomical theory*, and *orbital theory* among them. There is now a very extensive literature on this topic. Milankovitch (1941) summarizes his own work on the problem, starting in 1911. An excellent historical review is given by Berger (1988). The central prediction is that the pattern of insolation changes is driven by variations in three distinct orbital parameters. These are the *eccentricity* of the Earth's orbit, the *obliquity* of the Earth's axis, and the *precession* of the equinoxes (see Box 6.2). Gravitational calculations have succeeded in working out the changes in insolation caused by these astronomical elements for the last few million years (Berger and Loutre, 1991) and for the future (Berger *et al.*, 1991). For our purposes, the important result is that each of the three parameters is associated with a characteristic periodicity. In round figures, these are 100,000 years for the eccentricity, 40,000 years for the obliquity, and 20,000 years for the precession; collectively, these are generally referred to as *Milankovitch cyclicities*.

The topics discussed in this chapter are very diverse, ranging from astronomy to zoology, but always with an emphasis on geological time. In this regard, the interplay between polarity reversals, oxygen isotopes, and Milankovitch theory has been richly rewarding. This is nowhere better illustrated than in the outstanding contribution of Shackleton *et al.* (1990) wherein a convincing case for revising the GPTS itself is established on the basis of the synergy between the three methods. Ocean Drilling Program (ODP) site 677 (83°44′W, 1°12′N) provided them with excellent material for high-resolution isotope analysis, the results of which they matched to the pattern of insolation variations deduced from astronomical calculations. Knowledge of where the GPTS boundaries occur in marine sediments prompted them to suggest that the "currently adopted radiometric dates for the Matuyama/Brunhes boundary,

Box 6.2 Milankovitch Theory

Gravitational interactions with the other bodies in the solar system cause small, recurring changes in the geometry of the Earth in space, which lead to fluctuations in *insolation* (the amount and distribution of solar radiation reaching the Earth). It is possible to work out the details — at least for the last few million years — and it is now firmly established that there are three main ingredients, each related to a different orbital parameter. The slowest one (with a quasi-period of ~100,000 years) involves changes in the shape of the Earth's orbit (the *eccentricity e*), which varies between ~0 (almost circular) and ~0.05. Currently, the eccentricity is such that at *perihelion* (closest approach to the sun, on January 3), the top of the Earth's atmosphere receives ~3.5% more solar radiation than the annual average (and vice versa at *aphelion*, July 5). The seasons are controlled by the tilt (*obliquity t*) of the Earth's spin axis. Currently, this is ~23.5°, but it fluctuates between 21.8° and 24.4°, with a quasi-period of ~41,000 years. When the tilt increases, more radiation is received in polar regions during summer and less in winter. Finally, the spin axis undergoes a wobbling motion (*precession p*) like that of a spinning top, with a quasi-period of ~20,000 years (actually, the precession is characterized by two very similar quasi-periods of ~19,000 and ~23,000 years). This leads to a gradual change in the exact position in the Earth's orbit where the equinoxes and solstices occur. Right now, the northern (boreal) winter solstice occurs close to perihelion, but 11,500 years ago the northern summer solstice received the extra perihelion boost.

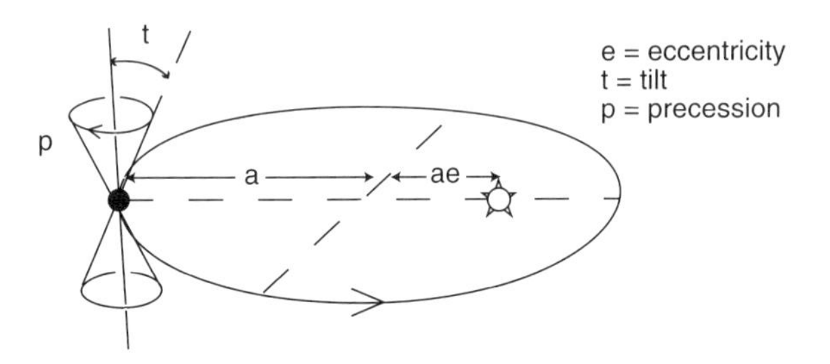

the Jaramillo and Olduvai subchrons, and the Gauss/Matuyama boundary underestimate their true astronomical ages by between 5 and 7%." This radical suggestion, which was dubbed the astronomical polarity timescale (APTS), was rapidly confirmed by a renewed effort to obtain high-quality radiometric dates on appropriate continental volcanic products (McDougall *et al.*, 1992; Spell and McDougall, 1992). What used to be the tail had succeeded in wagging the dog! As a result, using astronomical calculations in refining geological chronologies is now a widely adopted procedure.

The work of Hilgen (1991a,b) on the remarkable marl sequences in southern Italy represents an oustanding application of astrochronology to outcrops on land. He found that prominent cyclic sedimentation — mostly controlled by variations in carbonate content — could be convincingly tied to the pattern of insolation variations obtained by Laskar (1990). Individual strata could then be counted like tree rings and matched bed by bed to the astronomical fluctuations. This breakthrough has now provided timescales back to 12 million years that are more accurate and have a much higher resolution than previous geological timescales (Hilgen *et al.*, 1997). Sediments recovered in oceanic cores have also been found to provide excellent astrochronological archives. For example, the shipboard magnetic susceptibility profiles obtained at ODP site 926 correlate very well with the calculated insolation curves (Shackleton and Crowhurst, 1997) and provide a timescale accurate to the level of the individual precession cycle back to 13 million years (albeit with the occasional gap). This approach has now been extended to the Chinese Loess Plateau by Heslop *et al.* (2000), who have analyzed a number of data sets and carefully correlated them with the astronomical calculations of Laskar (1990). The data used are (1) the geomagnetic polarity stratigraphy at Baoji (Rutter *et al.*, 1991), (2) the detailed grain size profile at Baoji (Ding *et al.*, 1994), (3) the detailed magnetic susceptibility profile at Luochuan (Lu *et al.*, 1999a), and (4) the benthonic oxygen isotope profile from core ODP677 (Shackleton *et al.*, 1990). They succeed in linking together the fluctuations in global ice volume [reflected by (4)] and the variations in the strength of the summer and winter monsoons [reflected by (2) and (3)] with the whole scheme then tied to some 2.6 million years of geomagnetic polarity reversals. Thus are events in the solar system, the atmosphere, the hydrosphere, and the Earth's core brought neatly into a single chronology.

7

MAGNETOCLIMATOLOGY AND PAST GLOBAL CHANGE

7.1 INTRODUCTION

At first sight, it would seem that magnetism and climate are strange bedfellows. But the two were dramatically linked 30 years ago in a classic paper by Shackleton and Opdyke (1973), as we saw in Chapter 6. This early success of the magnetostratigraphic clock brought together two quite separate disciplines. At the beginning, the paleomagnetists involved restricted their attention to remanent magnetization because this is what carries the magnetostratigraphic signal. But, once involved, it was not long before they began to investigate other magnetic properties. In particular, magnetic susceptibility was often found to correlate with climatically controlled features (Kent, 1982). Subsequent work has expanded the magnetic parameters that can be used to decipher past global change; these now include laboratory-induced remanences (IRM and ARM), coercivity spectra, and the frequency dependence of susceptibility (Walden *et al.*, 1999). Furthermore, since the 1980s, a wide variety of geological archives have been found to render important magnetoclimatological information. These consist of sedimentary deposits preserved in several different settings—on the continents, in lakes, and in the oceans. In this chapter, we describe examples from each of these environments, paying particular attention to the geological record of the last few million years, for which the record is richest.

7.2 LOESS

7.2.1 Magnetic Enhancement

In the 1830s, the great British geologist Charles Lyell (1797–1875) traveled widely in Europe and America. Among his many other achievements, he can be credited with bringing the old German word "Löβ" into the English language, for which purpose he wrote it as "loess." Basically, it means "loose" and was used locally in the Rhine valley to describe the unconsolidated nature of certain geologically young unstratified "dirty yellowish-grey" deposits found there (Smalley *et al.*, 2001; Zöller and Semmel, 2001). Lyell himself described similar deposits in the Mississippi Valley, and we now know that Quaternary loess occurs throughout the globe (Pye, 1987). Particularly important are a discontinuous belt stretching from central Europe to China and the occurrences in Alaska and Argentina (Fig. 7.1). Loess is windblown dust winnowed off former periglacial areas and semiarid desert margins. It is produced by various processes in two types of source region: (1) glacial grinding, frost action, and fluvioglacial abrasion in glacial areas (Smalley, 1966) and (2) frost action and salt weathering in cold, dry deserts (Derbyshire, 1983). The resulting material consists of up to 90% quartz and feldspars with small amounts of clay and carbonate minerals. Grain size distributions are dominated by the silt fraction and generally peak in the vicinity of 30 μm (Fig. 7.2). During weathering, the feldspars are often replaced by clay minerals, particularly smectite and illite (Bronger and Heinkele, 1989). Loess invariably also contains minor amounts (on the order of 1% or less) of magnetic minerals (magnetite, maghemite, hematite, and goethite); because of the ease, speed, and nondestructive nature of most magnetic measurements, these have assumed a significance out of all proportion to the amounts actually present.

In north-central China, the blanket of loess commonly exceeds 100 meters (see Fig. 1.1) and in some places reaches 300 meters in thickness. During cold, arid glacial periods, the loess builds up steadily, but as interglacial conditions develop, increased precipitation and warmer temperatures promote *pedogenesis* (soil formation). This involves a complex cocktail of physical, chemical, and biological processes each of which is fascinating in its own right, as indicated in Chapter 5. Given the need to feed the world's ever-increasing population, the importance of understanding soil can

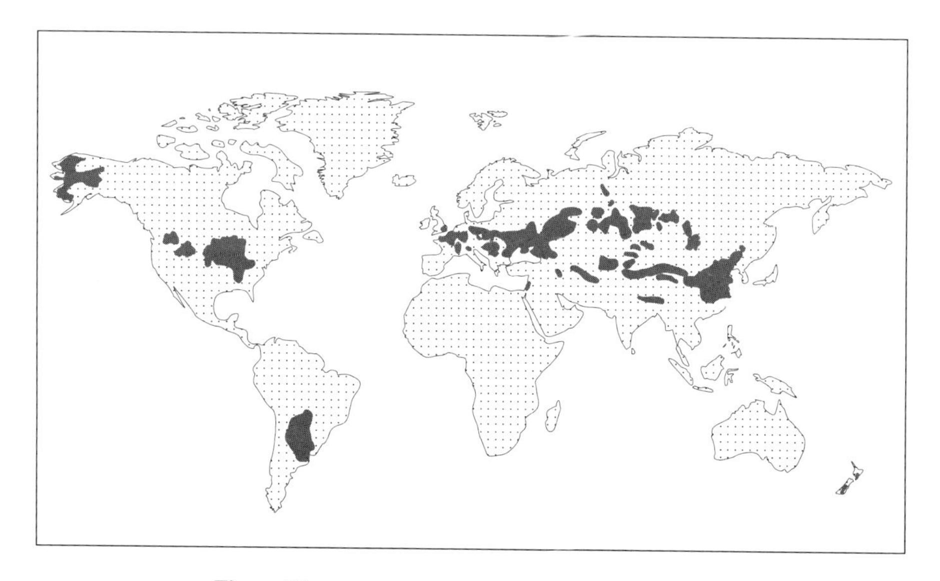

Figure 7.1 Global distribution of the major loess deposits.

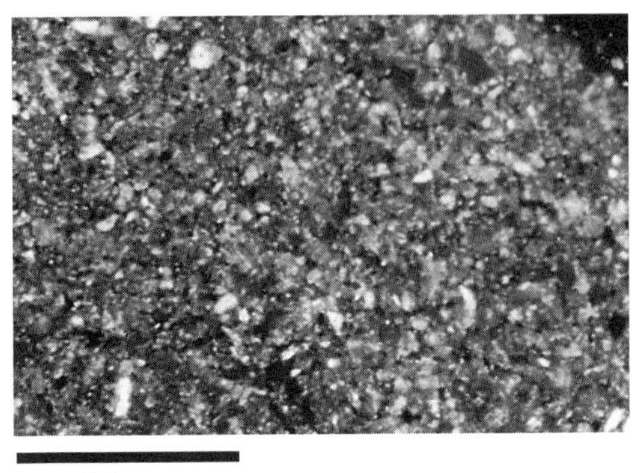

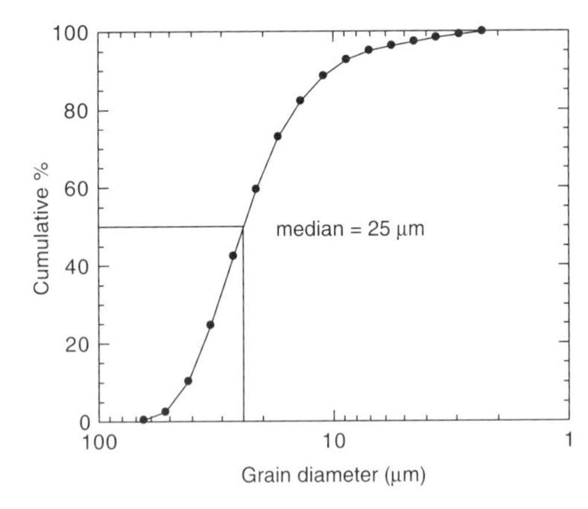

Figure 7.2 A closer look at typical loess (from China and the United States). The photomicrograph shows a sample from Xiagaoyuan in the Chinese Loess Plateau. The graph shows the cumulative grain size distribution of loess from Vicksburg, Mississippi (based on data compiled from Pye, 1987). See color plate.

hardly be overstated. Fitzpatrick (1986), for example, draws attention to this by speaking of the *pedosphere*—a thin, discontinuous global skin composed of air, water, minerals, and organisms that together constitute the multiple interface of atmosphere, hydrosphere, lithosphere, and biosphere.

The interglacial soils develop on preexisting loess and are, in turn, buried by loess brought in during subsequent glacial periods. Once buried, they become fossil soils, or *paleosols* (Fig. 1.1). An excellent example from Roxolany (30.4°E, 45.8°N) in the Black Sea area of Ukraine is given in Fig. 7.3; it has been thoroughly investigated by Tsatskin *et al.* (1998), who provide a highly informative paleopedological description and paleoenvironmental interpretation. Comparable studies have been carried out by several groups working on the magnificent outcrops so typical of the so-called Chinese Loess Plateau. These glacial/interglacial oscillations are clearly visible as alternating "yellow" and "red" strata. Indeed, the Chinese expression for loess is *huangtu* (yellow earth), and the Yellow River (Huang He) owes its color, and hence its name, to the great burden of eroded loess that it carries in suspension.

Impressive as they are, the loess/paleosol sequences of China stubbornly refused to yield up adequate evidence for their chronology to be worked out. All this changed

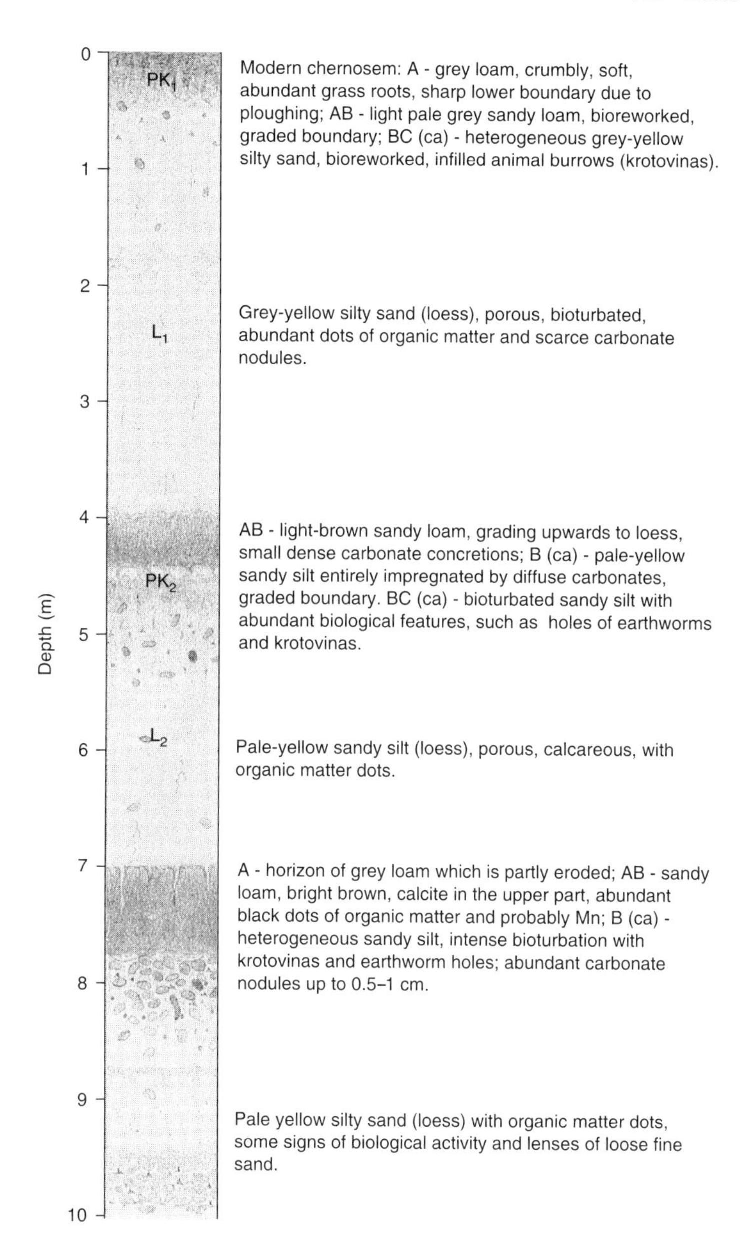

Figure 7.3 The uppermost 10 m of the loess/paleosol sequence at Roxolany (Ukraine). (Modified from Tsatskin *et al.* 1998.) © Elsevier Science, with permission of the publishers. See color plate.

about 20 years ago, when Heller and Liu (1982) succeeded in finding several major geomagnetic polarity reversals in a loess/paleosol sequence at the now famous site of Luochuan (109.2°E, 35.8°N). In addition to the Matuyama/Brunhes boundary

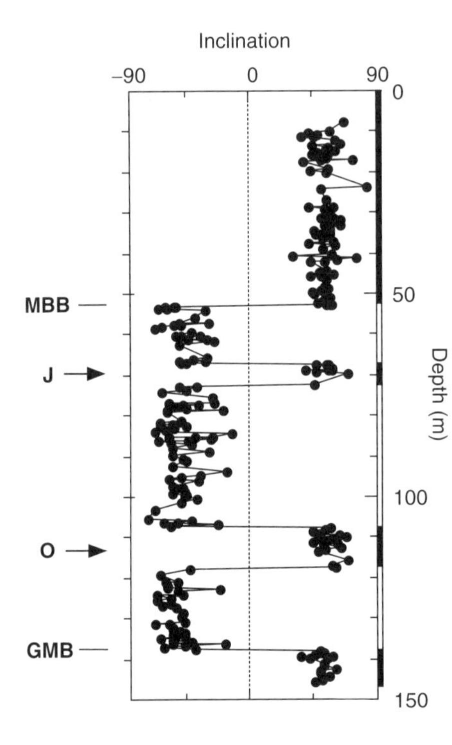

Figure 7.4 Geomagnetic polarity sequence at Luochuan, China. MBB = Matuyama/Brunhes boundary, GMB = Gauss/Matuyama boundary, J = Jaramillo subchron, O = Olduvai subchron. (Modified from Heller and Liu, 1982.) © Macmillan Magazines Limited. Reproduced with permission.

(MBB), their data revealed the Gauss/Matuyama boundary (GMB) and the Jaramillo and Olduvai subchrons (Fig. 7.4). Loess deposition in China was thus established as having been initiated more than 2.5 million years ago, with some 5 to 10 cm accumulating every thousand years. Subsequent investigations extended the record back to 7 million years (Ding *et al.*, 1999), and more recent work has pushed this back to 22 million years (Guo *et al.*, 2002).

Having set up the proper timescale by comparison with the geomagnetic polarity timescale (GPTS), Heller and Liu (1986) went on to measure the magnetic susceptibility. They found a close match with the marine oxygen isotope (MOI) stages and thus established the vital teleconnection between continental and oceanic proxy records of past global change. An excellent example is given in Fig. 7.5, which compares susceptibility data from another well-known site in China (Xifeng, 107.6°E, 35.7°N) with oxygen isotope results from samples cored from the equatorial Pacific Ocean [site 677 of the Ocean Drilling Program (ODP), 83°44′W, 1°12′N]. Also shown are the alternating loess and paleosol strata at Xifeng. The loess layers (labeled L_1, L_2..., starting from the top) exhibit a "background" susceptibility value (Forster *et al.*, 1994) as low as 25×10^{-8} m^3/kg, whereas the paleosols (labeled S_0, S_1...) are much more magnetic, with values exceeding 250×10^{-8} m^3/kg in the case of S_1. The correspondence between these two records, one related to the magnetism of dust in a continental setting, the other to the oxygen composition of microorganism skeletons in the deep ocean, is quite remarkable. It demonstrates that

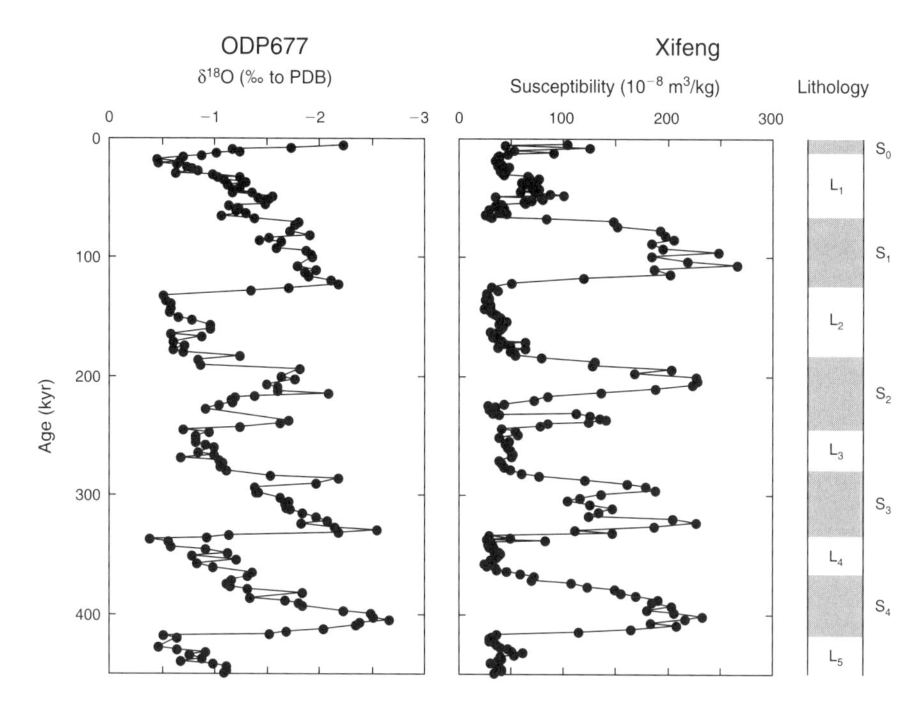

Figure 7.5 Magnetic susceptibility profile at Xifeng, China, compared with the oxygen isotope profile at ODP677. (PDB represents a standard reference material to which oxygen isotope measurements are compared.) The sequence of soils (S) and loess layers (L) at Xifeng is indicated on the right.

magnetoclimatology is a promising exercise, even though visual matching of "wiggly lines" is a tricky business.

As Edward Bullard (1907–1980) points out in his discussion of an earlier set of extremely important geophysical wiggly lines, the oceanic magnetic stripes, "similarity is not a simple idea, and is only partly metrical" (Bullard, 1968). He rejects a more objective approach using cross-correlation because slight variations in the spreading velocity of the ocean floor introduce enough mismatches to defeat the purpose. In sedimentary sequences, the same argument holds for the accumulation rate, with the added complication of possible stratigraphic breaks representing times of no deposition or even of erosion. Such problems can sometimes be overcome by a procedure known as *sequence slotting* (Thompson and Clark, 1989), which combines profiles together in an optimum way by minimizing the total path length obtained by summing all the (parameterized) distances between successive pairs in the combined sequence. Nevertheless, for the celebrated marine magnetic anomalies, Bullard himself opted for qualitative assessment, concluding that the agreement between the curves he was interested in was "so good that it seems impossible to ascribe it to chance." The same can be said of the curves illustrated in Fig. 7.5.

Regardless of the particular approach taken for the comparison of curves like those shown in Fig. 7.5, it is clear that pedogenesis in the Chinese loess has given rise to magnetic enhancement. Many sections throughout the Loess Plateau have now been studied in great detail, and the entire sequence of alternating loess and paleosols exhibits the same pattern (Fig. 7.6). It will be noted, however, that the degree of enhancement is by no means uniform. Some paleosols are represented by very strong

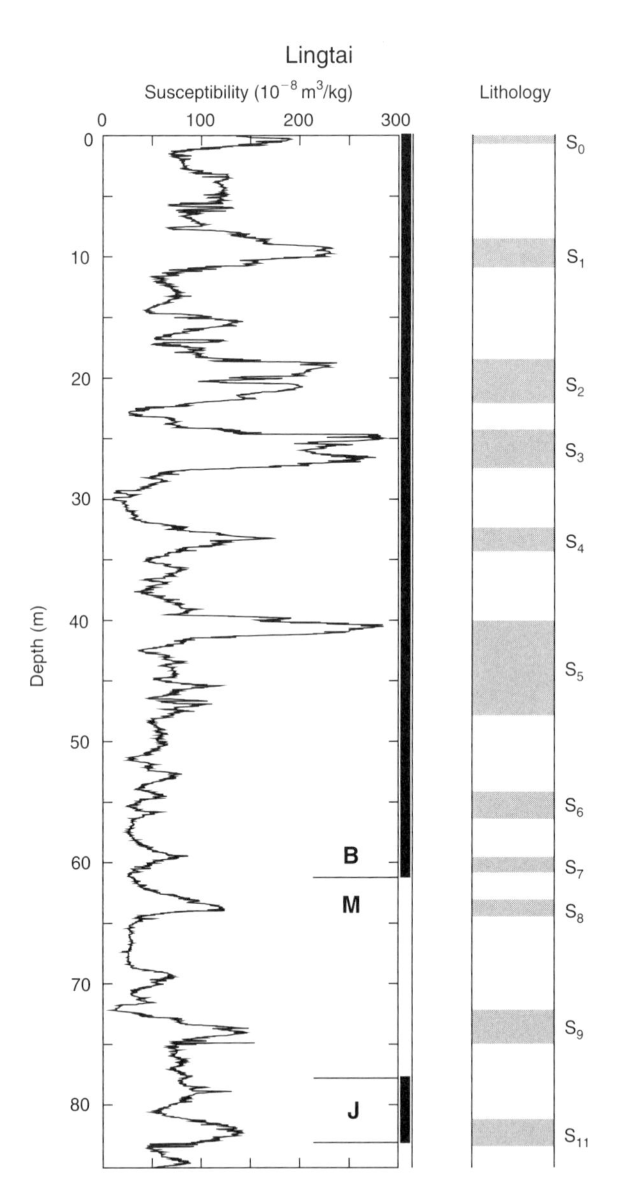

Figure 7.6 Magnetic susceptibility and lithological profiles at Lingtai, China. The susceptibility profile is based on more than 4000 samples representing continuous stratigraphic coverage. (Data kindly provided by Simo Spassov.)

susceptibility peaks, but others are quite subtle. Nevertheless, the overall pattern of magnetic enhancement in soils—albeit of variable intensity—is universally present over the whole plateau, an area in excess of 400,000 km^2. Taken completely in isolation, this already suggests that magnetic susceptibility is a useful climate proxy. But combined with the already established sequence of isotope stages, the notion becomes entirely convincing.

During glacial periods, the polar ice caps grow, sea level drops, and the oceans become isotopically heavier. In various continental areas, dust is blown off sparsely vegetated arid regions and deposited hundreds of kilometers downwind as loess. Some material may stay aloft much longer and travel farther afield. A well-documented example is reported by Begét *et al.* (1993), who identify mineral particles on the island of Hawaii that were evidently transported more than 10,000 km from Asia by late Quaternary wind storms. Similar material is deposited even farther afield in the oceans (Rea, 1994) or falls on—and is eventually incorporated into—ice sheets, as documented in the well-known ice cores from Vostok, Antarctica (Petit *et al.*, 1990), and GISP2, Greenland (Biscaye *et al.*, 1997). In the ensuing interglacial, the ice caps retreat, the ocean is restored to its original level and isotopic composition, and soils begin to form on the recently formed loess blanket. Soil formation is partly aided by increased atmospheric temperatures during interglacials, but, in China, the main driving force seems to be the behavior of the summer monsoon. Currently, the bulk of the mean annual precipitation over the Loess Plateau falls during July, August, and September when the summer monsoon brings its moisture-laden air masses from the Pacific Ocean, whereas the winter monsoon blows out of Siberia and is cold and dry. By contrast, when glacial conditions take over, the winter monsoon dominates for much of the year and the Loess Plateau becomes more arid, slowing down, or completely arresting, pedogenesis (Liu and Ding, 1998; Rutter and Ding, 1993; Fang *et al.*, 1999; Porter, 2001). The idea is illustrated diagramatically in Fig. 7.7.

Enhanced magnetism resulting from interglacial pedogenesis is now an indisputed observational fact, and because the Loess Plateau is the most striking example, it is now customary to speak of it in terms of a so-called Chinese model. As we shall see later, it is by no means the only model, but it does seem to operate in many parts of the globe. Figure 7.8 summarizes several important examples from Asia and Europe.

The preceding discussion focuses entirely on the last few million years of geological history, particularly the Quaternary. It is here that the record is best preserved and most easily interpreted, and for these reasons it is here that the vast majority of the relevant work has been carried out. Nevertheless, there are occasional reports of similar findings from geologically much older sequences. A case in point is a magnetic investigation of an upper Paleozoic loessite sequence (so called because the material involved is lithified and is therefore no longer a true unconsolidated loess) in the Eagle Basin, Colorado (Soreghan *et al.*, 1997). The section studied is 23 m thick and is characterized by fairly uniform susceptibility values between 2×10^{-8} and 4×10^{-8} m^3/kg interrupted by a few paleosol horizons yielding values of up to 17×10^{-8} m^3/kg. Petromagnetic analysis indicates that the observed enhancement

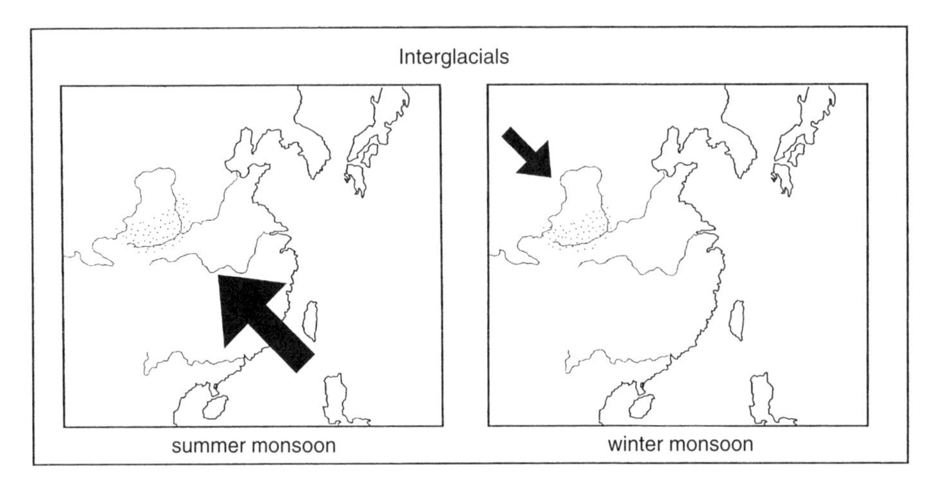

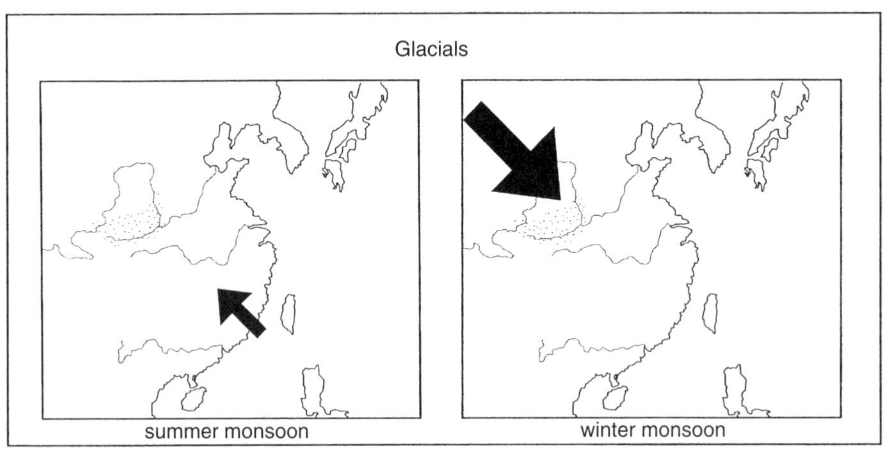

Figure 7.7 Schematic representation of summer and winter monsoon behavior over China during glacial and interglacial periods. The extent of the Chinese Loess Plateau is indicated by the dots.

parallels that found in the classic Chinese sites, and the authors therefore conclude that "highly resolved records of terrestrial paleoclimate" are potentially available from "very ancient rocks."

7.2.2 Minerals and Mechanisms

Given the striking correlation between magnetism and climate, the question of the underlying mechanism naturally arises. What is it about soil production on an existing ground surface consisting of loess that leads to enhanced magnetic susceptibility? The general aspects of this problem are discussed in Chapter 5 and a specific example, taken from the Chinese Loess Plateau, is illustrated in Box 5.4. Because the loess in China represents one of the most fully studied cases, it merits a closer look.

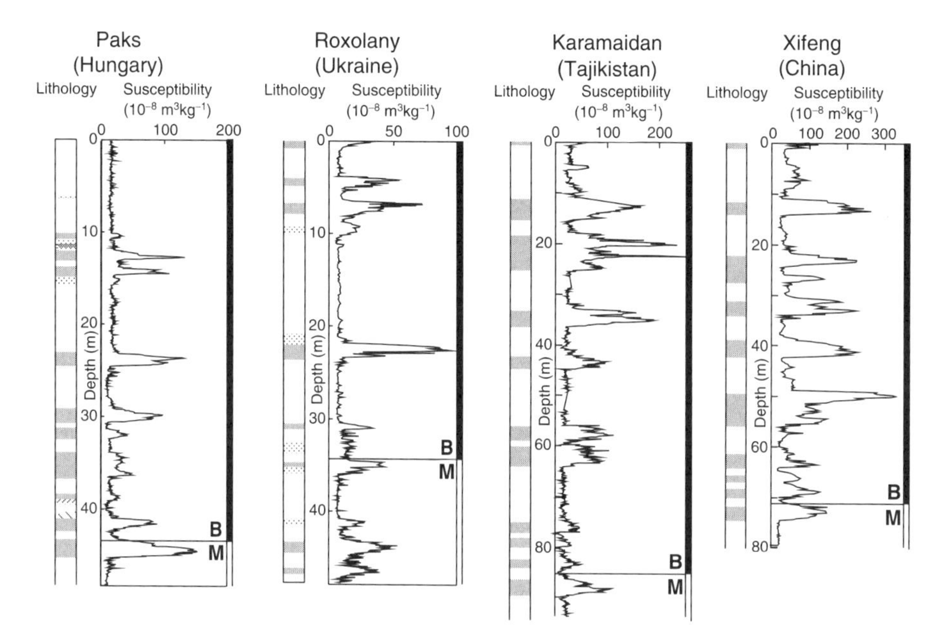

Figure 7.8 Representative magnetic susceptibility profiles at important sites in Europe, central Asia, and China. Paks (18.8°E, 46.6°N) (Sartori *et al.*, 1999), Roxolany (30.4°E, 45.8°N) (Tsatskin *et al.*, 1998), Karamaidan (69.4°E, 38.4°N) (Forster and Heller, 1994), Xifeng (107.6°E, 35.7°N) (Liu *et al.*, 1992).

The first important point to note is that the enhancement is not simply an increase in the amount of magnetic material. There is indeed a substantial increase due to the formation of new material; at Luochuan, for example, Heller and Liu (1984) find that the 159 soil samples they measured had an average susceptibility 2.4 times greater than the average obtained from 225 loess samples. But the increase is due to an ingredient which differs from that already present in the original loess. This can be deduced from a number of magnetic parameters but is particularly well demonstrated by the coercivity spectra obtained from isothermal remanence (IRM) experiments. Figure 7.9 shows the results obtained from the S_3/L_4 couplet at two sites in the Chinese Loess Plateau, one (Baicaoyuan) in the relatively cool, arid western area, the other (Luochuan) in the warmer, wetter central area where stronger pedogenesis has taken place. The original loess is the same at both sites, but the change due to soil formation is much greater at Luochuan. Notice also that at fields greater than 100 mT all four curves are essentially identical. This suggests that the magnetic properties can be explained by a two-component model. Component A comprises the magnetic material brought in as part of the original airfall loess; it is ubiquitous and is not modified by pedogenesis. Component B is the magnetic material created during soil formation; it is magnetically softer than component A, having no coercive forces beyond 100 mT.

This AB model provides a convenient framework for the discussion of magnetic enhancement (Evans and Heller, 1994; Mishima *et al.*, 2001), but we still have not

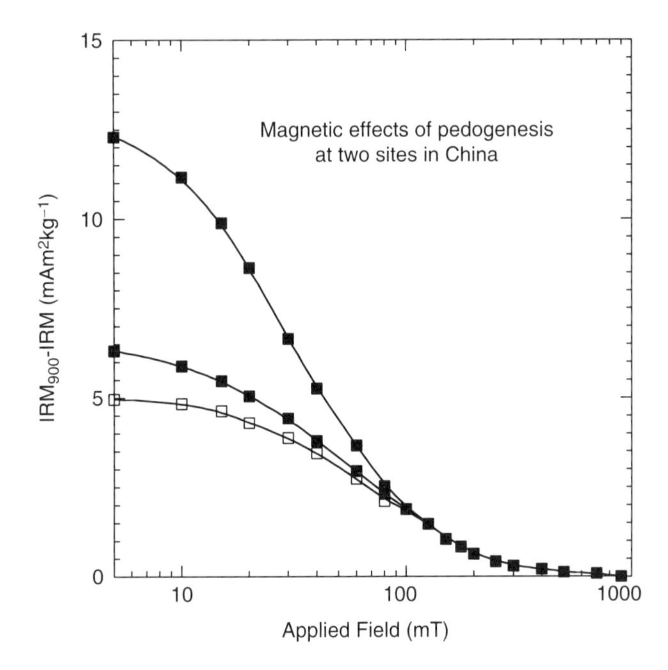

Figure 7.9 Coercivity spectra of IRM at two sites in the Chinese Loess Plateau. The upper and middle curves represent the S_3 paleosol at Luochuan and Baicaoyuan, respectively. The lower curve represents the parent L_4 loess at both sites. Note that all the curves coalesce at 100 mT with a common value of $\sim 2\times10^{-3}$ Am^2kg^{-1}. Present-day mean annual temperature and precipitation are 9°C and 620 mm for Luochuan and 6°C and 350 mm for Baicaoyuan. The IRM plotted is the difference between the IRM after exposure to a 900-mT field (the highest field available) and the IRM at the field in question. See also Boxes 7.1 and 7.2.

determined the actual minerals involved or their amounts. The way to do this is indicated in Boxes 7.1 and 7.2. The outcome is that the original airfall loess contains about 1% of ferromagnetic minerals, heavily dominated by hematite (α-Fe_2O_3) but with significant magnetic contributions from magnetite (Fe_3O_4) and maghemite (γ-Fe_2O_3). Pedogenesis leads to the production of ultrafine magnetite particles (<100 nm; about a thousand times smaller than the detrital windborne grains making up the pristine loess). This latter ingredient increases as the degree of soil development advances. At Baicaoyuan (present-day mean annual precipitation and temperature ~350 mm and ~6°C), the ferromagnetic content is enhanced by only 0.04% (by mass), but at Luochuan (~630 mm, ~9°C) four to five times as much is created. At a third site farther along the climatic gradient (Baoji: ~700 mm, ~13°C), a further doubling of the pedogenic addition takes place (Evans and Heller, 1994). These B-component estimates are based on measured susceptibilities, but essentially the same pattern emerges from the corresponding IRM curves, as indicated in Fig. 7.10.

Box 7.1 The AB Model (Component A)

Consider first the IRM of "pure" A material, $IRM^A = 5\times10^{-3}\,Am^2kg^{-1}$ (see Fig. 7.9). Data from standard reference samples indicate that hematite is the only likely contributor to the IRM curve beyond 100 mT, which amounts to $2\times10^{-3}\,Am^2kg^{-1}$. But hematite always has some "softer" coercivities, typically about 10%, below 100 mT. Thus we deduce a total hematite contribution of $2.2\times10^{-3}\,Am^2kg^{-1}$, which leaves $2.8\times10^{-3}\,Am^2kg^{-1}$ to be accounted for. For this, the most realistic candidates are magnetite and maghemite. Thermal demagnetization of the samples in question indicated the presence of both — in the ratio 70:30 (Evans and Heller, 1994). The saturation IRMs (= M_{rs} in Chapter 2) for hematite, magnetite, and maghemite can be estimated from corresponding typical values of M_{rs}/M_s of 0.5, 0.1, and 0.5 and the intrinsic M_s values of 0.48, 92, and $70\,Am^2kg^{-1}$, respectively. Thus the A-component mass fractions (m^A) of the three minerals are given by

$$m^A_{hematite} = 2.2\times10^{-3}\,Am^2kg^{-1}/0.24\,Am^2kg^{-1} = 0.92\%$$

$$m^A_{magnetite} = (0.7\times2.8\times10^{-3}\,Am^2kg^{-1})/9.2\,Am^2kg^{-1} = 0.021\%$$

$$m^A_{maghemite} = (0.3\times2.8\times10^{-3}\,Am^2kg^{-1})/35\,Am^2kg^{-1} = 0.0024\%$$

Considerable approximations are involved in these estimates, but they indicate that the magnetic fraction of the pristine loess is a mixture dominated by hematite and amounting to ~1% of the total mass. Although more hematite is present, do not forget that the other two minerals are much more magnetic per unit mass: of the total IRM^A ($5\times10^{-3}\,Am^2kg^{-1}$), hematite accounts for 44%, magnetite for 39%, and maghemite for 17%.

7.2.3 Milankovitch Cycles

As we saw in Chapter 6, the influence of Milutin Milankovitch on current ideas about global climatic changes is enormous. Even though he erred on many of the details, his contributions to the so-called astronomical, or orbital, theory remain at the very center of modern studies. For present purposes, the important result is that each of the three orbital parameters (eccentricity, obliquity, and precession) is associated with a characteristic periodicity. Strictly speaking, one should speak of these as quasi-periodicities because the gravitational interplay of the bodies in the solar system is sufficiently complicated that the resulting cyclicities are not exactly periodic in the true mathematical sense. Nevertheless, it has become common practice (at least in the initial stages of any study) to search for signals at ~100,000 years for eccentricity, at ~40,000 years for obliquity, and at ~20,000 years for precession.

Box 7.2 The AB Model (Component B)

Now let us look at component B — the magnetic material created by soil formation. The most important observations are that (1) thermal demagnetization curves are dominated by magnetite (Curie point near 580°C) and (2) the frequency dependence of susceptibility (~11%) indicates that very small grains near the SP/SSD threshold (~30 nm) are common. The latter point emphasizes the difference between components B and A (for which the measured frequency-dependent susceptibility is $<2\%$). Points (1) and (2) allow us to use the reference data for magnetite grains of known size (Maher, 1988), in particular a susceptibility value of $\sim 10^{-3}\,\mathrm{m^3/kg}$ for the mass-specific susceptibility. The samples from Baicaoyuan and Luochuan have measured susceptibilities of 0.65×10^{-6} and $1.97\times10^{-6}\,\mathrm{m^3/kg}$, respectively, but from these we must first subtract the measured susceptibility of the prepedogenic component A ($=0.29\times10^{-6}\,\mathrm{m^3/kg}$). Thus, we obtain B-component mass fractions (m^B) for the two sites,

$$m^{B}_{\mathrm{Baicaoyuan}} = 0.36\times10^{-6}\,\mathrm{m^3kg^{-1}}/10^{-3}\,\mathrm{m^3kg^{-1}} = 0.036\,\%$$

$$m^{B}_{\mathrm{Luochuan}} = 1.68\times10^{-6}\,\mathrm{m^3kg^{-1}}/10^{-3}\,\mathrm{m^3kg^{-1}} = 0.17\,\%$$

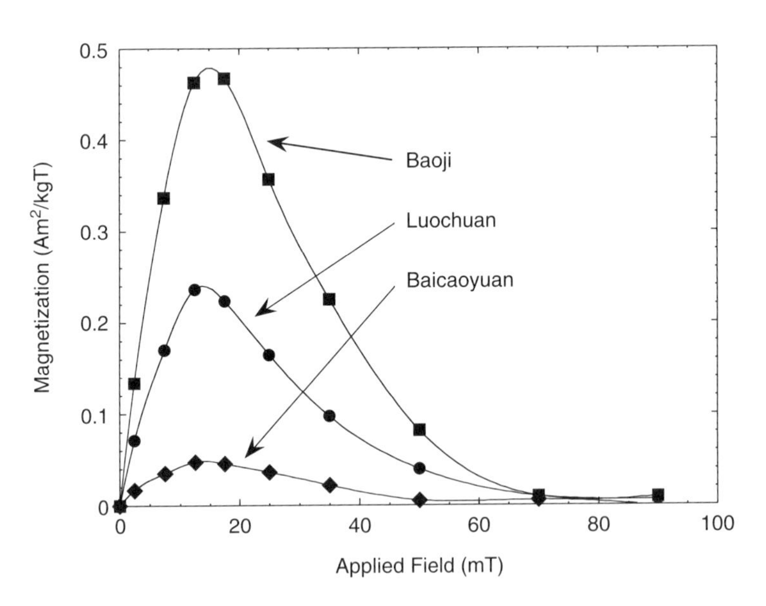

Figure 7.10 Differential IRM spectra at three sites in the Chinese Loess Plateau. These curves are the first derivative of standard IRM acquisition curves. All three curves have essentially the same shape but increase in magnitude along the climatic gradient from dry and cold (Baicaoyuan) to moist and warm (Baoji).

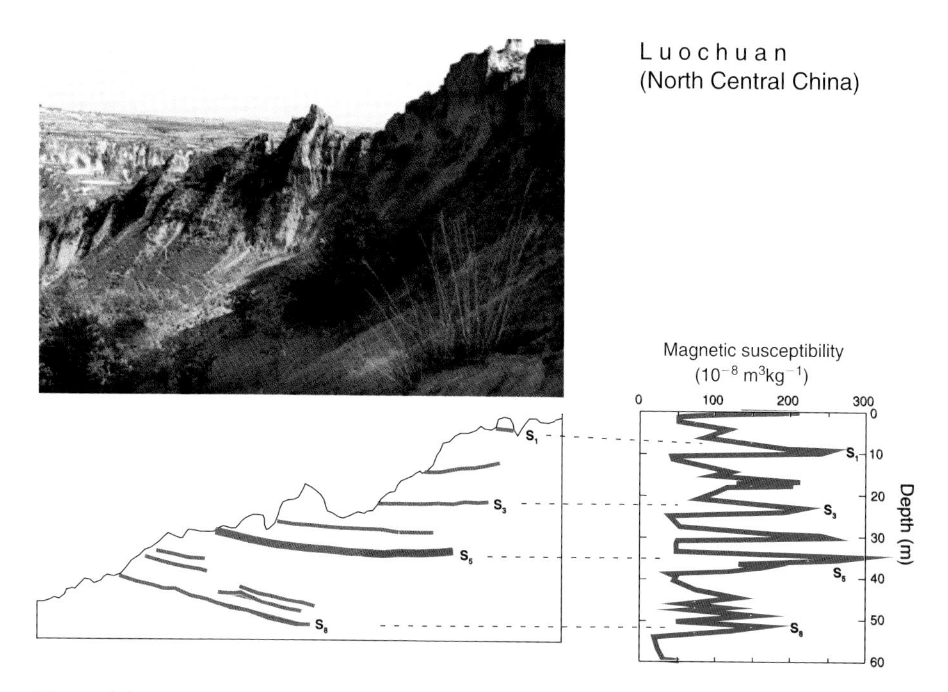

Figure 1.1 The famous sedimentary section at Luochuan, China. ***See page 3.***

Figure 6.5 The kiln at Pompeii studied archeomagnetically by Evans and Mareschal (1989) and Evans (1991). ***See page 122.***

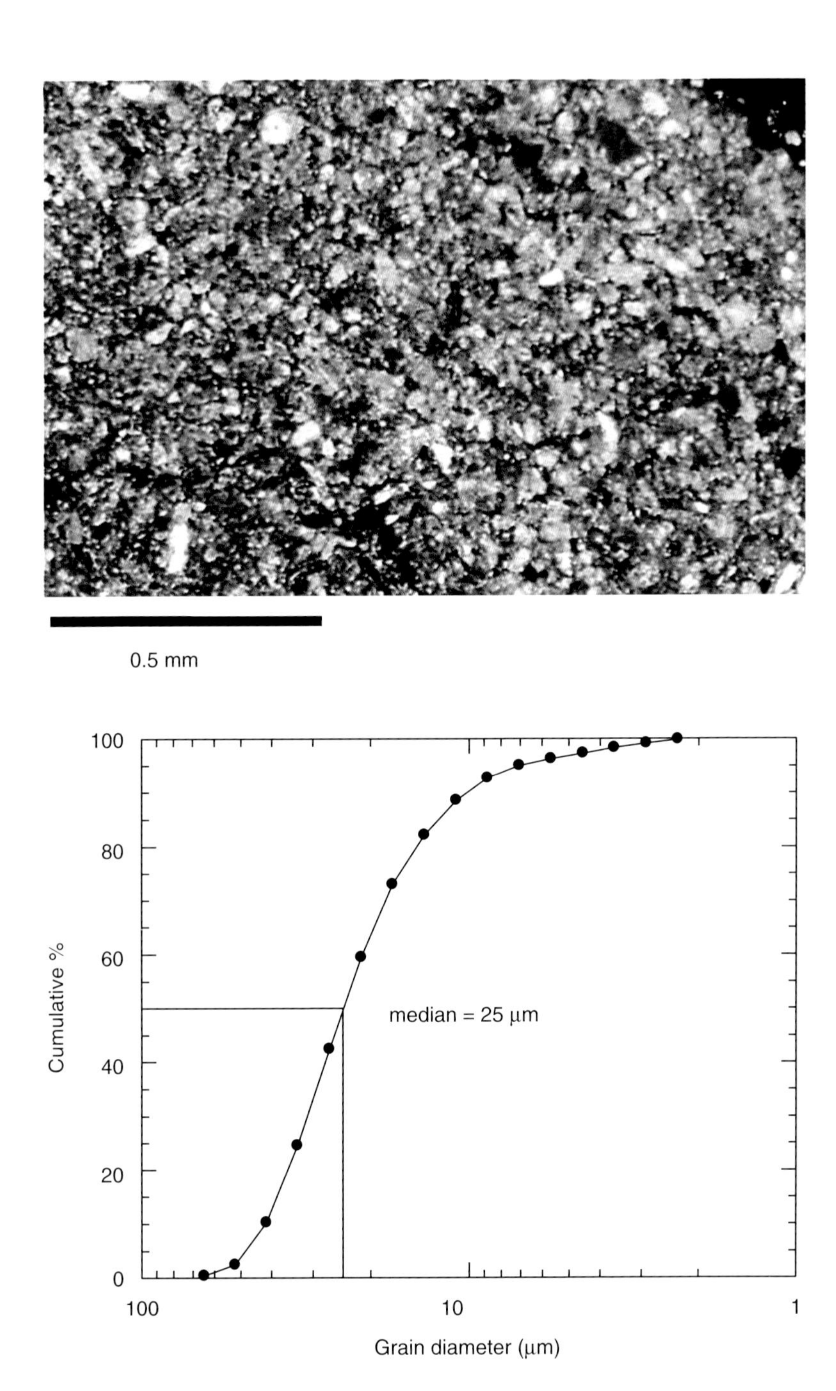

Figure 7.2 Typical loess. Photomicrograph of a sample from Xiagaoyuan, China. Plot of grain size distribution from Mississippi, United States. ***See page 138***.

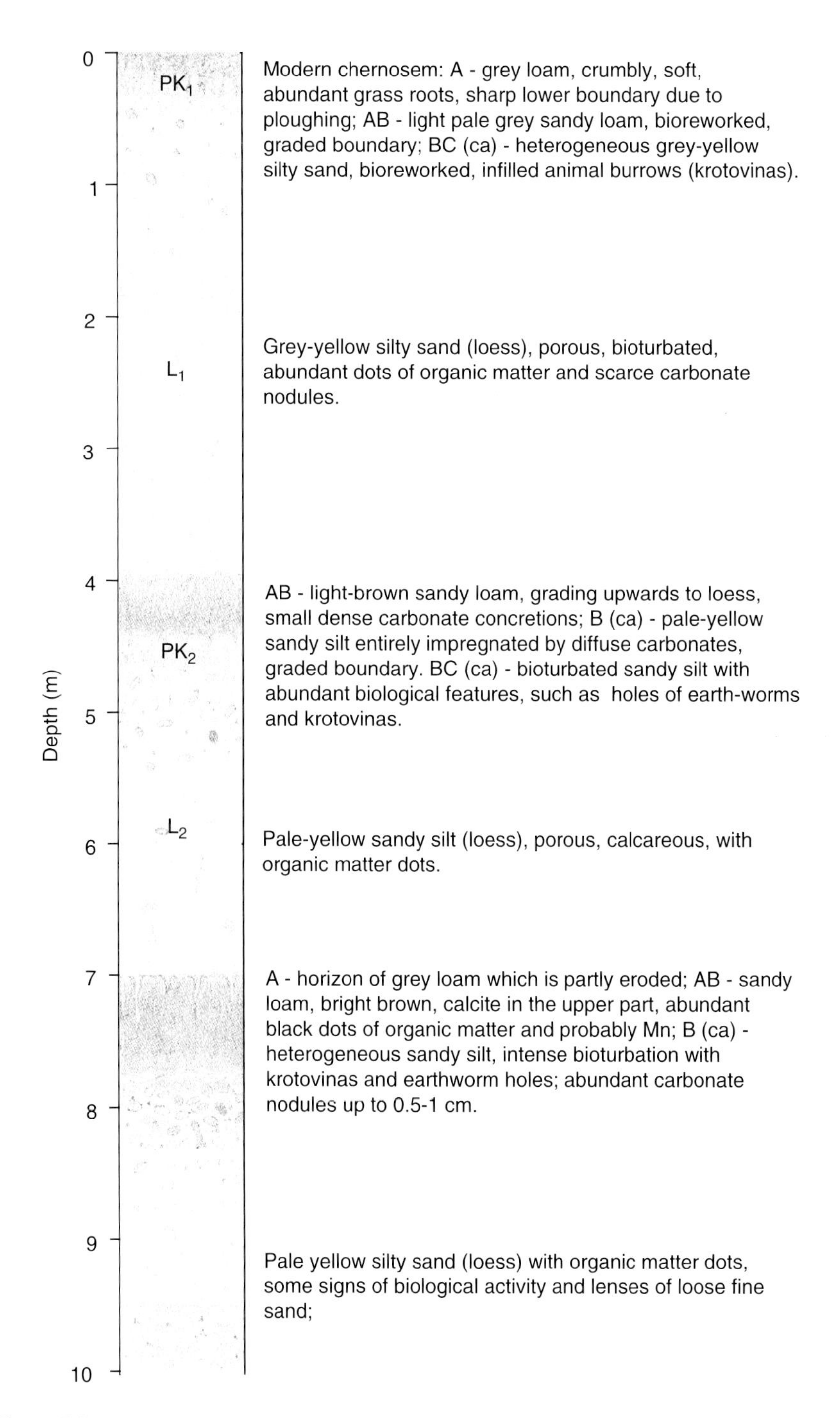

Figure 7.3 The uppermost 10 m of the loess/paleosol sequence at Roxolany (Ukraine). ***See page 139.***

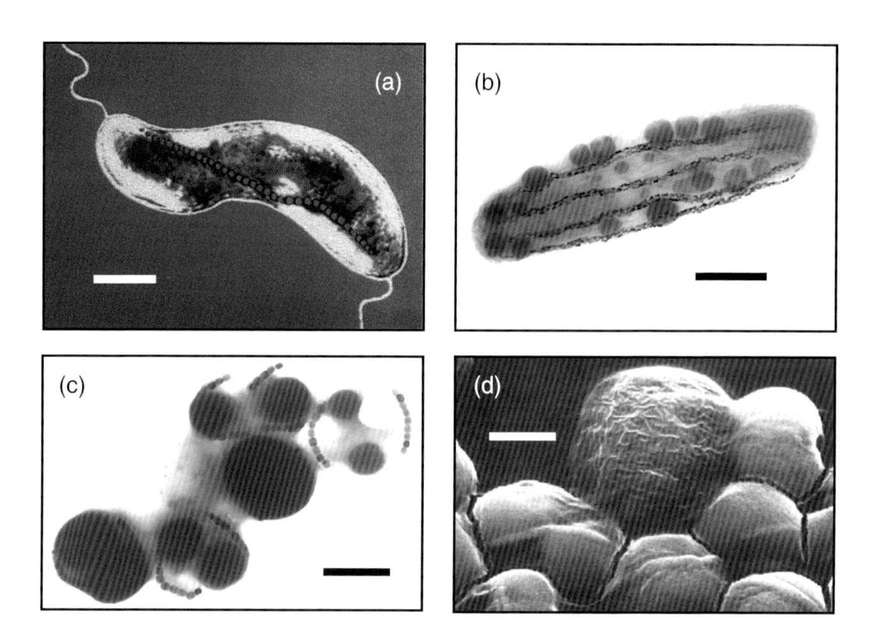

Figure 9.2 Electron micrographs of several types of magnetotactic bacteria. ***See page 191***.

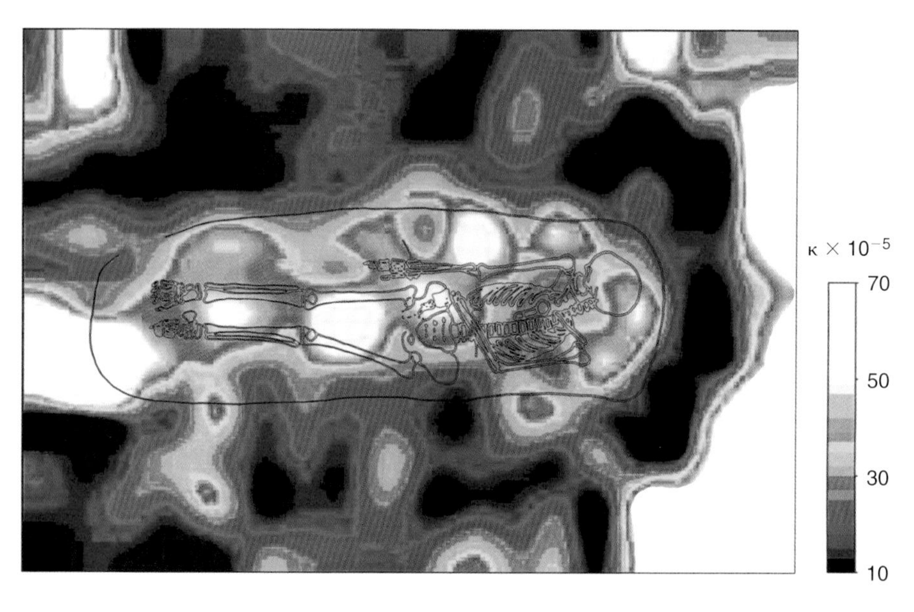

Figure 11.2 Magnetic susceptibility values recorded over an Anglo-Saxon grave at Lakenheath, England (0.6°E, 52.4°N). ***See page 235***.

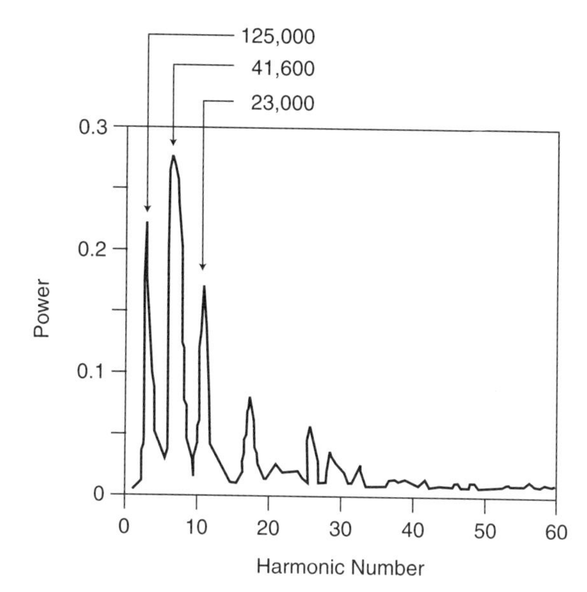

Figure 7.11 Milankovitch cyclicities revealed by Fourier spectral analysis of the magnetic susceptibility profile at Halfway House, Alaska. (Redrawn from Béget and Hawkins, 1989.) © Macmillan Magazines Limited. Reproduced with permission.

Using standard methods of spectral analysis, periodicities close to the expected Milankovitch predictions were reported by Begét and Hawkins (1989) from a loess/paleosol sequence near Fairbanks, Alaska (Fig. 7.11). The observed peaks in the power spectrum (125, 41.6, and 23 kyr) do not coincide exactly with the predictions of orbital theory. This is most likely due to the fact that the timescale associated with any geological sequence is rarely known very accurately; the approximate age is usually known from only a few "fixed" points (e.g., a handful of radiometric dates, a few geomagnetic polarity reversals, a key fossil horizon, or a distinctive volcanic ash layer — such as the Old Crow Tephra in this Alaskan case). Between these points, one is almost always forced to assume a constant rate of sedimentation. These geological timing problems must always be borne in mind. Nevertheless, the observations from Alaska were of great significance because they provided the first evidence that Milankovitch cycles are encoded in — and can be retrieved from — land-based sections. This important result was confirmed by Wang *et al.* (1990), who recognized the eccentricity and obliquity signals in a loess/paleosol section at Baoji (107°01′E, 34°20′N) in the Chinese Loess Plateau. Their sampling scheme was not dense enough to extract any possible precession forcing, but more detailed work by Florindo *et al.* (1999) at the nearby Duanjiapo section (109.2°E, 34.2°N) did reveal a strong peak in the frequency spectrum corresponding to a period of ~21,000 years.

7.2.4 Paleoprecipitation

In China, the correlation of high magnetic susceptibilities with warmer, wetter interglacials is now a well-established fact, and its interpretation in terms of the intensification of the summer monsoon is generally accepted as the most satisfactory explanation. Useful as this magnetoclimatological signal is, it is only a qualitative indicator. How much more useful it would be if reliable quantitative estimates of paleoprecipitation could be obtained. Some authors have suggested that this may, in fact, be possible. Unfortunately, they do not all agree!

The first attempt was by Heller *et al.* (1993) using the radioisotope ^{10}Be as a means of tracking the rate at which loess accumulates. The method relies on the fact that ^{10}Be provides a convenient tracer of mineral grains and aerosols in the atmosphere. It is produced by the interaction of cosmic radiation with atmospheric nitrogen and oxygen and is therefore referred to as *cosmogenic* ^{10}Be. Because the atmosphere is being constantly bombarded by incoming radiation, ^{10}Be is being created continuously (see Box 7.3). Many other isotopes are also produced, but ^{10}Be is the one with the longest half-life (1.5×10^6 yr), which makes it useful in geological applications. Once they have been created, ^{10}Be atoms are scavenged by aerosols, which are, in turn, removed from the atmosphere when it rains or snows. On average, a ^{10}Be atom may remain in the atmosphere for a year or two. The atoms that are precipitated onto the continents may become attached to mineral grains in the ground. Some of these may be reintroduced into the atmosphere as dust particles, which will eventually be returned to the surface and incorporated into sediments, such as the loess in China.

At Luochuan, in the central part of the Chinese Loess Plateau, it is found that the ^{10}Be content of the sediments closely parallels the observed variations in magnetic susceptibility (Fig. 7.12). Furthermore, a timescale can be obtained by matching individual features on the ^{10}Be curve to corresponding points on the δ^{18}O curve, which is itself tied to the absolute time frame obtained from the astronomical calculations behind Milankovitch theory (Imbrie *et al.*, 1984). Given this chronological

Box 7.3 Cosmic Beryllium

Cosmic rays consist of protons, alpha-particles, and heavier nuclei arriving at the top of the Earth's atmosphere from outer space. Collisions with the atmosphere may produce secondary particles, many of which have sufficient energy to cause further reactions leading to a particle cascade. If energies exceed $\sim$10 MeV, ^{10}Be can be produced by collisions with oxygen and nitrogen. Many other isotopes are produced (from target nuclei of argon and krypton as well as oxygen and nitrogen), but only a few have half-lives of more than a year [they are ^{3}He, which is stable, ^{26}Al (7.4×10^5 yr), ^{36}Cl (3.1×10^5 yr), ^{81}Kr (2.1×10^5 yr), ^{14}C (5730 yr), ^{32}Si (500 yr), ^{39}Ar (270 yr), ^{3}H (12.3 yr), and ^{32}Na (2.6 yr)]. Once they are produced, ^{10}Be atoms are easily scavenged by aerosols, particularly sulfates, and precipitated in rain and snow. They may then circulate with water or become attached to sediment particles (generally silicates with diameters of $\sim$0.1 to 10 μm), which, in turn, may reenter the atmosphere as dust. Because of its long half-life (1.5×10^6 yr), ^{10}Be has found wide application as a natural tracer in the earth sciences. An excellent review is given by McHargue and Damon (1991).

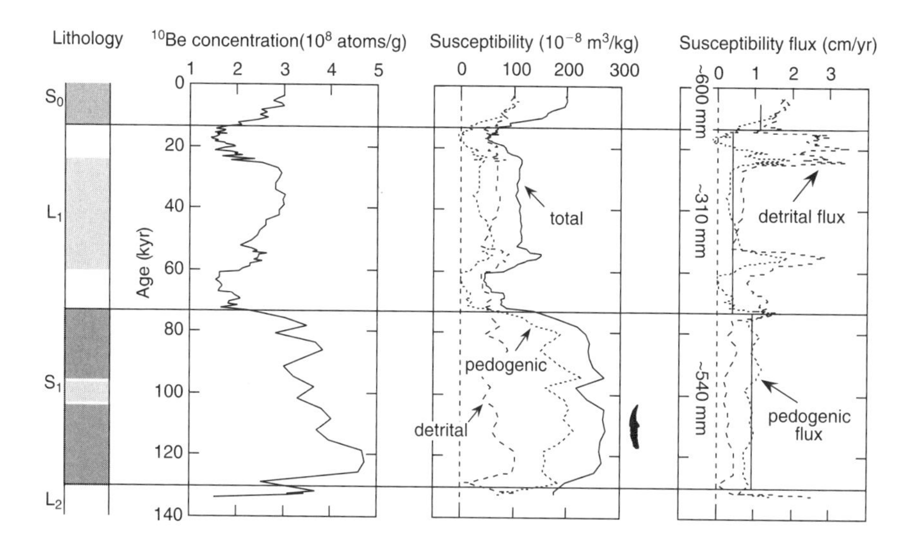

Figure 7.12 Lithology, ^{10}Be, magnetic susceptibility, and paleoprecipitation at Luochuan, China. As desrcibed in the text, the pedogenic and detrital components can be separated. It is found that the pedogenic component is essentially zero in MOI stages 2 and 4 ($\sim$18 and $\sim$70 kyr, respectively = uppermost and lowermost parts of L_1, repectively) and much higher in stages 1 and 5 (S_0 and S_1, respectively). A somewhat elevated contribution is seen in stage 3 (middle part of L_1). (Redrawn from Heller *et al.*, 1993.) © Elsevier Science, with permission of the publishers.

control (Beer *et al.*, 1993), one can calculate the rate at which the sediments accumulated. It is found that the ^{10}Be flux (F_B, in atoms/unit area/unit time) has a strong linear correlation ($R = 0.94$) with the sediment accumulation rate. However, the regression line does not pass through the origin. This is because the total beryllium flux consists of two components, a detrital flux (F_D) consisting of beryllium attached to dust particles (as described previously) and a so-called atmospheric flux (F_A) due to local precipitation. Thus, we have

$$F_B = F_A + F_D$$

The intercept of the regression line yields $F_A = 1.29 \times 10^6$ ^{10}Be atoms/cm^2/year, which is within the range found for modern rain in India (Somayajulu *et al.*, 1984).

The susceptibility signal can be treated similarly by writing

$$F_S = F_{D'} + F_X$$

where F_S is the total susceptibility, $F_{D'}$ represents the susceptibility of detrital origin, and F_X is the susceptibility enhancement due to pedogenesis. Beer *et al.* (1993) make the reasonable assumption that $F_{D'}$ and F_D are proportional to one another, that is, $F_{D'} = \alpha F_D$. The relative contribution of pedogenic susceptibility is therefore given by

$$F_X/F_S = 100 \times (1 - \alpha(F_B - F_A)/F_S)\%$$

By comparing the profile obtained at Luochuan with the susceptibility of the Holocene soil (S_0) and the present-day precipitation (630 mm/year), Heller *et al.* (1993) conclude that the mean annual precipitations during the formation of L_1 and S_1 were ~310 and ~540 mm, respectively.

A rather different approach was taken by Maher *et al.* (1994). They collected together susceptibility data on the modern soil at nine sites widely spread across the Loess Plateau and compared them with the current annual precipitation at those localities. There is a reasonable correlation, as indicated by the regression line in Fig. 7.13. The suggested procedure, therefore, is simply to determine the maximum susceptibility in a given paleosol after subtracting the detrital component due to the original eolian dust input. The paleoprecipitation is then estimated from the equation of the regression line, on the assumption that the enhancement of susceptibility is entirely due to the pedogenesis resulting from the increased moisture. During interglacial periods and in the early Holocene, Maher and her coauthors infer increased rainfall throughout central China but particularly pronounced in the western plateau. During glacial periods, rainfall was reduced across the whole loess plateau, especially in the southeast (Fig. 7.14).

Despite its promise, it must be admitted that quantitative paleoclimatology using magnetic susceptibility is not yet able to provide robust estimates of paleoprecipitation. General trends are agreed on by all authors, but actual rainfall values differ markedly from author to author. This is illustrated graphically in Fig. 7.15; in isotope stage 5, for example, estimates range from 540 to 740 mm/yr.

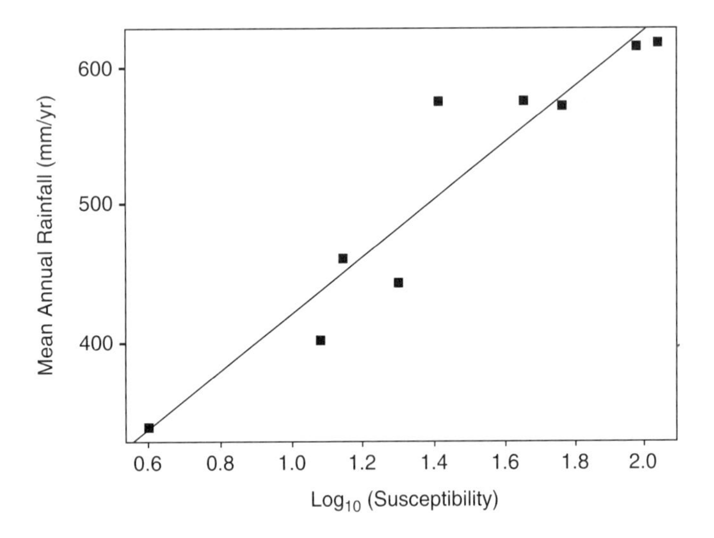

Figure 7.13 Present-day rainfall and magnetic susceptibility of the modern soil at nine sites in the Chinese Loess Plateau. Rainfall values (R, in mm/yr) represent 30-year averages (1951–1980). Magnetic values (χ, in $10^{-8}\,m^3kg^{-1}$) represent pedogenic susceptibility (i.e., the susceptibility of the B horizon minus that of the C horizon, see Box 5.3). The regression line is given by $R = 222 + 199\log_{10}(\chi)$. Correlation coefficient = 0.95. (Redrawn from Maher *et al.*, 1994.) © Elsevier Science, with permission of the publishers.

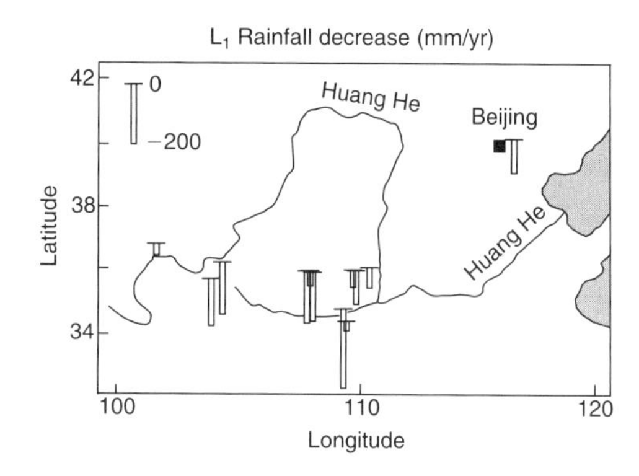

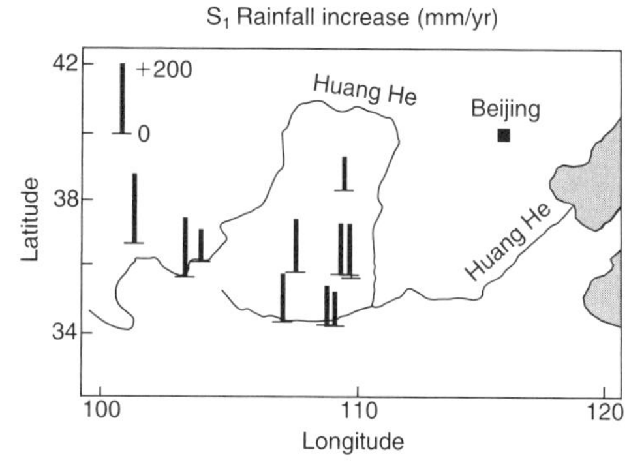

Figure 7.14 Map of the Yellow River (*Huang He* in Chinese) region showing changes in mean annual precipitation (mm/yr) deduced from magnetic susceptibility measurements on sediments of the Chinese Loess Plateau, following the relationship illustrated in Fig. 7.13. The upper panel shows rainfall decrease during the deposition of loess layer L_1; the lower panel shows rainfall increase during the formation of paleosol layer S_1 (both compared with the present-day rainfall at each site). (Generalized from Maher *et al.*, 1994.) © Elsevier Science, with permission of the publishers.

In China, the overall picture is one of alternating periods of dominance by either the winter monsoon or the summer monsoon: in glacial intervals, the dry northerly winter monsoon is stronger, whereas interglacials are characterized by the dominance of the moist southerly summer monsoon. Rainfall estimates of this kind represent an important step toward completing the overall paleoenviromental picture. It now appears that precipitation changes are just as important as temperature variations when assessing global change scenarios. Stine (1994), for example, has investigated relict tree stumps in California and Patagonia and reports that they indicate severe intervals of wetland desiccation during medieval times. He suggests that the cause was a reorientation of the midlatitude storm tracks due to a general contraction of the circumpolar vortices. Whatever the cause, however, he stresses that aberrant atmospheric circulation can produce far greater departures in precipitation than in temperature. He therefore makes the plea that such well-known terms as *medieval warm period* be replaced by neutral terms such as *medieval climatic anomaly*. This will help to avoid prejudice in future analyses.

Finally, it is important to note that even if a suitable procedure can be developed, magnetic data are by no means the only avenue being explored in the quest for quantitative paleoclimatic information. Oxygen isotopes in the polar ice sheets are perhaps the most successful case (Dansgaard *et al.*, 1993), but evidence from past

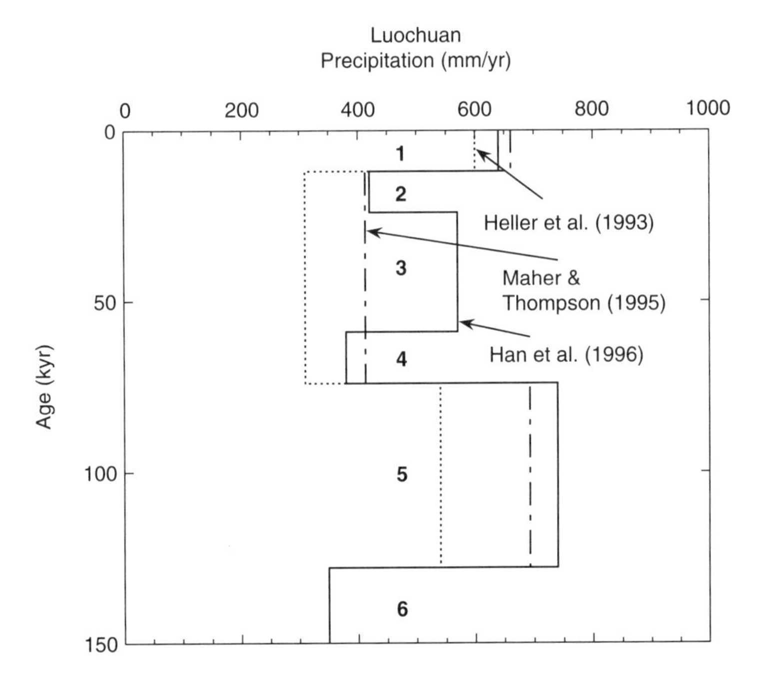

Figure 7.15 Paleoprecipitation at Luochuan (Chinese Loess Plateau) according to different authors. The bold numbers refer to the corresponding oxygen isotope stages. (From Evans and Heller, 2001.) © Elsevier Science, with permission of the publishers.

biota (pollen and tree rings, in particular) is also of prime importance. Another promising methodology for loess deposits, which is being developed in conjunction with magnetic measurements, involves statistical analysis of snail assemblages (Rousseau, 1991). This provides an objective assessment of biological diversity that can be related (qualitatively) to environmental factors such as mean annual precipitation and mean annual temperature. Joint susceptibility/snail (malacological) studies have been carried out at Eustis, Nebraska (Rousseau and Kukla, 1994), at Achenheim, France (Rousseau *et al.*, 1998), and at Luochuan, China (Rousseau and Wu, 1997). The data from the Peoria Loess at Eustis on the Great Plains of North America show a particularly convincing correspondence between magnetic susceptibility and biodiversity, as measured by the so-called Shannon index of diversity (Box 7.4; see also Rousseau and Kukla, 1994, Fig. 6). Over the lower part of the section (spanning the depth interval 18 to 9 m), the susceptibility falls steadily from 8.4×10^{-8} to 5.5×10^{-8} m^3kg^{-1} while the Shannon index rises from 1.0 to 2.5. Throughout the upper part of the section (from 9 to 1 m depth), susceptibility and diversity remain approximately constant (at about 5.0×10^{-8} m^3kg^{-1} and 2.5, respectively). The available chronological control suggests that these two intervals correspond to the advance and retreat of the Laurentide ice sheet.

Box 7.4 Biodiversity

In their combined study of magnetic susceptibility and snail assemblages in the loess section at Eustis, Nebraska, Rousseau and Kukla (1994) quantify biodiversity by means of the Shannon index. This is given by $H' = -\sum p_k \log_2 p_k$ (where p_k is the frequency of the kth species in a given assemblage). Suppose, for example, that only two species are present in an entire section. Diversity is maximum if each species represents 50 % of the assemblage, in which case $H' = 1.0$. As the proportions become more and more one sided, the index decreases until finally, when only one of the species is present, $H' = 0$, that is, there is no diversity at all. At Eustis, 16 species were identified, so the theoretical maximum diversity is 4.0. In practice, the highest observed diversity was 2.8. As a concrete example, consider the assemblage at 14.2 m depth where only four species are found (*Pupilla muscorum*, 205 individuals; *Vollonia gracilicosta*, 160; *Succinea grosvenori*, 69; and *Columella avara*, 11) with a resulting diversity index of 1.59. The stratigraphic variation of diversity is shown in the diagram along with the corresponding susceptibility profile (in units of $10^{-9}\, m^3 kg^{-1}$). A strong negative correlation clearly exists, suggesting that both parameters are responding to a common driving force. Rousseau and Kukla (1994) interpret this to be the climatic and ecological changes associated with the advance and retreat of the Laurentide ice sheet, which, at its greatest extent, came within about 100 km of the site.

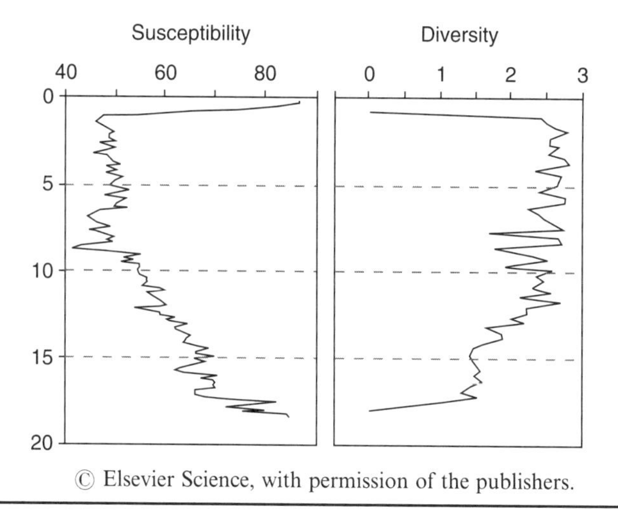

7.2.5 The Wind-Vigor Model

Mention has already been made of the important loess investigations in Alaska, particularly at the famous Halfway House site near Fairbanks. Apart from its significance as the first land-based record of Milankovitch cyclicities, this section also deserves our attention as a magnetoclimatological archive resulting from an entirely different model from that responsible for the Chinese records. There, the climate is encoded as magnetic susceptibility highs during interglacials, whereas at Halfway House, susceptibility lows correspond to interglacials when soils were formed. Begét *et al.* (1990) argued that this is due to the stormier nature of glacial periods. The essential point is that stronger winds are more able to pick up and transport denser mineral grains, and it so happens that the magnetic oxides of interest to us are about twice as dense as most common silicates (5190 compared with 2650 kg/m^3). It is like winnowing as employed in traditional agriculture; normally, the wind blows away the chaff as the denser kernels fall to the threshing floor. But if the wind gets up, everything is blown away.

A good example of the wind-vigor model is provided by data from a 34-m section at Kurtak in southern Siberia (Fig. 7.16). Susceptibility peaks in the

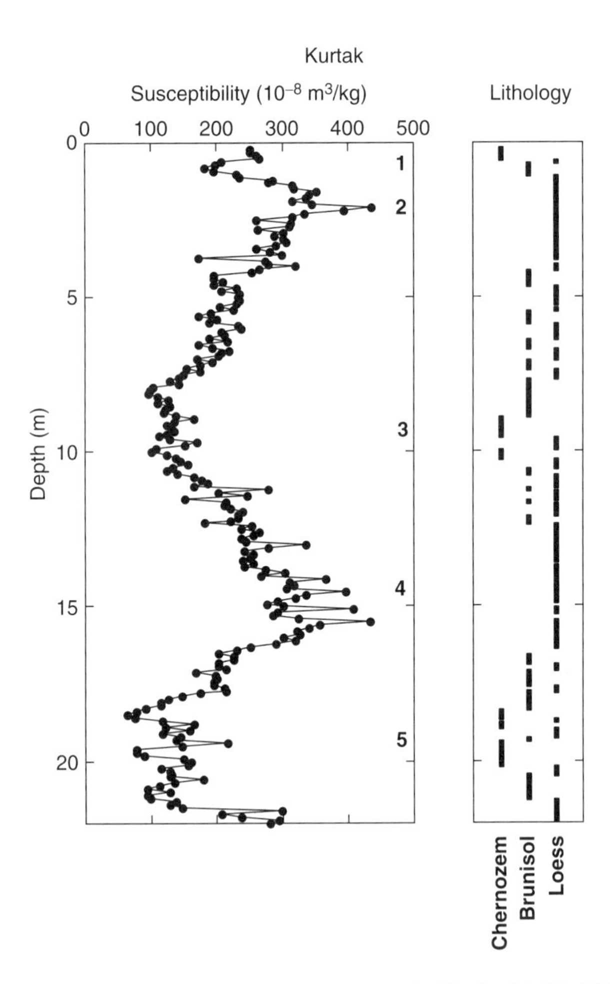

Figure 7.16 Magnetic susceptibility and lithology at Kurtak, Siberia (91.4°E, 55.1°N). The bold numbers indicate the corresponding oxygen isotope stages. (Redrawn from Chlachula *et al.*, 1998.)

loess layers exceed 4×10^{-6} m^3/kg (median = 2.7×10^{-6} m^3/kg), whereas paleosol values fall below 1×10^{-6} m^3/kg (median = 1.4×10^{-6} m^3/kg). These values are almost the mirror image of the corresponding observations in the Chinese Loess Plateau; at Baoji, for example, the most prominent soil (S_3) has a susceptibility of 3.6×10^{-6} m^3/kg while the most primitive loess (L_{15}) has values as low as 0.2×10^{-6} m^3/kg.

Stronger and more frequent winds are the basic cause of the susceptibility profiles seen in Siberia and Alaska, but it is also possible that changes in atmospheric circulation under different climatic conditions could lead to different source regions being abraded. The prevailing winds might deliver particles derived from a granitic bedrock, say, at one time, but these might be replaced later by material derived from a more magnetic volcanic province. Even if the bedrock remains constant, proximity to the source will influence the magnetic signal because of preferential fallout of the denser iron oxides, similar to that observed in the ash plume generated by the 1980 eruption of Mount St. Helens (Evans, 1999). Of course, this is not an eolian deposit in the same sense as the loess occurrences discussed before, but it does provide a useful natural experiment concerning the dispersal of atmospheric particulates. It was found that there is a strong exponential decrease of magnetic content in the ash as a function of distance from the vent. Such a gradient can also explain the magnetoclimatological patterns observed in Alaska and Siberia, as illustrated in Fig. 7.17.

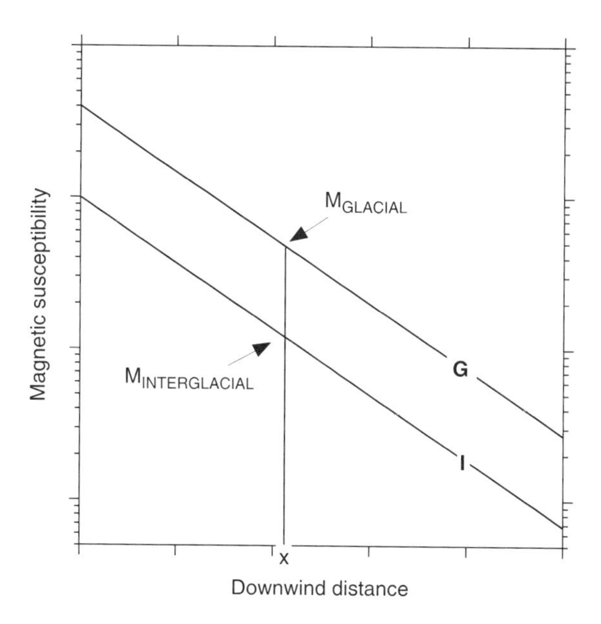

Figure 7.17 Schematic representation of the magnetic susceptibility signature of the wind-vigor model. Magnetic susceptibility is assumed to decrease exponentially downwind (note that the ordinate is logarithmic). During glacial intervals, the curve moves upward (because stronger winds entrain more dense magnetic oxide particles) and downwind (because the source region expands). The net result is that, at any given site (at a downwind distance x, say), the magnetic susceptibility (M) fluctuates between an upper bound (M_{GLACIAL}) and a lower bound ($M_{\text{INTERGLACIAL}}$). Evans (1999, 2001) gives further details.

The wind-vigor model thus gives rise to a magnetoclimatological signal opposite in sign to that produced by soil formation in China, but it is important to recall that under some circumstances pedogenesis can also lead to susceptibility minima in interglacial periods due to magnetic depletion caused by the pedological process of gleization. We now turn to a brief description of some specific examples.

7.2.6 Examples of Gleization

In Chapter 5, we saw that pedogenesis is a highly complex physical, chemical, and biological process that does not always lead to magnetic enhancement. In particular, if conditions lead to waterlogging, gleization or gleying takes place. Stagnant water becomes exhausted of oxygen, and minerals containing iron in the reduced state (Fe^{2+}) are formed. Magnetic susceptibility decreases, as illustrated by the example from the Orzechowce brickyard (22°46′E, 49°51′N) near Przemysl in southern Poland shown in Fig. 7.18. Similar results are also reported by Nawrocki *et al.* (1996) from a loess profile at Bojanice in the Ukraine: They observe an almost threefold susceptibility decrease between the glacial (stage 2) loess (381×10^{-6} SI) and the underlying interstadial (stage 3) gleyed paleosol (137×10^{-6} SI). At the Late Pleistocene periglacial site on the North American Great Plains discussed earlier (Eustis, Nebraska), Rousseau and Kukla (1994) also find that gleying causes significant susceptibility decreases (by about a factor of 2) that are superimposed on the broader regional effects of ice sheet dynamics. The possibility of such gleying requires careful scrutiny when undertaking paleoclimatic interpretations of magnetic data. Particular attention must be paid to site specificity: as Retallack (1990, p. 87) emphasizes, gleying "reflects local waterlogging rather

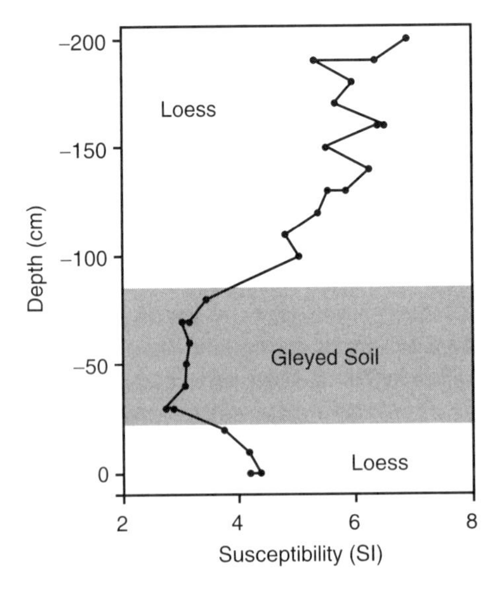

Figure 7.18 Gleyed soil in Poland. Measured in the Orzechowce brickyard by Jerzy Nawrocki and the authors, August 1995. Susceptibility (κ) is in 10^{-5} SI units.

than wider effects of climate and vegetation evident from other soil-forming processes."

7.3 LAKE SEDIMENTS

7.3.1 Early Work

The importance of lake sediment magnetism for paleoenvironmental studies was first recognized during the 1970s in Holocene sequences cored from British lakes. Much of the early lacustrine work was carried out on material collected with a pneumatic coring device (Mackereth, 1958), which was unable to penetrate more than a few meters below the water/sediment interface. In Britain, this typically represents a time interval on the order of about 10,000 years rather than the millions of years recorded in the Chinese loess sequences, for example. Nevertheless, investigations of this kind provided important data for the study of the secular variation of the geomagnetic field (Mackereth, 1971; Creer *et al.*, 1972), greatly extending the historical observatory record and improving on the geographical distribution of the archeomagnetic record. Similar studies were subsequently undertaken throughout Europe, in Africa, in North and South America, and in Australia, a useful representative selection being summarized in Table 7.1. In some cases, specific details of the secular variation patterns observed have been sufficiently well defined to provide chronological control, as described in Chapter 6.

It appears that the addition of magnetic susceptibility to the limnologist's arsenal grew out of the desire to correlate, rapidly and nondestructively, between cores

Table 7.1 Selected Lacustrine Paleosecular Variation Archives

Barombi Mbo	Cameroon	25,000 yr	Maley *et al.*, 1990
Fish	U.S.A	13,000 yr	Verosub *et al.*, 1986
Keilambete	Australia	10,000 yr	Barton and McElhinny, 1981
Lac de Joux	Switzerland	13,400 yr	Creer *et al.*, 1980
Mara	Canada	5,000 yr	Turner, 1987
Morenito	Argentina	14,000 yr	Creer *et al.*, 1983
Sea of Galilee	Israel	5,000 yr	Thompson *et al.*, 1985
St. Croix	U.S.A.	10,000 yr	Banerjee *et al.*, 1979
Superior	Canada	14,000 yr	Mothersill, 1979
Tanganyika	Zambia	25,000 yr	Williamson *et al.*, 1991
Trikhonis	Greece	6,000 yr	Creer *et al.*, 1981
Vatnsdalsvatn	Iceland	9,000 yr	Thompson and Turner, 1985
Vuokonjarvi	Finland	5,000 yr	Stober and Thompson, 1977
Windermere	England	10,000 yr	Turner and Thompson, 1981

collected at different parts of a given lake (Thompson, 1973; Thompson *et al.*, 1975). But, as so often happens, it was discovered that a procedure adopted for one reason yields useful information for an entirely different topic. In the lake sediment case, significant correlations were soon found between susceptibility and established indicators of sedimentary environment, such as grain size, organic carbon, and pollen content. The need for longer lacustrine sequences to extend the geomagnetic secular variation record further into the past led to the use of more powerful coring equipment, which, in turn, provided longer sequences for comparison with paleoenvironmental reconstructions inferred from loess and oceanic data. The sediments of Lac du Bouchet in France, for example, have now been penetrated to a depth of almost 50 m (Thouveny *et al.*, 1990). In what follows, we explore some of the main magnetic observations bearing on paleoclimate and paleoenvironmental reconstructions derived from representative studies of small lakes in Europe and from the world's largest and oldest body of freshwater, Lake Baikal in Siberia.

7.3.2 Italian and French Crater Lakes

Crater lakes form when water collects in the caldera of an extinct, or dormant, volcano. The drainage into such lakes is often limited to, or at least dominated by, runoff from the inner walls of the caldera, and the sediments they collect therefore consist mostly of detrital material derived from the volcanic products of the surrounding catchment area. Several such sequences have been studied under the auspices of two major international projects, PALICLAS (Palaeoenvironmental Analysis of Italian Crater Lake and Adriatic Sediments) (http://www.iii.to.cnr.it) (Oldfield, 1996) and EUROMAARS (European Maars) (Creer and Thouveny, 1996)—a *maar* being a volcanic crater that formed explosively (phreatomagmatically). Both projects involved a wide range of topics including magnetism, geochemistry, palynology, and lithostratigraphy and yielded very large databases. Here, we describe results germane to magnetoclimatology collected from Lago Albano in Italy (12.6°E, 41.8°N) and Lac du Bouchet in France (3.8°E, 45.0°N). The Albano cores yield a maximum of 14 m of sediment spanning the last 30,000 years, whereas the 46-m Bouchet record reaches back an order of magnitude further, to about 300,000 years.

The oldest Lago Albano sediments are from the base of core PALB 94-1E. The lowest 300 cm, representing 4 to 5 kyr, reveal three strong oscillations in magnetic susceptibility with minima suggesting an almost complete absence of magnetic minerals, well below the range expected from material derived from the catchment area (Fig. 7.19). These minima correlate well with indicators of biological productivity such as diatom and pollen concentration and are interpreted as zones wherein organically driven reductive (dissolution) diagenesis has strongly depleted the ferrimagnetic minerals (Rolph *et al.*, 1996). They bear witness to climatic amelioration during interglacials or interstadials. When these results are combined with those from other Albano cores (PALB 94-6A and B), there are proxy data for at least seven oscillations in climatically forced ecosystem responses during the period ~30 to ~15 kyr BP. They are regarded by Oldfield (1996) as lacustrine counterparts to the

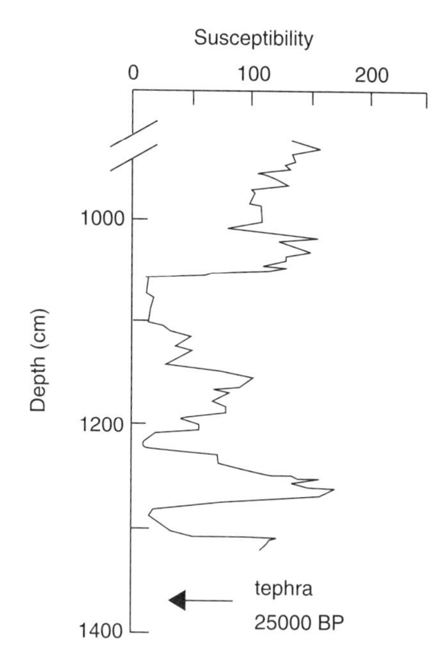

Figure 7.19 Magnetic susceptibility profile of the lowermost part of core PALB 94-1E from Lago Albano near Rome, Italy. (Redrawn from Oldfield, 1996.)

rapid (millennial-scale) changes indicated by the Greenland ice core (GRIP) oxygen isotope data (Dansgaard *et al.*, 1984; Bond *et al.*, 1993) sometimes referred to as Dansgaard–Oeschger cycles.

Between ~15 and ~10 kyr, the shift from an ice age earth to the interglacial we are currently enjoying is witnessed by a major increase in the amount of tree pollen as natural reforestation took place. This important paleoecological event is captured magnetically as a marked decrease in susceptibility, albeit with a superimposed short-lived peak corresponding to the so-called Younger Dryas (YD) cooling interval (~12,000 kyr BP) (Fig. 7.20). Concomitant PALICLAS oxygen isotope data obtained on foram shells from the Adriatic marine cores quantify this cooling as a drop of 5 to 6°C in sea surface temperature (SST) between the peak of the preceding Bölling/Alleröd interstadial and the YD minimum. Unlike the millennial-scale signals observed between 30 and 15 kyr, which begin and end abruptly, this rather leisurely glacial-to-interglacial susceptibility decrease is partly attributed to the lake being gradually starved of erosional influx as vegetation was progressively established. Magnetic remanences (ARM, SIRM) given to samples in the laboratory suggest that this is not the whole story, however. Were it just a matter of gradually cutting off the erosive supply, these parameters would decrease in parallel with the susceptibility, whereas a strong increase in the ARM/SIRM ratio is observed, well beyond the catchment envelope (Fig. 7.20). Rolph *et al.* (1996) attribute this to the increasing importance of bacterial magnetosomes as the lake becomes richer in nutrients, evolving from glacial oligotrophy to interglacial eutrophy (but not yet becoming so rich in organics as to suffer marked dissolution of the iron oxides).

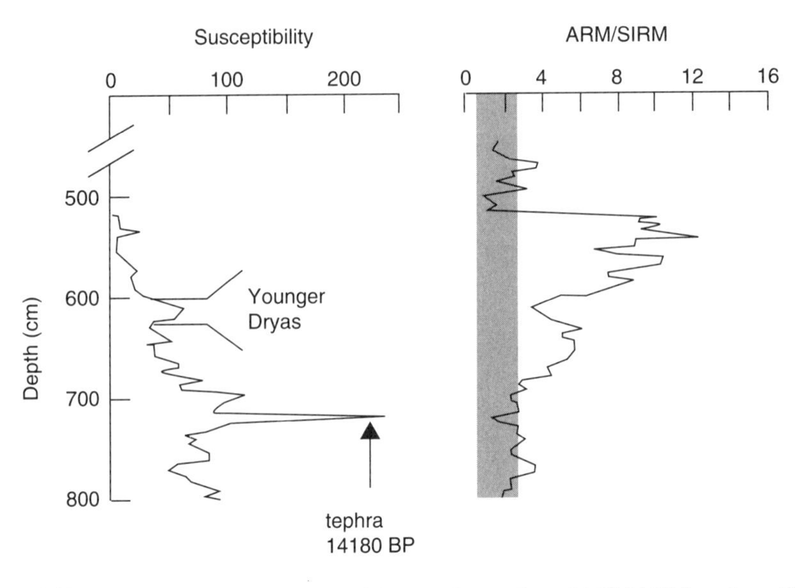

Figure 7.20 Magnetic susceptibility profile of the central part of core PALB 94-1E from Lago Albano near Rome, Italy. The shaded zone indicates the envelope of ARM/SIRM ratios obtained from parent materials in the catchment. (Redrawn from Oldfield, 1996.)

The sediments of the maar lake, Lac du Bouchet, in the Velay region of France contain a detrital magnetic fraction composed mainly of titanomagnetite grains close to magnetite in chemical composition (Williams *et al.*, 1996). The measured magnetic susceptibility of the sediments reflects the relative importance of various climate-sensitive processes. In glacial conditions, aridity and low temperatures restrict vegetation and soil development, while cryoclastic erosion produces rapid terrigenous sedimentation. The lake was oligotrophic, diagenetic change was restricted, and the sediments produced have high magnetic susceptibilities. During interglacial times, vegetation growth and soil formation—both favored by medium to high temperatures and increased moisture—restricted erosion, whereas organic productivity was enhanced. In the resulting diagenesis, the composition of the water–sediment mixture determines the degree of alteration of iron-bearing minerals, advanced eutrophic conditions favoring reduction leading to partial or complete dissolution of titanomagnetites and authigenesis of weakly ferromagnetic and paramagnetic minerals (sulfides, phosphates, and carbonates) as described in Chapter 5. The sediments produced have low magnetic susceptibilities. The important controlling processes are therefore essentially the same as those in Lago Albano and the sign of the magnetoclimatological signal in both lakes is therefore the same: glacials produce high magnetic susceptibility, interglacials low. This mimics the situation in soil profiles in which gleying has occurred but is opposite to the signal found at other sites (such as the Chinese Loess Plateau) where soils are magnetically enhanced.

Thouveny *et al.* (1994) report results from a 20-m-long susceptibility record at Lac du Bouchet representing some 140,000 years, which they successfully correlate with the oxygen isotope record obtained from the Greenland ice cap and with the North Atlantic microfossil and lithological records. Later work at Lac du Bouchet extended the coring to a depth of 50 m, corresponding to ~300 kyr according to $^{40}Ar/^{39}Ar$ laser dating of a volcanic ash layer at 42 m depth (Williams *et al.*, 1996). Furthermore, the same ash layer occurs in the upper part of a sequence sampled in nearby Praclaux Maar. Another tephra, at the base of the Praclaux sequence, has yielded an $^{40}Ar/^{39}Ar$ age of 606 ± 6 kyr (Roger *et al.*, 1999), offering the prospect that lacustrine records of this kind may (at least in some cases) eventually yield paleoclimatic records in excess of half a million years. But this remains as a future exercise.

The data currently available provide a striking example of the linkages it is possible to establish between continental and oceanic paleoclimatic records. The wide geographic distribution of these teleconnections can be appreciated from Fig. 7.21, where the 0- to 150-kyr part of the Lac du Bouchet record is compared with oxygen isotope profiles from the Pacific and Indian Oceans (Sowers *et al.*, 1993; Bassinot *et al.*, 1994) and with the temperatures deduced from oxygen isotope measurements on the Greenland ice cap (Dansgaard *et al.*, 1993). As Williams *et al.* (1996) point out, linkages of this kind demonstrate the wider significance of the Lac

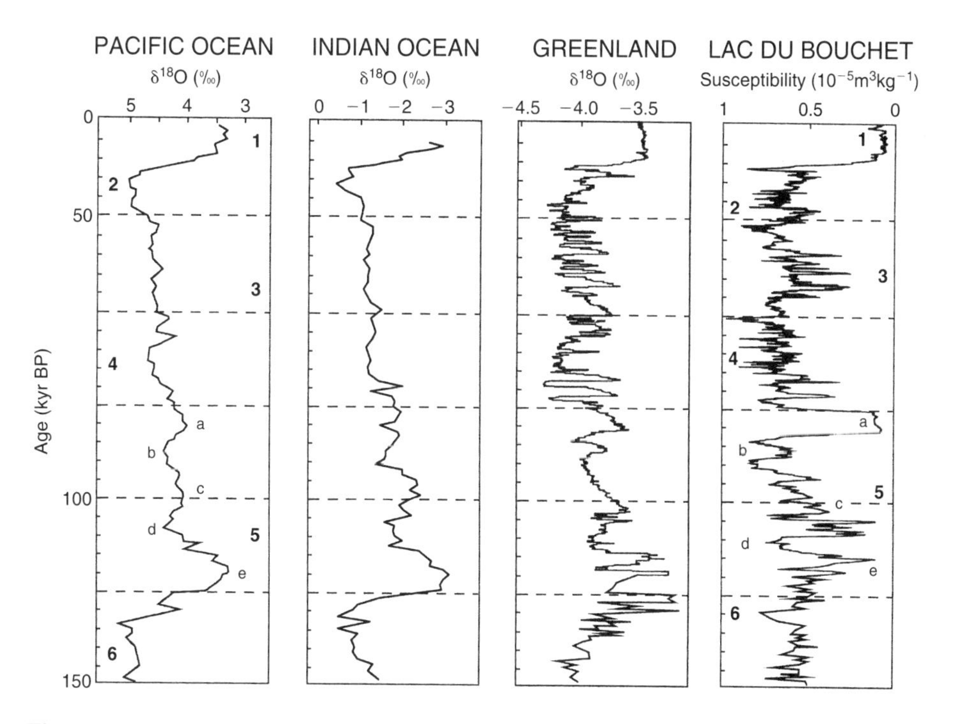

Figure 7.21 Lac du Bouchet (France) magnetic susceptibility profile (note reversed scale) compared with oxygen isotope records from the Pacific Ocean, the Indian Ocean, and the Greenland ice cap. (After Heller *et al.*, 1998a, compiled from data by Thouveny *et al.*, 1994, Sowers *et al.*, 1993, Bassinot *et al.*, 1994, and GRIP members, 1993.) © Elsevier Science and Macmillan Magazines Limited. Reproduced with permission.

du Bouchet record—it is one of the better continental records offering both high resolution and long time coverage. The short-wavelength features are consistent with the climatic variability associated with Dansgaard–Oeschger and Heinrich events, whereas the longer wavelength features can be correlated with the classic isotope stages and substages.

7.3.3 Lake Baikal

Lake Baikal is situated in Siberia about 1000 km east–southeast of the Kurtak site discussed in Section 7.2; it is the world's oldest (5 million years), largest (23,000 km^3), and deepest (1620 m) freshwater lake. As part of a joint Russian–American project, several cores were recovered from the Academician Ridge region of the lake. The longest of these penetrated more than 10 m of sediment that have been demonstrated to provide a quasi-continuous climatic record of the last quarter of a million years (Peck *et al.*, 1994). The magnetic properties reveal the influence of two distinct sources of sedimentation, (1) an eolian detrital influx blown in from near and far and (2) a biogenic input—consisting largely of diatom opal—created by organic activity within the lake itself. These two processes respond in opposite directions to climate forcing—when one goes up, the other goes down. During interglacials, the biogenic input increases due to enhanced organic activity but the eolian input decreases because of lower average wind vigor. Exactly the reverse takes place during glacial intervals. The net result is that the overall sedimentation rate remains approximately constant. Magnetically, however, the two processes reinforce each other. This is because the diatom opal is essentially nonmagnetic whereas the eolian influx is magnetic (Fig. 7.22). This means that during interglacials, decreased eolian input reduces the magnetism and increased biogenic content further dilutes the overall signal.

Peck *et al.* (1994) also show that the Lake Baikal magnetoclimate record can be successfully correlated with the SPECMAP oxygen isotope time series (Imbrie *et al.*, 1984) obtained from oceanic sediments and thereby demonstrate that their record reaches back more than 200 kyr. Spectral analysis of the magnetic susceptibility profile shows all three Milankovitch orbital periodicities. More recently, the Lake Baikal coring has been extended to much greater depths and the biogenic silica content used to continue the climate history back to 5 million years, the crucial chronological control being provided by an excellent magnetostratigraphic record (Williams *et al.*, 1997).

7.4 MARINE SEDIMENTS

7.4.1 Early Work

The advantages of magnetic measurements previously pointed out (speed, nondestructiveness) have proved particularly effective in oceanographic work. As Manley *et al.* (1993) indicate, the appropriate measurements can be made on board ship and the data can often be available within 24 hours of core retrieval. This is a great

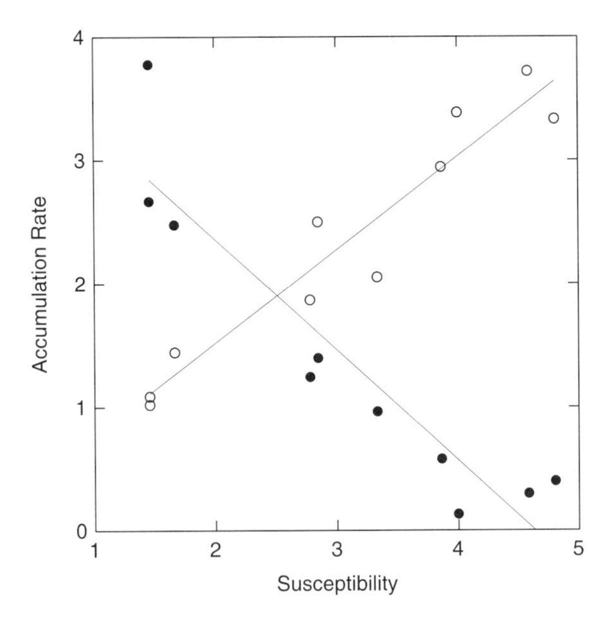

Figure 7.22 Basis of the Lake Baikal enviromagnetic signal. The terrigenous (eolian) influx (open circles: $R = 0.96$) and biogenic production (closed circles; $R = 0.93$) vary antithetically, as explained in the text. Accumulation rate for the terrigenous material is in 10^{-1} gcm^{-2}yr^{-1}, for the biogenic production it is 10^{-2} gcm^{-2}yr^{-1}. Susceptibility is in 10^{-4} SI units. (Compiled from Peck *et al.*, 1994.)

advantage to the shipboard scientists, even if no consideration is given to the origin of the susceptibility variations, and it was soon realized that the down-core profiles obtained in this way contained paleoclimatic information. Robinson (1986) convincingly demonstrated such a link. In cores of late Quaternary sediments from the North Atlantic, he found that glacial horizons had high magnetic susceptibility whereas interglacials were magnetically much weaker. Subsequently, it has emerged that other magnetic parameters, such as coercivity and laboratory-induced magnetizations (ARM and IRM), also provide climatic signatures caused by (1) variations in terrigenous influx, (2) variations in the productivity of magnetotactic bacteria, and (3) variations in postdepositional diagenesis. These three factors are the topics of the following three sections (*7.4.2–7.4.4*).

7.4.2 Terrigenous Magnetic Influx

Trace amounts of strongly magnetic Fe_3O_4 (magnetite) and related minerals (titanium-bearing iron oxides and iron sulfides) are delivered to the ocean basins as part of the continent-derived fraction of deep-sea sediments. They will be relatively diluted or concentrated depending on the other ingredients in the sediments, particularly biogenic carbonate and (to a lesser extent) silica. Many processes that are climatically controlled can therefore be monitored magnetically. In mid- to high latitudes, ice rafting is one such process; in lower latitudes, eolian dust flux is another.

Results from ODP Leg 117 provide a striking example from low latitudes (Bloemendal and deMenocal, 1989). The Owen Ridge, in the western Arabian Sea (~20°N), lies directly beneath the path of eolian dust carried from Africa and Arabia by southwest winds of the Asian summer monsoon. Plio-Pleistocene sediments deposited on the Owen Ridge consist predominantly of carbonate oozes diluted by varying amounts of terrigenous eolian material. Extraction of the continent-derived material from 94 representative samples indicates that there is a strong correlation between the total amount of this terrigenous material present and the measured susceptibility of each sample ($R = 0.98$), thus permitting the whole-core magnetic susceptibility logs—which contain vastly more susceptibility data—to be used to construct a proxy record of eolian deposition. The nature of the actual minerals responsible for the magnetic signal, however, is complicated; Hounslow and Maher (1999) argue that different ingredients dominate at different depths, depending on the relative significance of carbonate production and source area aridity (reflecting fluctuations in global ice volume and reduction diagenesis leading to postdepositional loss of ferrimagnetic minerals, as discussed in Chapter 5). Despite these complexities, the Owen Ridge sediments provide an excellent magnetoclimatological archive. A particularly convincing interval, spanning some 8×10^5 years, is illustrated in Fig. 7.23. Spectral analysis indicates that the record is dominated by ~22 kyr cyclicity that can be attributed to the precession of the Earth's spin axis. This is consistent with the results of

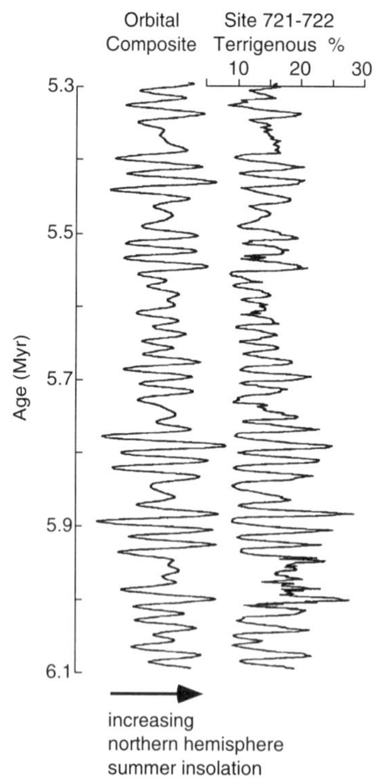

Figure 7.23 Terrigenous input into the Owen Ridge area (northwest Indian Ocean) derived from magnetic susceptibility data compared with northern hemisphere insolation derived from astronomical calculations. (Modified from deMenocal and Bloemendal, 1995.) © Yale University Press, with permission of the publishers.

atmospheric general circulation models (GCMs), which demonstrate that precession causes strong variations in low-latitude summer insolation to which the monsoons are "extremely responsive" (Bloemendal and deMenocal, 1989). When the summer solstice coincides with perihelion, summer insolation is maximized, the monsoons are intensified, and more windborne material from the African continent is delivered to the Arabian Sea.

An extremely important demonstration of magnetic monitoring of ice-rafted detritus (IRD) is provided by the work of Robinson *et al.* (1995) on a set of North Atlantic cores. Here, a major source of IRD is material derived from North American igneous rocks containing significant amounts of ferromagnetic particles. Figure 7.24 indicates that magnetic susceptibility correlates strongly with the coarse-grained (>150 μm) IRD fraction and thereby reveals individual ice-rafting episodes, the so-called *Heinrich events* (Heinrich, 1988; Broecker *et al.*, 1992). Robinson *et al.* (1995) argue that the magnitude of individual Heinrich events, as well as that of the last glacial maximum (LGM) itself, can be deduced directly from the magnetic susceptibility data by taking the ratio of the peak ("glacial") magnetic susceptibility value to

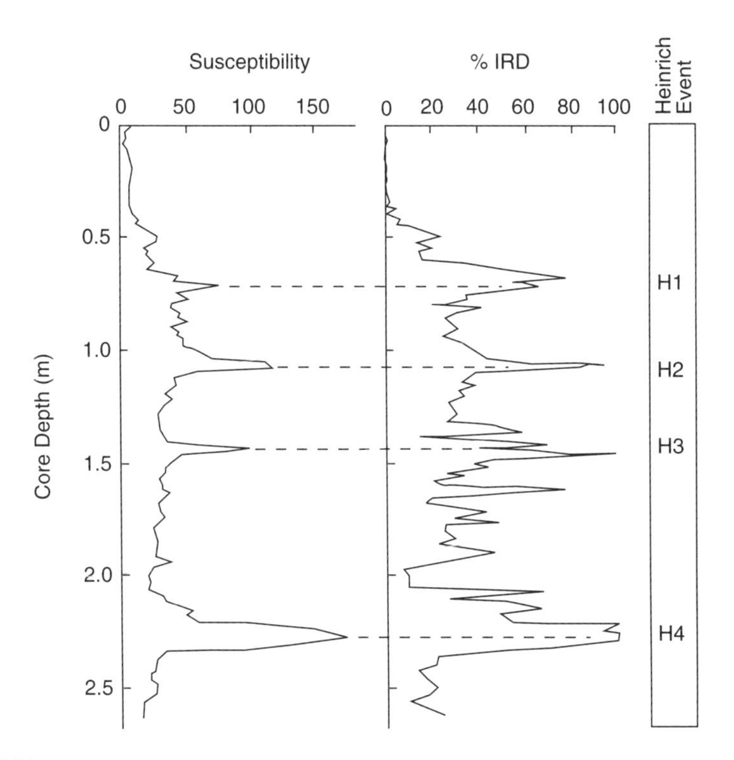

Figure 7.24 Magnetic susceptibility and ice-rafted detrital (IRD) influx maxima resulting from the so-called Heinrich events identified in a North Atlantic core studied by Robinson *et al.* (1995). Susceptibility is in 10^{-5} SI units; %IRD denotes abundance of lithogenic grains $>150\times10^{-6}$ m. Heinrich events 1, 2, 3, and 4 occurred at ~14.3, 21.0, 27.0, and 35.5 ka, respectively. © American Geophysical Union. Reproduced by permission of American Geophysical Union.

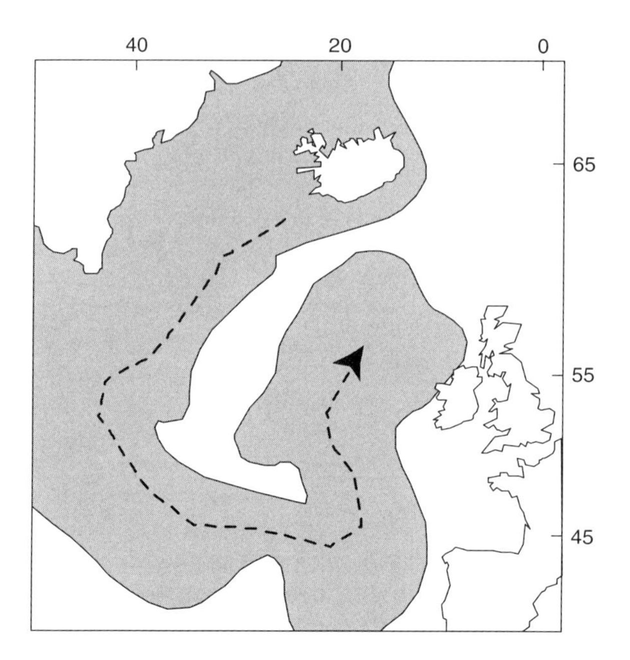

Figure 7.25 Magnetic susceptibility map of North Atlantic deep-sea sediments during the last glacial maximum (LGM: 18,000–19,000 years ago). The shaded region indicates where the ratio of the magnetic susceptibility of LGM sediments to that of the present-day sediments at the same site exceeds unity. The cyclonic pattern of the so-called Central North Atlantic Gyre is evident. (Redrawn from Robinson *et al.*, 1995.) © American Geophysical Union. Modified by permission of American Geophysical Union.

the present-day value from the same core. This is necessary in order to allow for geographic variations due to other influences, such as nearby volcanic sources. In this way, they were able to map the distribution of deep-sea sediments in the North Atlantic during the LGM (18,000–19,000 years ago) and thus to reveal the cyclonic pattern of paleoceanographic surface currents, particularly the so-called *central North Atlantic gyre* (Fig. 7.25).

Whole-core magnetic susceptibility scans have also been used convincingly to map the geographic spread of specific Heinrich events emanating from the Laurentide ice sheet (Fig. 7.26) (Dowdeswell *et al.*, 1995). Thouveny *et al.* (2000) have succeeded in magnetically identifying many of the Heinrich layers as far east as the Portuguese margin. Furthermore, climatic fluctuations that correlate with the Heinrich events have now been recognized magnetically in marine sediments off the coast of the Antarctic peninsula (Sagnotti *et al.*, 2001). In this case, however, the enviromagnetic signal does not immediately stem from changes in detrital magnetic input but rather from changes in detrital organic input. Cooling leads to extension of the ice cover, which reduces organic input. This, in turn, inhibits the authigenesis of iron sulfides. The final outcome is that these cooling events are represented by a magnetic mineralogy consisting solely of detrital magnetite, whereas sediments from other time intervals contain a mixture of magnetite and pyrrhotite. This means that the cold intervals

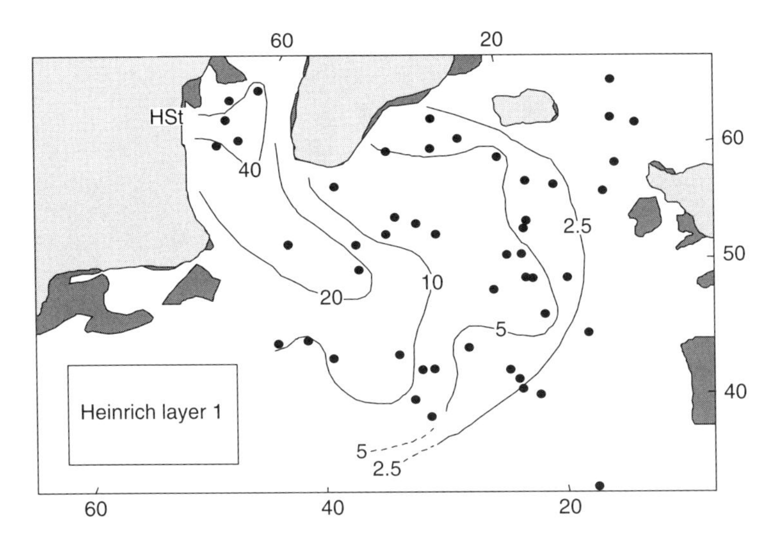

Figure 7.26 Heinrich layer 1 (~14,300 years BP). The contours show the thickness of this layer (in cm) deduced from whole-core scans of magnetic susceptibility. The Hudson Strait (HSt) is thought to have been the major ice stream through which detrital material from the Laurentide ice sheet was supplied to the North Atlantic. Light gray represents the main ice sheets, darker gray represents exposed land, and dots indicate the core locations. (Redrawn from Dowdeswell *et al.* 1995.)

are characterized by coercivity minima, the recognition of which enables Sagnotti and his coauthors to make the important step of putting forward an interhemispheric correlation of climatic changes.

A similar investigation by Kissel *et al.* (1999) involving seven deep-sea cores ranging over a wide span of latitude, from the Nordic seas (67°N) as far south as Bermuda (33°N), provides evidence of changes in North Atlantic Deep Water (NADW) circulation paralleling the Dansgaard–Oeschger cycles during MOI stage 3. The data from one of the cores (MD95-2010, from the Norwegian Sea) are used to great effect in a report that compares the magnetic results with several other climate proxy records, including (among others) oxygen isotope measurements on Greenland ice, forams in the Santa Barbara basin, organic content in cores from the Pakistani Margin, and pollen abundances in lake sediments in Italy (Trins workshop participants, 2000).

Stoner *et al.* (1995b) studied sediments cored in the Labrador Basin off southwest Greenland. They use petromagnetic measurements (particularly susceptibility and ARM susceptibility) to investigate the last two glacial–interglacial transitions. In both cases, the magnetic signals mark the arrival of continental detritus carried by meltwater from Greenland. The older deglaciation (Termination II) is found to coincide with the 6/5 oxygen isotope stage boundary, but the younger one (Termination I) was delayed until several thousand years after the 2/1 boundary. The different magnetic signatures of the two deglaciations are in keeping with earlier interpretations of oxygen isotope data (Duplessy *et al.*, 1986; Fairbanks, 1989), which suggest that Termination II was a rapid, single process, whereas Termination

I took place in two stages. Only the second stage (Ib) is observed by Stoner *et al.* (1995b), implying that the source of the so-called *meltwater pulse Ia* was not Greenland. The geography of deglaciation events is therefore starting to be revealed, although many more cores will need to be investigated if the overall picture is to be brought into focus.

An investigation of an Eocene/Oligocene sequence of glaciomarine sediments from the Ross Sea has yielded important information concerning the onset of continent-wide glaciation in Antarctica, "one of the most important paleoclimatic events in the Cenozoic" (Sagnotti *et al.*, 1998). The lower half of a 700-m core (the CIROS-1 core, drilled at 164.6°E, 77.1°S from a sea-ice platform 12 km offshore in 197 m of water) was studied in detail and revealed an alternating series of five stratigraphic zones of high and low magnetic susceptibility (Fig. 7.27). Petromagnetic measurements indicate that magnetite is the dominant magnetic mineral in both high- and low-susceptibility zones, although the former also contain significant amounts of maghemite. Observed variations in clay mineralogy demonstrate that smectite content covaries with magnetite. The interpretation is that subaerial chemical weathering of the dolerites of the Ferrar Group delivered increased magnetite and smectite fluxes into this part of the Victoria Land Basin. At such times, Antarctica was relatively warm and humid and even had sufficient soil cover to provide the observed maghemite. (The word "relatively" is important — the actual climate was probably similar to that of Tierra del Fuego today.) Sagnotti and his co-authors conclude that a persistent continent-wide ice sheet was not established over Antarctica until after the termination of the younger high-susceptibility zone. This requires a significant delay in glacial onset (~3 million years — from the middle/late Eocene boundary to the Eocene/

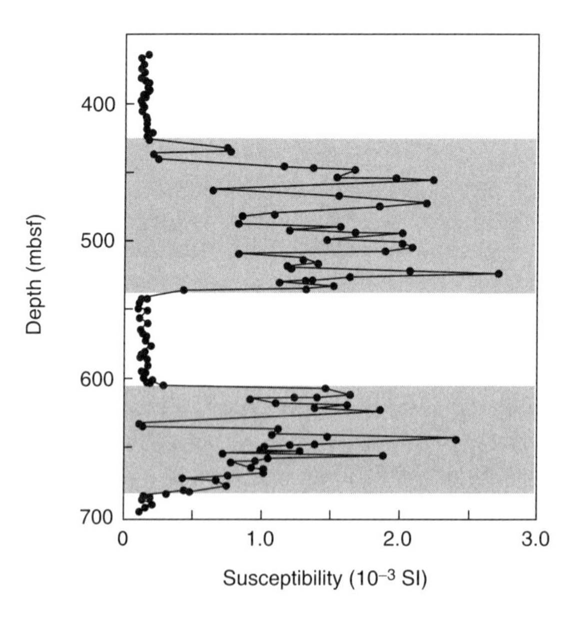

Figure 7.27 Eocene magnetoclimatology in Antarctica. The high-susceptibility zones (shaded) are interpreted as relatively warm intervals during which (at least) some of the Antarctic landmass was ice free (mbsf = meters below sea floor). (Redrawn from Sagnotti *et al.*, 1998.) © Blackwell Publishing, with permission of the publishers.

Oligocene boundary) compared with previous interpretations based on paleontological evidence (Berggren and Prothero, 1992). A fascinating discussion of this problem is given by Wilson *et al.* (1998), who provide a synthesis of paleogeography, tectonics, faunal changes, oxygen isotope data, sea level fluctuations, and oceanic currents leading to the development of what they refer to as the *Antarctic cryosphere*.

7.4.3 Biogenic Magnetite

Bacterial biomineralization is an important source of fine-grained magnetite in marine sediments (Petersen *et al.*, 1986; Chang and Kirschvink, 1989; McNeill, 1990). The magnetite grains, which are used by magnetotactic bacteria to orient themselves (see Chapter 9), are grown within an organic matrix. They generally consist of pure magnetite of ultrafine size (~100 nm) resolvable only by means of high-resolution transmission electron microscopy (HRTEM).

A possible bacterial magnetite component in cores from the eastern equatorial Atlantic was detected by Bloemendal *et al.* (1992), who showed that peaks in its concentration occurred during warm oxygen isotopic stages exhibiting a distinct 23-kyr periodicity. However, identification of the magnetic component was not confirmed by HRTEM. On the other hand, Hesse (1994) used HRTEM to determine the morphology of bacterial magnetosomes in sediments from the eastern Tasman Sea (between Australia and New Zealand), which showed fluctuations of magnetic properties throughout the Brunhes chron. Three main types of magnetosome were recognized—equant, prismatic, and elongate—and their down-core relative abundances were determined. The different shapes are thought to represent different species, and changes in the overall bacterial assemblage were therefore attributed to paleoecological fluctuations (see discussion in Chapter 5). Hesse (1994) was also able to show that increases in stable single-domain bacterial magnetite were associated with warm oxygen isotope stages. He concluded that changing oxygenation of the sediments was the most likely control of the productivity and species composition of the magnetotactic bacterial populations and that these records thus reflect the influences of factors such as bottom-water oxygen concentration and sedimentation rate, both of which are climatically modulated.

In a study of a core taken from the Chatham Rise east of New Zealand, Lean and McCave (1998) give further evidence of the importance of bacterial magnetite. They observe susceptibility peaks $>20\times10^{-8} m^3/kg$ in isotope stages 1 and 5, with low values ($\sim 10\times10^{-8}\ m^3/kg$) in stages 2, 3, 4, and 6. Furthermore, they obtained excellent electron microscope images of distinctive chains of 20- to 200-nm magnetite grains of bacterial origin. Measurements of the size and axial ratios of such grains indicate that they fall in the single-domain (SD) field (Fig. 7.28), which is essential for magnetotaxis to be effective (see Chapters 2 and 9).

7.4.4 Postdepositional Diagenesis

The loss or transformation of magnetic mineral grains through geochemical processes poses a very real threat to the ultimate survival of geomagnetic polarity records

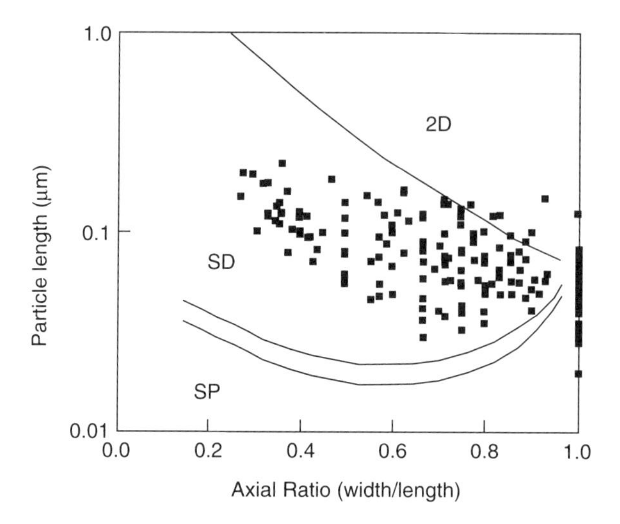

Figure 7.28 Domain state of magnetosomes identified in deep-sea sediments near New Zealand. Three fields predicted by theoretical calculations are indicated (SP = superparamagnetic, SD = single domained, and 2D = two-domained particles) (Butler and Banerjee, 1975). See also Chapters 2 and 9. (Redrawn from Lean and McCave, 1998.) © Elsevier Science, with permission of the publishers.

in a sedimentary sequence (see Chapter 5). Because such records have played a major role in working out suitable chronologies for many important marine records, it is only natural that considerable attention has been devoted to this problem. Tarduno (1994) draws attention to the fact that in cores from the equatorial western Pacific (Leg 130 of the Ocean Drilling Program) there is a broad correspondence between the water depth at the drilling site and the maximum subbottom depth to which an interpretable polarity zonation is preserved. As the water depth goes from 2521 to 3862 m, the depth to which a measurable polarity signal can be retrieved goes from 11 to 48 mbsf (meters below sea floor). At these depths in each core, the strength of the magnetic remanence drops by several orders of magnitude, leading to the loss of meaningful magnetostratigraphy due to dissolution of magnetite. Small grains should be preferentially removed by chemical dissolution, leading to a net coarsening that might be detectable by magnetic hysteresis measurements because of the known inverse relationship between grain size and coercive force (Stacey and Banerjee, 1974; Heider *et al.*, 1987). This is, in fact, what Tarduno (1994) finds. At site 806A, between depths of 1 and 8 m, he observes three sharp coercivity minima that imply a doubling of the average grain size. Furthermore, when matched to the timescale, these are found to occur at 100-kyr intervals, prompting the obvious correlation with orbitally forced glacial/interglacial fluctuations. When plotted together with the corresponding oxygen isotope data, the expected correlation is found, but with a 30- to 40-cm offset (Fig. 7.29). This is because the dissolution-controlled coercivity data reflect redox boundary processes at depth, not at the water/sediment interface. Increased organic carbon supply during glacials promotes this enhanced dissolution of magnetite—it not only leads to a reduction in the measured coercive force (due

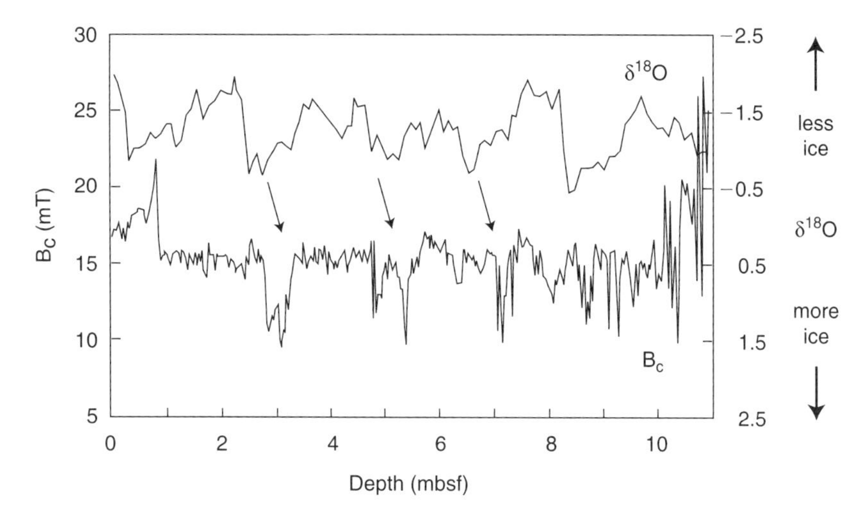

Figure 7.29 Magnetic signal of oceanic paleoproductivity. The three magnetic coercivity minima indicated are correlated by Tarduno (1994) to marine oxygen isotope stages 6, 8, and 10, reflecting increased dissolution of small magnetite particles (mbsf = meters below sea floor). © Elsevier Science, with permission of the publishers.

to preferential removal of small grains) but also raises the redox boundary from ~80 cm, as found in the present-day interglacial, to 30–40 cm. Tarduno (1994) thus draws the important conclusion that magnetic hysteresis profiling has potential as a means of gauging paleoproductivity of the oceans.

8

MASS TRANSPORT

8.1 INTRODUCTION

Environmental phenomena cannot be understood in isolation: by their very nature they are always part of a broader network of interacting processes. It is therefore a rather arbitrary undertaking to devote a separate chapter to the discussion of mass transport and its study by magnetic methods. After all, most topics discussed in this book involve movement of material. Without it, no enviromagnetic archives would be assembled in the first place! Nevertheless, it is sometimes useful to focus on the flux of material itself and its immediate outcome. For example, meteorologists may wish to monitor the amount of dust in the atmosphere, geomorphologists may wish to assess erosion in a river's catchment, stratigraphers may wish to discover the rate of accumulation of a particular set of strata, economic geologists may wish to identify the effects of fluids flowing through a sediment package, or paleoceanographers may wish to know more about oceanic currents in the past. It turns out that magnetic data have yielded useful information concerning aspects of each of these. In this chapter, we describe a few examples chosen to illustrate the kinds of problems that can be addressed by means of magnetic measurements. These involve transport of material in the atmosphere, in the hydrosphere, and in the lithosphere. Most of them follow the central theme of this book by focusing on the relatively recent past, but in order to demonstrate the wide diversity of problems that can be attacked magnetically, a few cases from more remote geological times are included.

8.2 DUST FLUX AND CLIMATE

Atmospheric dust plays many important roles, such as neutralizing acid rain, delivering nutrients to marine and terrestrial ecosystems, and acting in complex feedbacks with incoming and outgoing radiation. It may even be on the threshold of becoming an exportable, money-making, commodity (Saydam and Senyuva, 2002). Nor is its influence purely local: long-distance transport of dust was recognized at least as early as 1833 by Charles Darwin (1809–1882) during the famous voyage of the *Beagle*. He concluded, correctly, that the dust that fell on the ship as it traversed the central Atlantic came from the Sahara Desert. In fact, we now know that this area is one of two major dust sources operating today, the other being the Gobi Desert. Modern estimates indicate that ~0.6 and ~0.3 Pg (Pg = petagram = 10^{15} grams) of mineral dust are transported annually from northern Africa and central Asia, respectively (Bergametti, 1992; Prospero *et al.*, 1989). These, and other smaller sources, provide much of the input into marine sediments remote from continental margins. For example, more than 75% of the sediments in the central North Pacific is eolian material derived mostly from Asian sources: it is the marine counterpart of the loess in China—which is how magnetism enters the picture.

With the first geomagnetic polarity dating of the Chinese Loess (Heller and Liu, 1982), it immediately became clear that not only had deposition started much earlier than previously thought but also the accumulation rate had not been constant. In the lower part of their now classic section at Luochuan the rate was found to be 4.6 cm/kyr, whereas in the upper part (above the Jaramillo subchron), it increases to 7.3 cm/kyr. These broad outlines of depositional history, established by means of polarity stratigraphy based on magnetic remanence, can be refined by appealing to another magnetic parameter—susceptibility. A good example of the procedure is provided by data from loess in the Tadjik depression (central Asia) covering the last million years (Forster and Heller, 1994; Shackleton *et al.*, 1995; Ding *et al.*, 2002). By establishing tie lines between the susceptibility profile at Karamaidan (see Fig. 7.8) and the MOI timescale (which is itself well dated by astronomical tuning, see Fig. 6.12), Forster and Heller were able to show that the sediment accumulation rate (SAR) fluctuates up and down between an average value of 19 cm/kyr for loess layers and 7.4 cm/kyr for paleosols (Fig. 8.1). In the same way, at Luochuan itself, Porter (2001) concludes that the SAR was >30 cm/kyr during the last glacial interval but <10 cm/kyr during the preceding interglacial. All of this is in keeping with reduced atmospheric dust loading during interglacials due to weaker and less frequent winds and reduced ablation in the source areas as a result of more extensive vegetation cover.

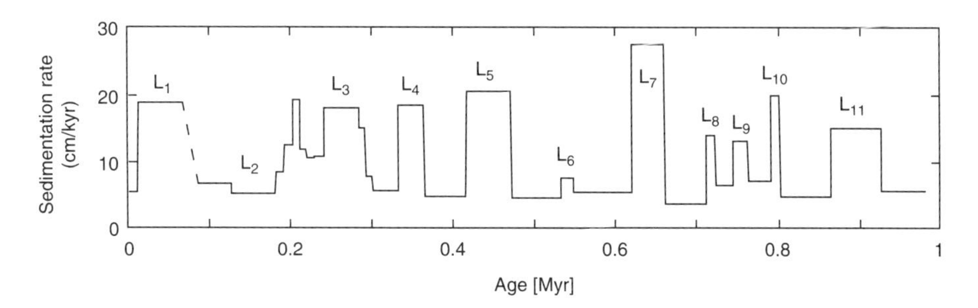

Figure 8.1 Sediment accumulation rates in the loess/paleosol section at Karamaidan, Tadjikistan. The anomalously low value for L_2 is attributed to an erosional boundary observed in the field. (Modified from Forster and Heller, 1994.) © Elsevier Science, with permission of the publishers.

Sediment accumulation rate (SAR) can be converted to mass accumulation rate (MAR) by multiplying by the bulk density [(mass/unit surface area/unit time) = (stratigraphic thickness/unit time) × (mass/volume)]. Ding *et al.* (1994) have reported SAR values for the Chinese loess site at Baoji, which Kohfeld and Harrison (2001) convert to MAR values assuming an average density: they obtain values of up to $100\,g/m^2/yr$ in warm MOI stages 3, 5, 7, 9, and 11, and 400 to $800\,g/m^2/yr$ for the intervening even-numbered cold stages. Sun *et al.* (2000) have produced MAR maps of the loess in China. During MOI stage 5 (the last interglacial), MARs were between 20 and $200\,g/m^2/yr$ at most sites but were significantly higher (50 to $>1000\,g/m^2/yr$) during stage 2 (the last glacial maximum, LGM). Even if the entire modern output of dust from the Gobi plume ($\sim 3\times 10^{14}\,g/yr$, see earlier) were to be delivered to the Chinese Loess Plateau, it would provide no more than $\sim 500\,g/m^2/yr$. Because much of the dust must actually travel farther downwind and out over the Pacific, it is clear that the supply must have been considerably greater at various times in the past. Combined with dust flux estimates obtained from ice cores and marine sediments, these loess MARs give an indication of how the atmospheric dust burden has fluctuated in the past. Thus, they provide significant input into current efforts to understand global climate change as part of projects MAGIC (Mineral Aerosol and Glacial–Interglacial Cycles; Harrison *et al.*, 2001) and DIRTMAP (Dust Indicators and Records of Terrestrial and Marine Environments; Kohfeld and Harrison, 2001) (see www.bgc-jena.mpg.de/bgc_prentice/projects). These international projects are still in their early stages, but the data compiled so far already indicate that during the LGM atmospheric dust loading was as much as an order of magnitude higher than today. Not only was the Earth colder, it was also dirtier. This is important information for Quaternary geologists and climate modelers.

8.3 EROSION AND SEDIMENT YIELD

In one of the very earliest contributions to environmental magnetism, Thompson *et al.* (1975) observed a clear correlation between magnetic susceptibility and the amount of grass pollen in sediment cores from Lough Neagh in Northern Ireland. They concluded that high susceptibility indicated increased flux of titanomagnetite grains (derived from the surrounding basaltic bedrock) as part of the material washed into the lake from soils in the catchment during times of forest clearance and soil disturbance resulting from farming. Other examples of anthropogenic disturbance of the landscape have been reported from Loch Lomond in Scotland (Thompson and Morton, 1979) and Lac d'Annecy in France (Dearing, 1979). In the United States, land-use changes have been suggested as a possible cause of magnetic increases seen in cores collected from lakes in Pennsylvania, although increased pollution from fossil fuel burning is also thought to be a possible contributor (Kodama *et al.*, 1997) (see Chapter 10).

In addition to these applications, magnetic measurements have been used to study sediment flux in estuarine and glacial settings. Oldfield *et al.* (1989) report a study of 11 cores recovered from the estuary of the Potomac River downstream from

Washington, DC. At depths of typically a meter, they observe a rapid change of mineral magnetic properties. Above this level, χ_{fd}, ARM, and SIRM increase while B_{cr} and SIRM/ARM decrease. These changes indicate a shift from an earlier assemblage dominated by hematite and/or goethite to a later one dominated by magnetite and/or maghemite with more SD and SP grains. They are consistent with an increased flux of magnetically enhanced topsoil material (see Chapter 5) resulting from land clearance and agricultural intensification since the early 19th century. These findings support the earlier conclusions of Oldfield *et al.* (1985c), who undertook a similar mineral magnetic study of the Rhode River estuary in nearby Maryland.

An example of the application of environmental magnetism to glaciation is provided by the work of Rosenbaum and Reynolds (2002), who studied a 12.8-m core from Upper Klamath Lake in southern Oregon (122.0°W, 42.4°N). They found that glacial sediments are about an order of magnitude more magnetic than postglacial sediments [mean values: magnetic susceptibility, 4.4×10^{-6} compared with $0.62\times10^{-6}\,m^3/kg$; ARM (peak field = 100 mT, bias field = 100 nT), 3.6×10^{-3} compared with $0.30\times10^{-3}\,Am^2/kg$; IRM (1.2 T), 7.5×10^{-2} compared with $1.0\times10^{-2}\,Am^2/kg$]. Using these data in conjunction with other magnetic and grain size measurements, a high-resolution history of glaciation for the eastern Cascade Range was worked out from the observed downcore magnetic variations.

These examples concerning the hydrosphere and cryosphere yield useful *qualitative* information about changes in sediment input into various depositional environments, but it has also proved possible in other cases to obtain *quantitative* estimates of material flux. The concept is straightforward: simply measure the thickness of sediments in cores taken at several spots in a lake, tie the measurements to a suitable chronology, and calculate the volume of material (or mass, if the density is known) entering the lake per unit time. In practice, it is not so easy. One of the persistent difficulties is that of correlating from core to core in order to build up a complete picture of the sediment architecture. It was shown by Thompson (1973) that magnetic susceptibility scanning of sediment cores provides a reliable, rapid, and nondestructive means of doing this. Many studies followed, throughout Britain (Bloemendal *et al.*, 1979; Dearing *et al.*, 1981; Hutchinson, 1995), in Sweden (Dearing *et al.*, 1987), and in Papua New Guinea (Oldfield *et al.*, 1985a). A particularly instructive example is that reported by Snowball and Thompson (1992) for the Welsh lake Llyn Geirionydd (3.8°W, 53.1°N). A combination of susceptibility, SIRM, pollen analysis, and ^{14}C dating on 16 cores was used to compile the sedimentation history throughout the Holocene (the last 10,000 years), and these were converted to average sediment yields in tonnes per hectare per year. An increase is observed from 0.02 t/ha/yr (= $2\,g/m^2/yr$) in the early Holocene to 0.05 t/ha/yr over the last 4000 years. The authors attribute this change to postglacial climatic warming and Neolithic forest clearance and agriculture. In a similar study using magnetic susceptibility measurements of 52 cores from the delta of the River Rhone where it enters Lake Geneva, Loizeau *et al.* (1997) discovered that, since 1961, there has been an annual deficit of 250,000 tonnes of sediment entering the lake. These "missing" sediments are all trapped upstream and now reside in several hydroelectric reservoirs constructed over the last few decades.

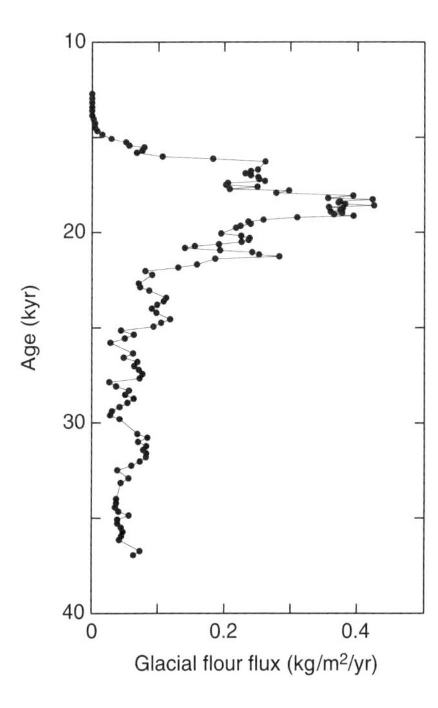

Figure 8.2 Flux of glacial flour entering Upper Klamath Lake, Oregon. (Data kindly provided by Joe Rosenbaum.)

In the case of the glacial history of Oregon described earlier, Rosenbaum and Reynolds (2002) deduce that the flux of glacial flour scraped off the highly magnetic volcanic bedrock and ground up by glacial action peaked about 18,500 years ago at a value of $\sim$400 g/m^2/yr, an increase by a factor of 4 to 8 over the preglacial background (Fig. 8.2). Then, as postglacial conditions set in (after 14,000 years BP), the glacial flour input fell to zero.

In the marine realm, Shackleton and Crowhurst (1997, see also Chapter 6) have made remarkable progress in using orbitally tuned magnetic susceptibility sequences to estimate sediment fluxes into the western equatorial Atlantic. For example, in the interval studied (5 to 14 Ma) they obtain terrigenous input varying between $\sim$5 and $\sim$20 g/m^2/yr (with similar, out-of-phase, variations in carbonate content). Bleil and von Dobeneck (2002) have investigated the mineral magnetic properties (IRM, ARM, frequency-dependent susceptibility) of much shorter cores ($<$9 m compared with $>$300 m) collected from the same area (the Ceará Rise, off the coast of Brazil, $\sim$42°W, $\sim$4°N). First they obtain a robust chronology from parallel oxygen isotope results that correlate very convincingly with the standard SPECMAP stack (see Chapter 6) back to 200 ka BP. This provides the critical time control necessary for estimates of material flux to be attempted. Next, they interpret their magnetic downcore profiles in terms of different mineral ingredients (Frederichs *et al.*, 1999; von Dobeneck, 1998), their main interest being the magnetite/hematite ratio. This is found to fluctuate (by a factor of about 2) between lows in glacial times and highs in the intervening warm interglacials. They suggest that this pattern arises from the interplay of two distinct sources, one involving transport by water (both river discharge and ocean currents) from the Amazon hinterland, the other consisting of material brought in by winds carrying the Saharan dust plume. (See http://visibleearth.nasa.gov for images of Saharan dust storms—and many others.) Bleil and von Dobeneck speculate that the African material is relatively poor in magnetite compared with its South American counterpart, so that during glacial times (when the Sarahan dust plume was enhanced), the overall magnetite/hematite ratio of the combined material received by the Ceará Rise decreases. In this way, they arrive at peak flux estimates of South American hematite and magnetite of ~ 750 and $\sim 12\,g/m^2/ka$, respectively. Corresponding values for the African source are ~ 600 and $\sim 2\,g/m^2/ka$, respectively. Climatic changes lead to fluctuations in the balance between these two inputs. At this present early stage, such estimates must be used with caution. However, the notion of not only gauging the amount but also magnetically identifying the provenance of the material accumulating in a depocenter is an important one with considerable promise for assessing environmental inventories.

8.4 PERMEATING FLUIDS

Numerous magnetic studies have been undertaken to determine the effect of fluid penetration through sediments in a variety of settings and on a wide range of time and length scales. The many chemical effects arising sometimes leave a magnetic legacy that can be rapidly and nondestructively monitored. For example, on the small scale, landfill sites have been exploited as natural laboratories responding to the upward flux of methane and the downward flux of meteoric water, atmospheric oxygen, and CO_2. The microbially catalyzed iron (and other) reactions are controlled by the redox conditions as described in Chapter 5. Ellwood and Burkart (1996) pioneered the use of magnetic susceptibility measurements to follow the evolution of conditions within landfill sites in Texas that have been covered for different lengths of time (in their case, 1, 10, and 20 years). After 1 year, a general decrease of magnetic susceptibility is observed (mean values for 1-m cores are 5.8×10^{-8} and $3.6\times10^{-8}\,m^3/kg$ for the control sample and the 1-year samples, respectively). After 10 years, the mean has risen beyond the starting value (to $7.5\times10^{-8}\,m^3/kg$), a trend that continues for the 20-year material ($21.6\times10^{-8}\,m^3/kg$). Furthermore, at 10 years, a distinct layering (with higher values of susceptibility in the lower half of the cores) begins to form, and this continues to develop for the next 10 years, to the point where a peak value of $163\times10^{-8}\,m^3/kg$ is observed at 60 cm depth (Fig. 8.3). Ellwood and Burkart attribute the eventual magnetic enhancement to the creation of maghemite by a process that seems to take a few years to complete (under the environmental conditions existing in Texas). They argue that insoluble Fe^{3+} in the overlying soil is reduced to soluble Fe^{2+}, which then infiltrates downward and is reprecipitated as an iron oxyhydroxide precursor to maghemite. Whatever the underlying mechanism might be, this type of data obviously has the potential to monitor the hydrological and (bio-)chemical evolution of landfill sites at little cost.

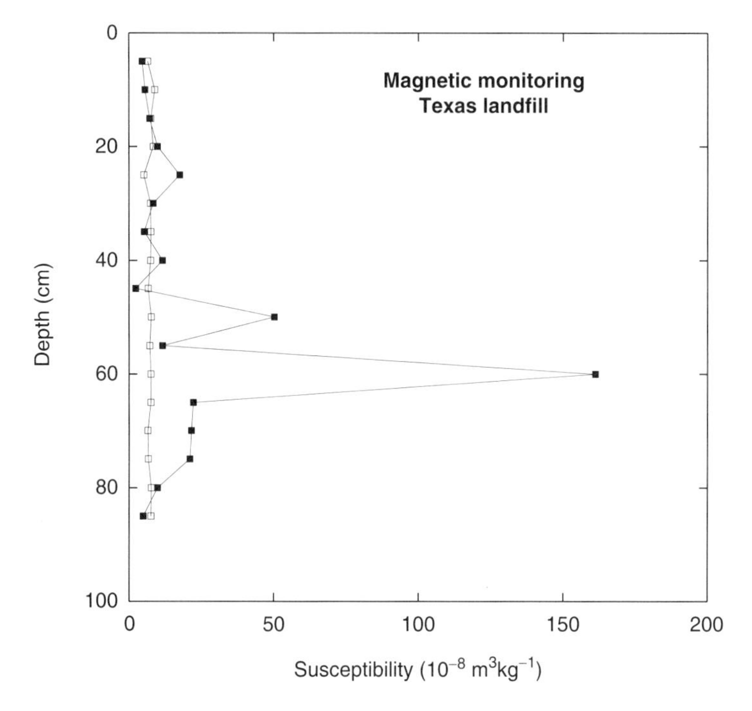

Figure 8.3 Evolution of magnetic susceptibility in a Texas landfill. Open squares represent the starting profile, closed squares show the situation after 20 years of upward-fluxing methane and downward-seeping meteoric water, atmospheric oxygen, and CO_2. (Modified from Ellwood and Burkart, 1996.) AAPG © 1996, reprinted by permission of the AAPG, whose permission is required for further use.

On a much larger spatial scale and longer timescale, many examples are available from the classic paleomagnetic literature because precipitation from permeating fluids is a potent source of remagnetization via the CRM mechanism (see Box 5.2.). If they escape proper identification, the paleomagnetic poles deduced from such results may cause considerable confusion to those engaged in the construction of apparent polar wandering paths (APWPs) and the paleogeographic reconstruction of tectonic plates. A discussion of plate tectonics is beyond the scope of this book, but a brief look at a couple of examples illustrates how magnetic responses can be used to detect pervasive fluid flow or, as McCabe and Elmore (1989) graphically put it, to probe the "ancient plumbing of sedimentary basins."

The basic idea is to look at the other side of the coin. If, for example, the APWP for a particular continent is already well known, then a magnetic overprint can be dated by reference to it. In many cases, it appears that the fluids responsible for the magnetic overprints ultimately derive from orogenic activity. It is supposed that during foreland basin evolution, sedimentary fluids migrate laterally as they are expelled toward the craton. The fluid expulsion may be due to (1) the development of overpressure during rapid sedimentation in the foreland basin environment

(Sharp, 1978), (2) the lateral compression due to the movement of thrust sheets (Oliver, 1986), or (3) gravitational flow from the mountain highlands (Bethke, 1986). A particularly illuminating example is reported by McCabe *et al.* (1989) for Devonian carbonates in the Appalachian Basin. Their investigation took the form of an east–west transect across New York State sampling the Onondaga Limestone. All the samples were shown to carry a late Paleozoic magnetization residing primarily in magnetite. Furthermore, a strong correlation was observed between magnetite content (as deduced from low-field and anhysteretic susceptibility) and the degree of illitization of detrital smectite in a bentonite layer (the Tioga Bentonite) within the Onondaga carbonates. Because illitization requires potassium and may release iron, McCabe and his coauthors argue that the illitization and the magnetite authigenesis result from the same process, namely the introduction of potassium in exotic brines expelled from the orogenic zone to the southeast.

An example of more economic interest concerns the emplacement of hydrocarbon resources into host formations in Canada (Lewchuk *et al.*, 1998). From cores drilled through Mississippian carbonates in southwestern Alberta, they obtained a magnetic overprint direction whose pole falls on the North American APWP close to the Cretaceous/Tertiary boundary (van der Voo, 1990, 1993). Lewchuk and his coauthors therefore conclude that the remagnetization was acquired during the Laramide Orogeny and reflects the migration of basinal fluids — including the hydrocarbons — into the traps they currently occupy. In other words, the host carbonates sat around for almost 300 million years before the natural gas showed up. Gillen *et al.* (1999) studied cores from a site ~350 km north of the area investigated by Lewchuck and his coworkers and found the same "migration" magnetic overprint direction. Similar studies concerning the Western Canada Sedimentary Basin have been reported by Enkin *et al.* (1997) and Cioppa *et al.* (2000, 2001). These demonstrate the robustness of the methodology and the widespread nature of the event.

The connection between paleomagnetic poles and hydrocarbon migration comes as no surprise. Close associations between magnetite and oil have been known for many years (see Elmore *et al.*, 1987 and McCabe *et al.*, 1987). The actual mechanism behind this association is not entirely clear, but one possibility involves microbial attack of the hydrocarbons. Machel (1995) summarizes this complex topic and concludes that microbial activity is likely to be important in surface and near-surface environments but that inorganic processes dominate at depth. The invasion of hydrocarbons into a sediment package lowers the redox potential and "almost invariably results in diagenetic remagnetization" (Machel, 1995, p. 9). The possible outcomes, however, range over the whole gamut from increased magnetization (due to the creation of magnetic minerals) to decreased magnetization (due to the destruction of those that were already there). What happens in any individual case depends on the exact chemical and biological conditions prevailing, as we saw in Chapter 5. One final point concerns the suggestion that the magnetic changes likely to occur in conjunction with hydrocarbon seepages may lead to detectable aeromagnetic anomalies and hence be important in the search for oil. This notion goes back to Donovan *et al.* (1979), but few undisputed cases have been forthcoming. One example (Reynolds *et al.*, 1991) involving biogenic greigite (Fe_3S_4) has been reported and is discussed in Chapter 9.

Finally, we note that obtaining chronological control by matching CRM paleopoles to APWPs is not restricted to hydrocarbons: ore bodies, particularly lead–zinc deposits, have also been successfully studied in this way (McCabe and Elmore, 1989; Pan *et al.*, 1990; Symons *et al.*, 1993). An excellent example from a Pb-Zn-Ba-F mining area in southern France has been described by Henry *et al.* (2001). They find a characteristic remanent overprint resulting from Early–Middle Eocene fluid migration related to the uplift of the Pyrénées mountain chain. In this way, they argue that the fluids were driven from the south, not from the east as previously believed.

8.5 OCEANIC AND ATMOSPHERIC CIRCULATION

The thermohaline circulation of water in the oceans plays a vital part in the overall climate system of the Earth (Broecker, 1991). Kissel *et al.* (1997) have shown how magnetic data might be used to reveal broad changes in water circulation in the oceans. Their suggestion is that a prevailing flow pattern can be imparted to the ocean-bottom sediments and preserved as a distinctive magnetic fabric in terms of the shape and orientation of the susceptibility tensor, which can be represented as a triaxial ellipsoid (see Chapters 4 and 5). For example, if the degree of anisotropy of magnetic susceptibility (AMS) observed in a core changes from one depth to another, it is supposed that the degree of alignment of the particles has changed accordingly. The susceptibility anisotropy is determined only by the magnetic particles, but grains of all compositions will respond similarly to the forces acting. The AMS ellipsoid therefore serves as a measure of the overall internal structure. It should be noted that, in general, the departure from isotropy is not great: the difference between maximum and minimum susceptibilities rarely exceeds 10%. In undisturbed sediments, the dominant feature is almost always the vertical alignment of the axis of minimum susceptibility (Tarling and Hrouda, 1993). The AMS ellipsoid thus has the shape of a slightly flattened sphere (i.e., an oblate spheroid). The intermediate and maximum axes thus lie in the horizontal (bedding) plane and may (or may not) show a meaningful pattern. A significant clustering of maxima is usually taken as evidence of increased particle alignment due to fluid flow. If the flow is weak ($<\sim 1$ cm/s), the maxima lie parallel to the direction of flow, but at higher velocities this switches to the perpendicular direction.

These ideas are by no means new: they go back at least as far as the work of Granar (1958) on varved sediments in Sweden. A useful summary of the early work is given by Hamilton and Rees (1970). Subsequently, several authors have extended AMS investigations to oceanic sediments, notably Ellwood and Ledbetter (1977) and deMenocal *et al.* (1988). In the specific example first mentioned, Kissel *et al.* (1997) collected samples every 5 cm from the top 11.5 m of a core (SU90-33) taken south of Iceland (22.1°W, 60.6°N) in a water depth of 2400 m. The site lies in the path of the so-called Iceland–Scotland Overflow Water (ISOW), a branch of the North Atlantic Deep Water (NADW). They find a clear pattern of the AMS footprint as a function of time, with higher degrees of anisotropy corresponding to interglacial periods (MOI stages 1, 3, and 5; AMS = 1.3, 1.2, and 1.8%, respectively) and lower values corresponding to glacial periods (MOI stages 2, 4, and 6; AMS = 0.3, 0.4, and 0.3%, respectively). These results are compatible with earlier geochemical data and modeling results that have also been interpreted in terms of reduced vigor, or complete cessation, of NADW circulation during glacial intervals.

Another aspect of oceanic circulation that has come under scrutiny by environmental magnetists concerns coastal upwelling. Not only is the upward flux of water an important part of the overall dynamics of the ocean, it also plays a central role in large-scale biogeochemical cycles, particularly that of carbon. This is because the deeper, colder water is rich in nutrients. Upwelling efficiency thus controls biological activity and CO_2 production (for an overview, see Thiede and Suess, 1983). This, in turn, influences the creation and/or destruction of magnetic minerals and the environmental information encoded therein. There are five main geographic areas of coastal upwelling in the world: two in the Pacific (United States/Mexico in the northern hemisphere; Peru/Ecuador in the southern), two in the Atlantic (Mauritania in the northern hemisphere; Namibia in the southern), and one in the northwest Indian Ocean (Somalia/Oman). In each of these, the interplay of oceanic currents, prevailing winds, topography, and the Coriolis force deflects warm, light water away from the coast and causes upwelling of colder water from below.

Haag *et al.* (2002) have investigated the magnetic properties of three cores collected off the coast of Mauritania between latitudes 25.0° and 21.5°N and longitudes 16.5° and 18.0°W as part of the SEDORQUA program (Sedimentation Organique marine et changement globaux au cours du Quaternaire). The sites cored lie under the path of the southerly flowing Canary Current before it turns into the North Equatorial Current flowing westward into the Atlantic. Their results for oxygen isotope stages 2 and 3 are summarized in Fig. 8.4. In the early part of stage 3 (t_1: about 45,000 years ago), NRM is stronger in the southern and central parts of the area and very weak in the north. By the end of stage 3 (t_2: about 25,000 years ago), the profile flattens out before eventually reversing its trend during the first half of stage 2 (t_3: about 18,000 years ago). Haag *et al.* (2002) interpret these observations in terms of variations in upwelling. The increased nutrient flux caused by stronger upwelling leads to greater activity of iron- and sulfate-reducing bacteria. As a result, more magnetite and greigite are produced and stronger NRMs are created. On the basis of other mineral magnetic measurements (ARM, IRM, susceptibility, coercivity spectra, high- and low-temperature properties) as well as microscope and geochemical investigations, Haag and her coworkers conclude that the observed NRMs are predominantly CRMs (see Box 5.2). Finally, they interpret the "seesaw" pattern apparent in Figure 8.4 in terms of spatial and temporal fluctuations of the Mauritanian upwelling. As yet, these are rather preliminary findings, but it appears that magnetic properties of marine sediments have some promise as a proxy for coastal upwelling and may provide useful input for the modeling of the fluxes involved in the global carbon cycle.

It is reasonable to suppose that prevailing winds in the atmosphere will affect eolian sediments (such as loess) in the same way that currents affect water-lain sediments. This has been demonstrated to be so by means of various nonmagnetic techniques such as photomicrographic analysis of preferred grain alignment,

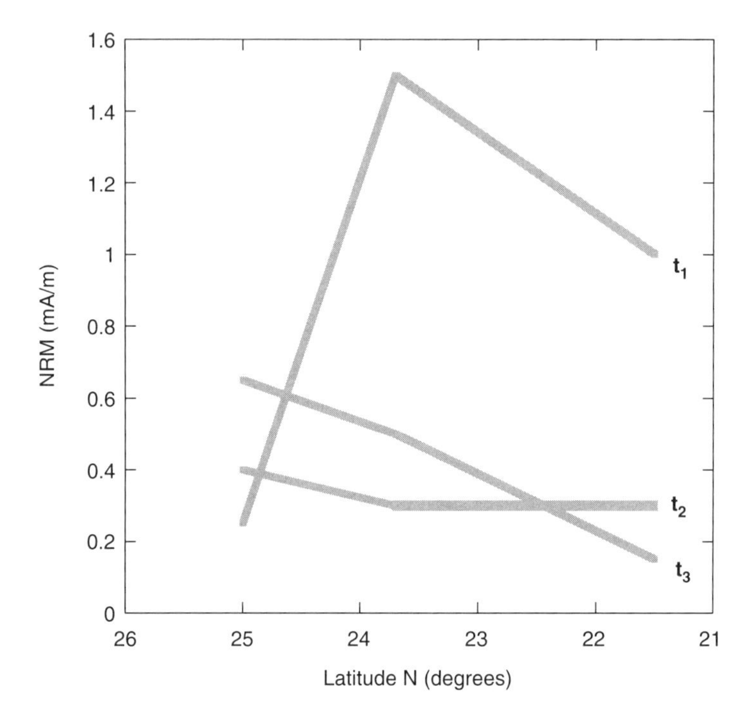

Figure 8.4 Natural remanent magnetization as a proxy measure of coastal upwelling for three time intervals (t_1 = early isotope stage 3, ~45 kyr ago; t_2 = end of isotope stage 3, ~25 kyr ago; t_3 = first half of isotope stage 2, ~18 kyr ago). The data are generalized from Haag *et al.* (2002), who studied three sediment cores collected off the coast of Mauritania (NW Africa).

determination of dielectric anisotropy, and identification of systematic geographic variations in bed thickness, grain size, and geochemistry. In particular, by comparing the regional grain size variations in the Chinese loess with that in the underlying eolian strata (the so-called Red Clay), Ding *et al.* (2000) deduce a shift in the atmospheric circulation pattern over China about 2.6 million years ago. Lagroix and Banerjee (2002) report AMS data for the famous loess/paleosol section at Halfway House in Alaska (148°27′W, 64°43′N), some 50 km west of Fairbanks. They find that the degree of anisotropy is ~4% throughout, with no systematic differences between loess and paleosol. However, the axes of maximum susceptibility are more strongly aligned in the loess beds than in the soils. This may be partly due to an originally stronger preferred alignment of grains caused by stronger winds transporting loess during glacial intervals and partly due to the degradation of any original magnetic fabric in the soils by the bioturbation associated with pedogenesis. Nevertheless, Lagroix and Banerjee conclude that the prevailing winds in this area were oriented northwest–southeast during the glacial periods corresponding to oxygen isotope stages 4 and 6.

9

MAGNETISM IN THE BIOSPHERE

9.1 INTRODUCTION

In a famous series of experiments, the Italian physiologist Luigi Galvani (1737–1798) discovered that the nervous system in animals is essentially a specialized form of electric circuit. Coupled with the discovery in 1820, by the Danish physicist Hans Christian Oersted (1770–1851), that a current-carrying wire produces a magnetic field, this implied that living organisms should be magnetic. This is, in fact, the case. But the fields involved are extremely weak and were not detected directly until about 40 years ago (Baule and McFee, 1963). The first successful experiments were carried out with induction coils, the sensitivity of which severely limited what could be achieved. In medicine and environmental health, the advent of the SQUID (superconducting quantum interferometer device) magnetometer (see Chapter 4) revolutionized the subject and made it possible to monitor the feeble magnetic fields associated with brain and heart activity in human subjects without the need for invasive surgery (Cohen *et al.*, 1970). A comprehensive analysis of modern instrumentation is given by Wikswo (1996), who points out that the sensors involved are now sufficiently small that spatial resolution is better than a millimeter, making it possible to speak of the SQUID microscope. The brief description in Box 9.1 indicates that the extremely small signals involved in these applications were theoretically predictable all along — it was simply a matter of waiting for technology to catch up so that they became measurable *in vivo*. Nowadays, *neuromagnetism* and *cardiomagnetism* are well-established physiological and clinical techniques: advanced medical facilities now offer *magnetoencephalography* (MEG) and *magnetocardiography* (MCG) services in addition to the well-known electroencephalography (EEG) procedures. For collections of relevant papers, see Williamson *et al.* (1989) and Weinberg *et al.* (1985). For a comprehensive, fully illustrated survey of the whole field refer to Malmivuo and Plonsey (1995) — or check out the current state of the art at http://biomag2000.hut.fi/tutorial.html. A graphic example, involving a human subject, is given in Fig. 9.1, which shows a magnetic map of the side of the skull following what the authors nonchalantly refer to as "stimulation" of one of the teeth.

Box 9.1 Neuromagnetic Signals

As a simple model, consider a motor neuron connecting a muscle fiber to the brain to be a long, straight wire carrying a current, I. The tangential magnetic field, B, is given by Ampère's law,

$$B = \mu_0 I / 2\pi r$$

where μ_0 is the permeability of free space [$4\pi \times 10^{-7}$ Vs/Am (=Tm / A) see also Appendix], and r is the perpendicular distance from the wire. A biologically plausible current of 1 μA produces a field of 2×10^{-11} T (20 pT) at a distance of 1 cm (Wikswo, 1989). This model is an oversimplification, but it yields the correct order of magnitude. The nerves actually carry an *action potential* caused by the movement of Na and K ions. The action potential propagates along the neuron at ~100 m/s and gives rise to a positive magnetic peak closely followed by a trough. The strongest fields in humans are associated with the heart, for which maximum values of 50 pT have been measured.

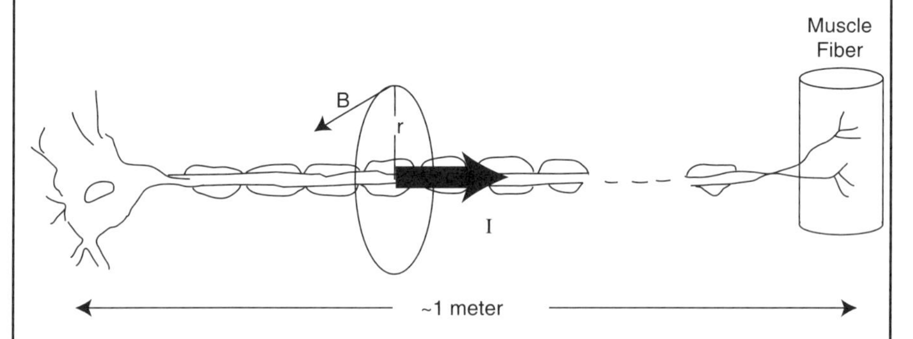

Prior to the development of the cryogenic (SQUID) magnetometer, the most sensitive instrument available was the optically pumped (cesium or rubidium) magnetometer, which typically has a sensitivity of 10^{-11} T (the earlier proton precession and fluxgate magnetometers were one or two orders of magnitude less sensitive). By comparison, SQUID configurations used in modern medical applications are able to measure fields as low as (in special cases, even lower than) 10^{-13} T. This means, for example, that the magnetic field associated with voluntary eye blinking (3–4 pT) is readily detected (Antervo *et al.*, 1985).

In contrast to the preceding applications of *biomagnetism*, which use the internal magnetic fields produced by organisms to explore various biological functions, it is possible to look at matters the other way around—in other words, to ask how organisms respond to externally applied magnetic fields. To emphasize this distinction, the study of the biological effects of external magnetic fields is sometimes referred to as *magnetobiology* rather than biomagnetism, but this terminology has not been adopted by all authors. Indeed, one of the most recent advances involves

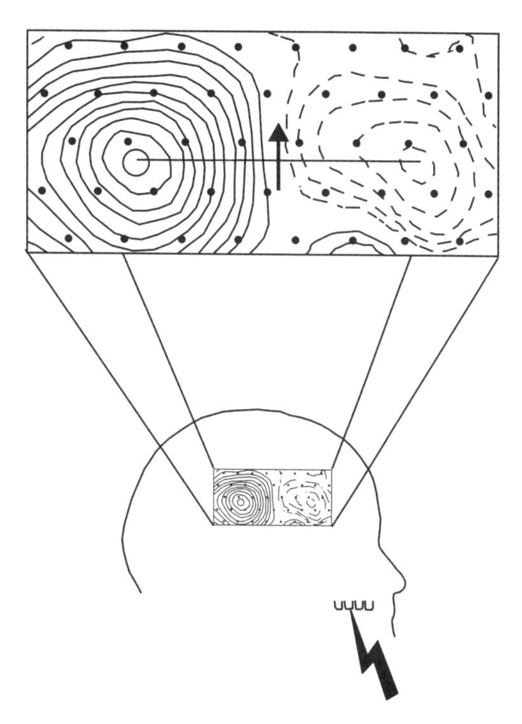

Figure 9.1 Magnetic signal produced by "stimulation" of a tooth. The small dots indicate the measurement positions, continuous (dashed) lines represent flux directed out of (into) the skull. The contour interval is 40 fT and the arrow indicates the equivalent electric current. (Redrawn from Hari *et al.*, 1985 with permission of the authors.)

what is called transcranial magnetic stimulation (TMS). In one report, an $\sim$1.5 T field pulse was applied to a small area of the subject's skull and the neuronal activity evoked was monitored by 25 standard EEG electrodes (Ilmoniemi *et al.*, 1997). It was found that the resulting maps enable connections between different areas of the brain to be worked out. Repetitive transcranial magnetic stimulation (rTMS) is proving to be an effective tool in the study of brain damage (Hilgetag *et al.*, 2001). These new clinical techniques employ external magnetic fields, yet they are generally included under the rubric of biomagnetism.

Much of the earlier work carried out in magnetobiology involved time-varying electromagnetic radiation rather than static magnetic fields, as in the ongoing debate on the possible carcinogenic effects of high-voltage power lines (Preece *et al.*, 2000). Vast sums of money have been spent on research into this touchy subject and the final word has perhaps not yet been heard. However, the U.S. National Institute of Environmental Health Sciences has come to the conclusion that the evidence for any causal link is "weak" (see http://www.niehs.nih.gov/emfrapid/home.htm). Now the debate has been extended to include the possible effects of the widespread use of mobile telephones. This is being actively investigated in many countries, but it seems to be too early for any consensus to have emerged. Indeed, some investigators argue that it will require years of tracking to assess fully the possible cumulative effects. Nevertheless, a balanced view strongly suggests that "the existing evidence for a causal relationship between RF radiation from cell phones and cancer is found to be weak to nonexistent" (Moulder *et al.*, 1999).

In addition to the possible effects arising from cultural sources, such as power lines and mobile telephones, there have been suggestions from time to time that natural variations in the ambient geomagnetic field, arising from the interaction of the solar wind with the earth's magnetosphere (see Chapter 12), may have biological effects (Dubrov, 1978). These are often based on spurious correlations and are invariably rather speculative: to our knowledge, no hard evidence has been forthcoming.

As far as steady magnetic fields are concerned, a variety of experiments have been carried out. Among the effects discovered are reduced hemoglobin content in the blood system of rabbits, modified heart function in monkeys, and the deflection of roots in certain plants (*magnetotropism*). However, these experiments always employ magnetic fields two to four orders of magnitude stronger than the geomagnetic field, so their significance to organisms living under normal environmental conditions is unclear. A comprehensive summary, giving many examples and also addressing the fundamental underpinning of the observed effects, is given by Wadas (1991).

Setting aside these fascinating—and often important—applications, we focus for the rest of this chapter on topics that involve the actual occurrence of magnetic material associated with biological organisms. It is convenient to group the topics of interest into two main categories—biomineralization and contamination—bearing in mind that, in some cases, the distinction may become a bit fuzzy. Where contamination is involved, one is mainly concerned with the intake of dust and metal aerosols, particularly into the lungs. This affects us all, especially in urban or industrial environments, but it is particularly significant in certain occupations such as welding and mining. Because such material is potentially harmful, there is obviously a link to medicine—as a monitoring technique, for example. For many purposes, however, it is simply a form of pollution, using organisms as passive collectors. We therefore include it as part of Chapter 10, under the heading *pneumomagnetism*. By its very nature, contamination is essentially accidental, whereas biomineralization is always the outcome of some specific biological purpose.

9.2 BIOMINERALIZATION

Biomineralization is a vast subject. In a nutshell, it is the process by which organisms convert ions in solution into solid minerals such as bones and shells, but it also includes mineral waste products resulting from ordinary metabolism. Some organisms are so effective at managing the microarchitecture of certain structures that serious efforts are being made to copy nature's processes to create new nanometer-scale technological materials *(biomimetics)* (Mann, 1993). In general, the most common cation involved is Ca, but Fe is the second most common metal (Simkiss and Wilbur, 1989). It has even been proposed that iron biomineralization may have played an important role in the very origin of life (Williams, 1990).

For our purpose, the most important iron biominerals are magnetite (Fe_3O_4), greigite (Fe_3S_4), and ferrihydrite ($5Fe_2O_3 \cdot 9H_2O$), but others do occur [e.g., goethite

(α-FeOOH), lepidocrocite (γ-FeOOH), pyrite (FeS_2), pyrrhotite (Fe_7S_8), siderite ($FeCO_3$), and vivianite [$Fe_3(PO_4)_2$]. Magnetite, in particular, has been identified in several species including fish, birds, insects, and bacteria. For many of these, it is thought that the creatures involved use the geomagnetic field for directional guidance (*magnetoreception*), but this is by no means proved in all cases and the whole subject is really in its infancy. There is one notable exception: the investigation of magnetic bacteria is now at an advanced stage and many aspects are reasonably well understood (for an excellent review, see Bazylinski and Moskowitz, 1997). Furthermore, their wide geographic distribution and numerous ecological habitats make these humble creatures one of the few likely sources capable of leaving a magnetic record of environmental change. This is made very clear in a summary by Konhauser (1998), who points out that bacteria inhabit every conceivable environment, including extremely harsh surroundings such as petroleum reservoirs, hypersaline lakes, black smokers in the deep sea, highly polluted groundwater, acid mine drainage, and even the core of a nuclear reactor! Indeed, their ubiquitous presence and biomineralizing activity mean that bacteria are "extremely important agents in driving both modern and ancient geochemical cycles" (Konhauser, 1998, p. 91).

9.3 BACTERIAL MAGNETISM

In 1975, while still a graduate student at the University of Massachusetts, Richard Blakemore serendipitously noticed that certain bacteria he was observing in a drop of muddy water under the microscope behaved in a very remarkable way. All the individuals swam in the same direction, like soldiers on some aquatic parade ground. His first guess was that they were somehow influenced by the direction in which the light fell on the microscope slide. By covering the microscope with a box and by moving to different rooms, he immediately ruled out this option. He then hit upon the idea that the Earth's magnetic field was the culprit and quickly confirmed it by bringing a small magnet nearby (Blakemore, 1975; Blakemore and Frankel, 1981). In Blakemore's own words "To my astonishment, the hundreds of swimming cells instantly turned and rushed away from the end of the magnet!" (Blakemore, 1982, p. 219). Experiments with controlled, uniform magnetic fields quickly followed, and the observed sensitivity of these creatures became known as *magnetotaxis* (as opposed to light sensitivity, *phototaxis* and chemical sensitivity, *chemotaxis*). Two important questions immediately arose. What makes these organisms magnetic? And, does their magnetism serve any useful biological function?

The source of bacterial magnetism was soon discovered to be the presence within them of tiny crystals of pure magnetite (Fe_3O_4) that they synthesize from iron in their environment, for example, in the species *Aquaspirillum magnetotacticum* (strain MS-1). There is now an extensive literature on the biochemistry and biophysics of magnetotaxis (Frankel and Blakemore, 1990; Moskowitz, 1995; Konhauser, 1998) covering such important problems as the means by which pure magnetite is synthesized and its precise crystal structure and morphology (Mann *et al.*, 1984, 1990b;

Matsunaga *et al.*, 1991; Meldrum *et al.*, 1992). The process by which the intracellular magnetic particles are synthesized is often referred to as *biologically organized mineralization*, or *BOM* for short (see Box 9.2). [Be warned, however, that some authors prefer to use the acronym *BOB—boundary organized biomineralization*. This is because the synthesis of the particles is controlled by some type of biological structure or surface. Other authors use yet another term—*biologically controlled mineralization (BCM)*. In this book, we stick with BOM.] The BOM particles themselves generally occur in a restricted size range (generally 20 to 120 nm, see Devouard *et al.*, 1998) and each individual bacterium typically possesses 10 to 50 of them, often

Box 9.2 BOM and BIM

Iron oxides (particularly magnetite) act in a number of ways in bacterial physiology—as an energy source, as an iron storage repository, and as a means of detoxification. Magnetite is produced both inside and outside the organism. The intracellular type is strictly controlled by processes within the cell that are collectively referred to as *biologically organized mineralization* (BOM). The most thoroughly investigated species is *Aquaspirillum magnetotacticum*, in which the composition, crystallography, grain size, and orientation are highly regulated. The resulting magnetite crystals are usually arranged in chains in which each one occupies its own cytoplasmic compartment, the whole thing being called a *magnetosome*. The magnetosome membrane isolates each compartment from the rest of the cell and controls the environment in which the magnetite is formed. Ferrous iron traverses the magnetosome membrane and is oxidized into a low-density hydrous ferric oxide, which in turn is dehydrated to high-density ferrihydrite. Finally, one third of the Fe^{3+} ions are reduced and further dehydration takes place to yield magnetite.

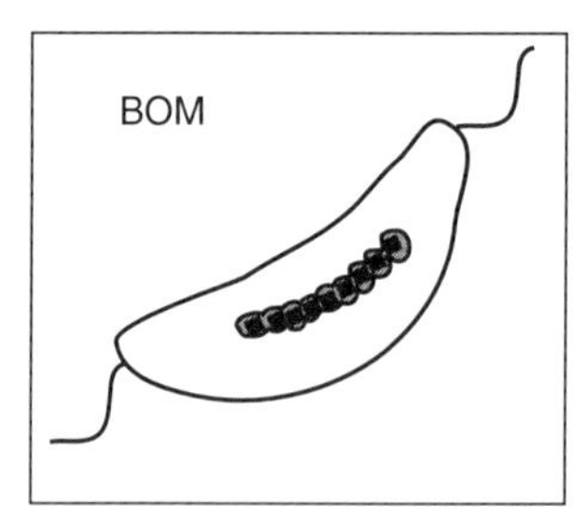

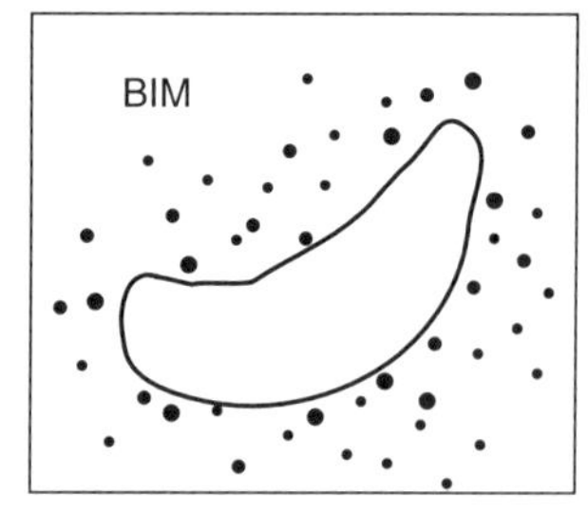

Some bacteria (such as *Geobacter metallireducens*) respire with oxidized iron (Fe^{3+}) in the form of amorphous ferric oxyhydroxide and secrete reduced iron (Fe^{2+}), which subsequently reacts with excess ferric oxyhydroxide in the environment to form magnetite. The mineral grains produced are generally poorly crystallized, irregular in shape, and often have a broad size distribution ranging well below the SP/SSD boundary. The whole process lacks the strict control of BOM and is referred to as *biologically induced mineralization, BIM*.

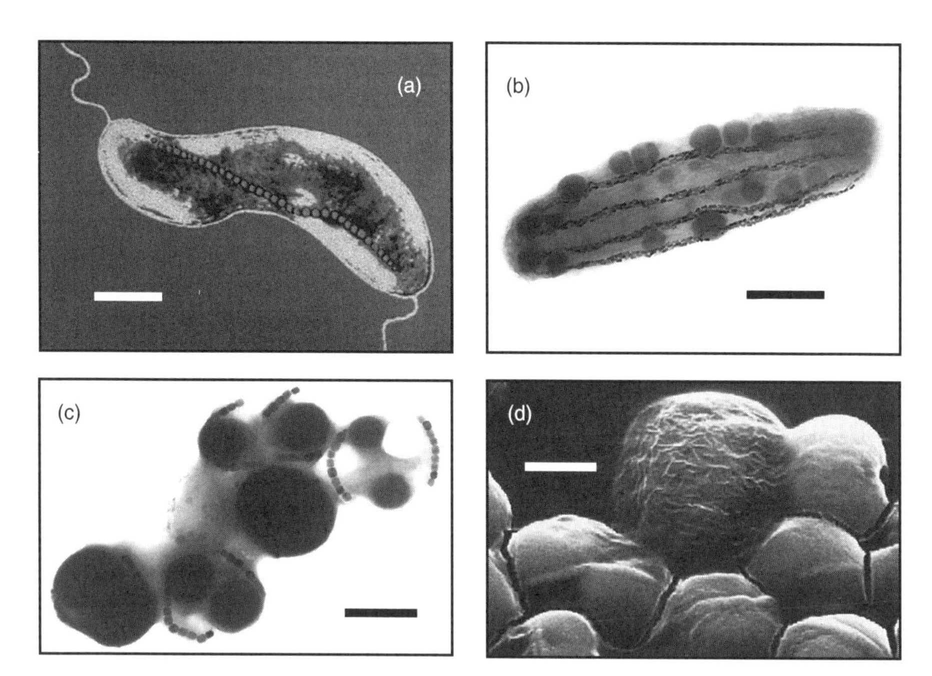

Figure 9.2 Electron micrographs of several types of magnetotactic bacteria. (a) False color electron micrograph of a soil bacterium similar to *Magnetobacterium bavaricum* with two flagellae and a single chain of 36 magnetosomes (orange) [from Williams (1990) by courtesy of Hojatollah Vali]. (b) *Magnetobacterium bavaricum* is one of the largest magnetotactic bacteria and can grow up to 12 μm in length. The rod-shaped cells contain 2–5 long chains which—again sitting near the outer membrane—are each made up of 2–3 braided subchains. The hook- or claw-shaped magnetosomes consist of magnetite. They have a size of about 100 nm and may be considered of single-domain size. Round sulfur bodies of variable color indicate different degrees of energy consumption. (c) *Coccus* bacteria from recent sediments of the Chiemsee (Bavaria) with two chains of magnetosomes. According to Hanzlik (1999), they consist of 5 to 28 cubic-octahedral to slightly prismatic magnetite single crystals of 40–110 nm size with slightly rounded edges. The large and dark spherical bodies are sulfur concentrations, which play an important metabolic role. (d) The chains of *Coccus* are located near the external cell membrane, thus providing mechanical stabilization of the body, high magnetic moments, and effective torque for fast motion (Hanzlik *et al.*, 1996b). Micrographs in (b,c) from Hanzlik (1999), kindly provided by Nikolai Petersen. Scale bars = 1 μm. © Macmillan Magazines Limited and Elsevier Science. Reprinted with permission of the publishers. See color plate.

arranged in one or more linear chains (Fig. 9.2). The magnetite chains are surrounded by a membrane consisting of a lipid bilayer admixed with proteins. Collectively, the membrane and its enclosed Fe_3O_4 crystals are known as a *magnetosome*. The restricted size range is compelling evidence of strong biological control during synthesis, and the reason for it is clear. Each magnetosome crystal is a stable single-domain particle (see Box 2.4). These creatures have thus contrived to maximize their magnetic moment per unit mass of magnetite. They are, in effect, biological dipoles that will be rotated—passively—into alignment with the local geomagnetic field. The organism then swims—actively—along the magnetic field line at speeds on the order of 100 μm/s (see Box 9.3).

Box 9.3 Magnetotaxis

Navigation in magnetotactic bacteria involves two steps. First, the organism is passively rotated in response to the torque exerted by the ambient magnetic field. Then it actively swims along the field direction. In a population of bacteria, thermal energy disrupts perfect alignment, and a statistical balance is achieved on the basis of magnetic potential energy and thermal energy. The degree of alignment is given by the Langevin function [$L(a) = \coth(a) - 1/a$], where $a = MB/kT$ (M being the magnetic moment of the organism, B the magnetic field, T the absolute temperature, and k Boltzmann's constant) (see Kalmijn, 1981). As an example, let us look at a population of identical bacteria, each containing N cubic magnetite crystals with 50-nm sides. The volume of each crystal is $1.25 \times 10^{-23}\,\mathrm{m}^3$ and the magnetic moment is $6 \times 10^{-17}\,\mathrm{Am}^2$, the spontaneous magnetic moment of magnetite being 480 kA/m. In a field of 50 μT, the magnetic potential energy is $3 \times 10^{-21}\,\mathrm{TAm}^2 = 3 \times 10^{-21}\,\mathrm{J}$ (because the tesla has the dimensions of $\mathrm{kg/As}^2$). At 300 K, the thermal energy is $4.1 \times 10^{-21}\,\mathrm{J}$. We can now plot a graph of the fractional alignment of the bacterial population as a function of N. About 20 crystals per individual suffice to get the population 90% aligned. If the crystals are larger, even fewer are needed.

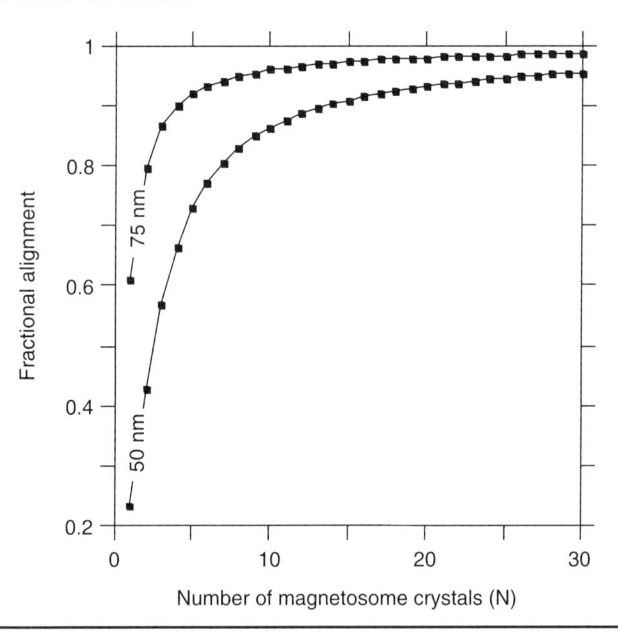

The biological function of this navigational system is to keep the organism in the proper oxygen environment. Being microaerophilic, they wish to avoid aerobic surface waters. In the northern hemisphere, not only do they swim northward, they also swim downward because of the sign of the vertical component of the geomagnetic field (see Box 12.1). Thus, if the mud in which they live is stirred up, they immediately endeavor to return to their preferred habitat. Bacteria living in the southern hemisphere (where the vertical component of the field points upward) are oppositely magnetized. Hence, not only do they swim southward but—like their counterparts in the northern hemisphere—they also swim downward, to safety.

By counting the number of magnetosomes in three *Magnetobacterium bavaricum* individuals, Steinberger *et al.* (1994) deduce magnetic moments of 16, 22, and 51×10^{-15} Am^2. This calculation assumes that all the magnetic material is single-domain magnetite in perfect alignment. The validity of the whole procedure was brilliantly confirmed by an ingenious experiment in which the same individuals were filmed swimming in a rotating magnetic field in an apparatus amusingly referred to as a "bacteriodrome" (Petersen *et al.*, 1989). A field of 160 μT (about three times the geomagnetic field in southern Germany, where these organisms live) rotating at 0.1 Hz caused the bacteria to swim in circles of ~ 25 μm radius, easily recorded with a video camera attached to an ordinary light microscope. The size of the circles is determined by the balance between the magnetic torque rotating the bacterium and the viscosity of the water resisting it. In this way, Steinberger *et al.* (1994) obtain magnetic moments of 13 ± 3, 19 ± 4, and $64 \pm 13 \times 10^{-15}$ Am^2, in excellent agreement with the estimates obtained from the electron microscope observations. Of course, dead individuals also respond to the forces acting on them, but they simply line up with the field and rotate passively with no forward swimming motion (Fig. 9.3).

Living populations of magnetotactic bacteria have been found in many different environments including soil, microbial mats, lakes, rivers, estuaries, and marine habitats (Blakemore *et al.*, 1979; Stolz *et al.*, 1986, 1989; Fassbinder *et al.*, 1990; Petermann and Bleil, 1993). For some years after their discovery, it was believed that these organisms required microaerobic conditions in order to synthesize Fe_3O_4 magnetosomes (Blakemore *et al.*, 1985), but it emerged later that some species were able to do so anaerobically, in the total absence of oxygen (Bazylinski *et al.*, 1988; Sakaguchi *et al.*, 1993). Further study then revealed that yet other species produce greigite (Fe_3S_4) anaerobically (Mann *et al.*, 1990a). All three types are potentially important to the paleoenvironmental and paleomagnetic records.

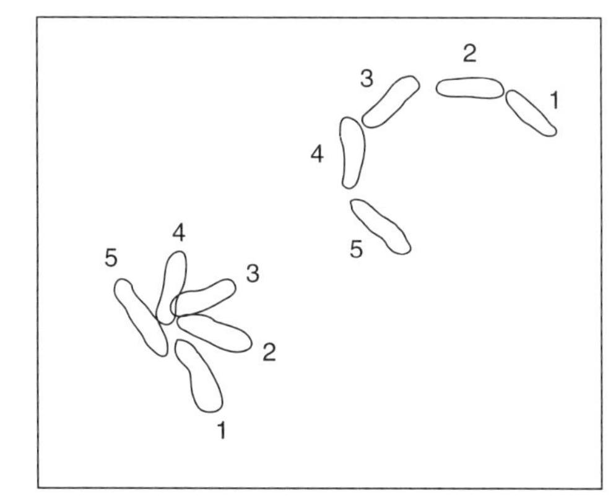

Figure 9.3 Living (upper right) and dead (lower left) bacteria (*Magnetobacterium bavaricum*) in a counterclockwise-rotating magnetic field. Traced from video images at successive times (1–5). (Modified from Steinberger *et al.*, 1994.)

After an individual bacterium dies and decays, its magnetosome crystals remain as *magnetofossils* that may make a significant contribution to the magnetic properties of the sediments in which they occur (Chang and Kirschvink, 1989; Petersen *et al.*, 1989; Stolz *et al.*, 1990). They have been reported from many different environments including:

- Holocene lake sediments in Sweden (Snowball, 1994; Snowball *et al.*, 1999)
- Holocene sediments in Lake Baikal, Siberia (Peck and King, 1996)
- Holocene hemipelagic sediments off the coast of California (Stolz *et al.*, 1986)
- Holocene carbonate sediments of the Great Bahama Bank (McNeill, 1990)
- Quaternary to Eocene deep-ocean sediments from the South Atlantic (Petersen *et al.*, 1986)
- Quaternary sediments from the Tasman Sea (Hesse, 1994) and the Chatham Rise (Lean and McCave, 1998), both in the southwest Pacific Ocean
- Cambrian limestones from Siberia (Chang *et al.*, 1987)
- Cretaceous chalk sequences in southern England (Montgomery *et al.*, 1998)
- Precambrian, Cambrian, and Tertiary stromatolitic sediments from around the world (Chang *et al.*, 1989)
- Quaternary Chinese loess (Jia *et al.*, 1996; Peng *et al.*, 2000)

Several distinct morphologies have been recognized — cubes, octahedra, elongated hexagonal prisms, bullet shapes, teardrops, and arrowheads (Fig. 9.4). These are thought to correspond to particular species (Hesse, 1994; Devouard *et al.*, 1998), although individuals containing mixed morphologies have also been observed.

In addition to the BOM magnetism responsible for magnetotaxis, there is another — potentially more important — bacterial source of magnetism. As a result of their surface properties and metabolic processes, certain species modify their local microenvironment in such a way that magnetic (and other) minerals are precipitated extracellularly. These bacteria are termed *dissimilatory*, to distinguish them from *assimilatory* bacteria, which reduce iron and incorporate it into the cell material. The type of process involved in extracellular production is known as *biologically induced mineralization*, or *BIM* (see Box 9.2 again). In this case, however, the organism exercises no control over the size and morphology of the end product. The result generally seems to be a distribution of grain sizes lying mostly within the superparamagnetic range. At the normal growth pH (5 to 8), polymers in the cell wall of bacteria are negatively charged and therefore attract and bind metal cations (such as Fe) to their surface. To do this, the bacteria do not even have to be alive. Once bound, these metals can become involved in a wide variety of subsequent reactions controlled largely by the chemical composition of the surrounding water. One common result is the production of ferrihydrite ($5Fe_2O_3 \cdot 9H_2O$) that can serve as a precursor to more stable iron oxides, such as goethite (Fig. 9.5) and hematite. Hanzlik *et al.* (1996a) find evidence for small ($\sim$10 nm) BIM particles with compositions intermediate between magnetite (Fe_3O_4) and maghemite (γ-Fe_2O_3) (Fig. 9.6). Lovley *et al.* (1987) describe the production of large quantities of magnetite by the dissimilatory iron-reducing bacterium *Geobacter metallireducens* (strain GS-15), which is not magnetotactic. The magnetite produced extracellularly is the end product of an

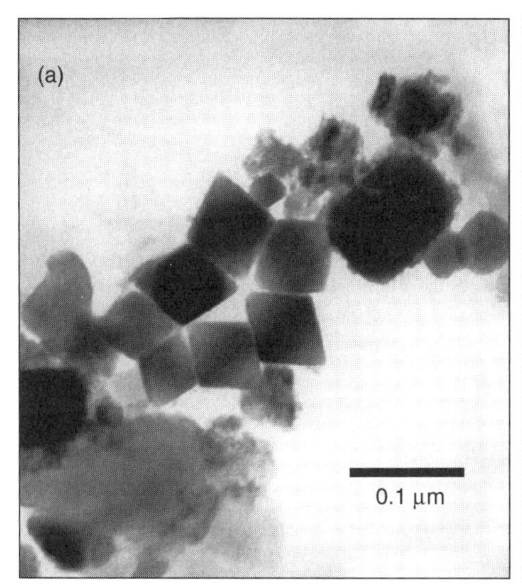

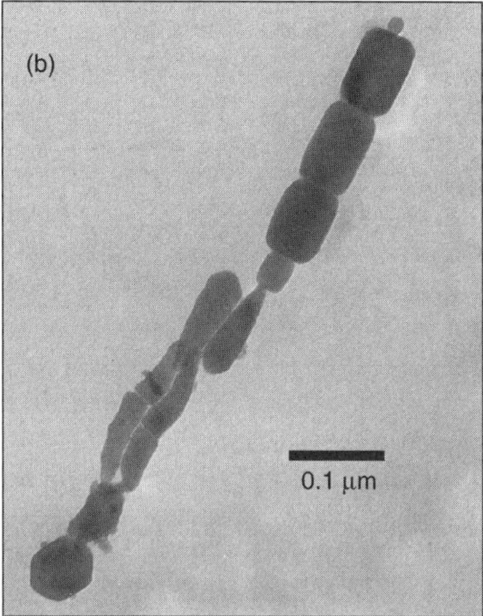

Figure 9.4 TEM images of different shapes of bacterial magnetites extracted at depths of (a) 46 cm and (b) 617 cm in core E39.72 from the Tasman Sea (154°36′E, 40°37′S, 4520 m water depth). The preference of different species (producing different shapes of magnetosome particles) to inhabit different depths (i.e., different chemical environments) is discussed by Hesse (1994) and also in Chapter 5 (e.g., see Fig. 5.5). (Images kindly provided by Paul Hesse.)

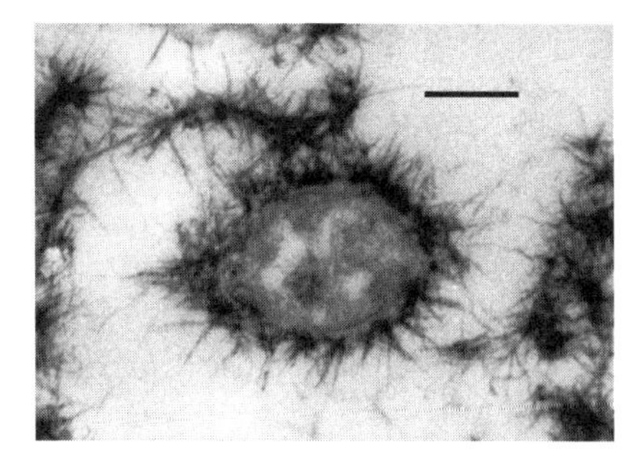

Figure 9.5 TEM image of a stained bacterial cell with crystalline, acicular (BIM) goethite. The scale bar is 200 nm long. (Image kindly provided by Kurt Konhauser.)

energy-generating metabolism that typically allows *Geobacter* to produce thousands of times more magnetite than an equivalent biomass of magnetotactic bacteria. For this reason, BIM is generally supposed to be quantitatively more important in environmental magnetism than is BOM.

Apart from their significance in magnetic studies, the metal-binding properties of BIM bacteria have been pressed into service for biorecovery of economically important metals (Basnakova and Macaskie, 1997) and for bioremediation of toxic metals and radionuclides (White *et al.*, 1995; Yong and Macaskie, 1997). It has also been suggested that bacterial oxidation of ferrous iron in the Precambrian ocean may

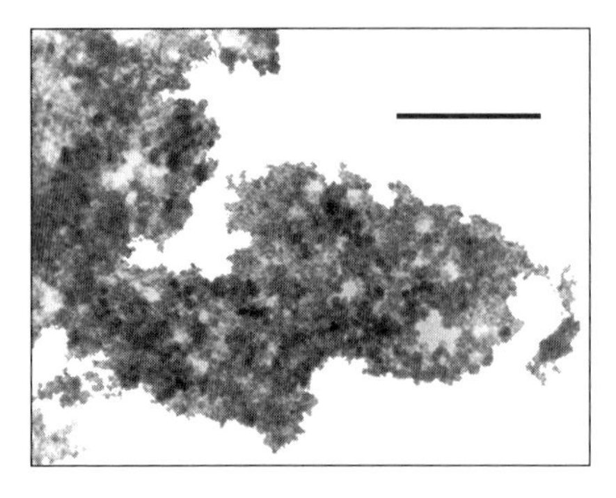

Figure 9.6 TEM image of BIM particles with compositions intermediate between Fe_3O_4 and γ-Fe_2O_3. Scale bar = 200 nm. (Image from Hanzlik, 1999, kindly provided by Nikolai Petersen.)

account for the origin of banded iron formations, the source of 90% of all iron mined today (Isley, 1995).

9.3.1 Two BOM Examples

Snowball (1994) describes a magnetic study of sediments collected from a lake (Pajep Njakajaure) in northern Sweden (18°50′E, 68°20′N). It is a small lake (380 m long and 220 m wide) with a maximum depth of 19 m. The region is situated far from potential sources of airborne pollution, and industrial magnetite (see Chapter 10) has not been found in the area. There is no significant inflow channel. The expected source of magnetic input is therefore restricted to the immediate catchment (Fig. 9.7a), which consists of gentle slopes developed on a bedrock of schist and amphibolite covered with glacial deposits. Soils — up to 25 cm thick — are podzolic, sometimes gleyed. Vegetation is birch forest and herbs. Sediments from the lake bottom were collected during winter by piston coring from the frozen surface of the lake. Soil samples from the catchment area were also studied to assess their role in the observed magnetic properties. Figure 9.7b shows the SIRM profiles for two cores. Between 370 and 200 cm, values are very low, but at shallower depths a steady rise takes place to a peak value of 28 mAm^2kg^{-1} at a depth of 35 cm. By contrast, 58 soil samples representing the entire catchment have a maximum SIRM of only 5.12 mAm^2kg^{-1}, with an average below 2 mAm^2kg^{-1}. The shortfall is even more serious when the diluting effect of the organic content of the lake sediments is taken into account. Snowball estimates that suitably adjusted maximum SIRMs would be over 40 mAm^2kg^{-1}, which is comparable to typical basalt! He concludes that bacterial magnetosomes are responsible. This is supported by other magnetic data and by direct electron microscope identification. Furthermore, the χ_{ARM}/χ diagnostic test used by Oldfield (1994) is positive, with values of 50 being observed — well above Oldfield's suggested threshold of 40 (actually, this is only part of Oldfield's test — see later for a full discussion). Snowball thus argues for the *in situ* postdepositional formation of bacterial magnetite. However, he goes on to investigate the downcore decrease in magnetism and concludes that most of the fossil magnetosomes are eventually dissolved by reductive diagenesis. As we saw in Chapters 5 and 7, this

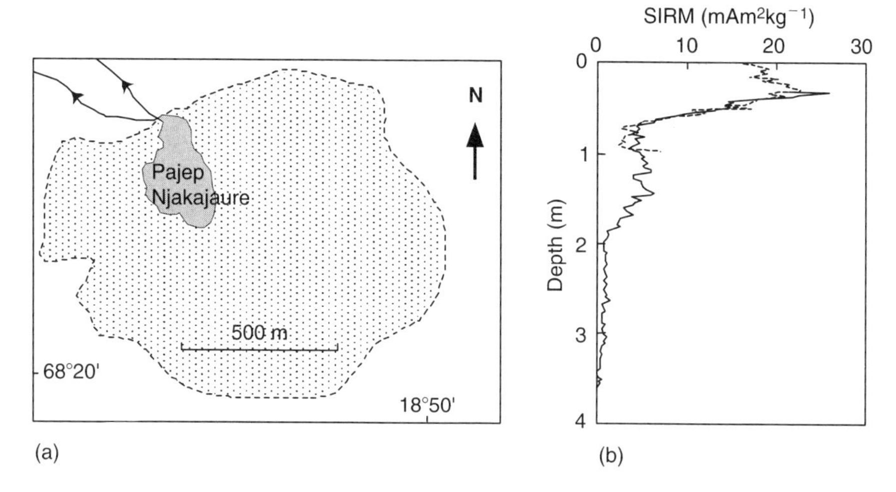

Figure 9.7 (a) The Swedish lake Pajep Njakajaure and its well-defined catchment (shaded). There are no significant inflow channels, but two outflow streams exit toward the northwest. (b) Depth profiles of saturation isothermal remanence (SIRM) in two cores from Pajep Njakajaure, Sweden. (Redrawn from Snowball, 1994.) © Elsevier Science, with permission of the publishers.

again demonstrates the potential complexities involved in interpreting environmental magnetic signals.

McNeill (1990) reports magnetic data from grab samples collected from the top 10 cm of uncemented carbonate sediments on Great Bahama Bank (78°W, 25°N), an area devoid of terrigenous sedimentation. Coercivity spectra obtained from IRM acquisition experiments yield values between 5 mT and slightly greater than 100 mT, consistent with biogenic magnetite that has undergone slight surface oxidation to maghemite (Vali and Kirschvink, 1989). McNeill also claims that comparison of the AF demagnetization characteristics of saturation isothermal remanent magnetization (SIRM) and anhysteretic remanent magnetization (ARM) (the so-called Lowrie–Fuller test, see Lowrie and Fuller, 1971) indicates that the remanence is carried by single-domain magnetite. But take care! More recent scrutiny of this test has cast considerable doubt on its ability to discriminate between single-domain and multi-domain magnetite assemblages (for a comprehensive assessment of our current understanding, see Dunlop and Özdemir, 1997). In addition to his magnetic experiments, McNeill undertook a thorough electron microscope investigation of magnetic separates. He found only fine-grained magnetite grains ranging in diameter from ~40 to ~110 nm and exhibiting various crystal morphologies (hexagonal, prismatic/cuboidal, octahedral, oval, and elongate). Multigrain chains were commonly observed, often with "progressively smaller grains toward the end, similar to the formation of new crystals in the magnetosome of magnetic bacteria" (McNeill, 1990, p. 4364). The physical dimensions of 102 grains were determined and they all fall in the stable single-domain field when plotted on a length versus axial ratio diagram (Butler and Banerjee, 1975; see also Fig. 7.28).

9.3.2 Two BIM Examples

Hanesch and Petersen (1999) describe an important example of BIM magnetism from a soil in southern Germany. Susceptibilities are $\sim 0.1 \times 10^{-6}$ m^{3}kg^{-1} in the C-horizon parent material and about twice this in the B-horizon, although with considerable scatter. The topsoil (A-horizon) generally has intermediate values (Fig. 9.8). This latter point is unusual — most soils show maximum magnetic enhancement in the A-horizon. Hanesch and Petersen were not able to detect any BOM-type magnetosomes, but industrial fly-ash magnetic spherules (see Chapter 10) with "orange peel" surfaces and typical diameters between 0.5 and 5 μm were common in the A_p-horizon. By mixing soil samples with the appropriate growth medium (i.e., by adding food), they convincingly demonstrated the presence of iron-reducing bacteria producing BIM magnetite. The magnetic susceptibility of the A-horizon material increased exponentially in the first 100 days of the experiment (by two orders of magnitude), after which the rate of increase gradually slackened toward saturation in about 200 days. Over the same period, B-horizon material increased in susceptibility by about one order of magnitude but showed no tendency to saturate. Material from the C-horizon showed no susceptibility increase over the entire time of the experiment. The enriched material (A- and B-horizons) was examined by transmission electron microscopy and it was found that the magnetic material consisted of particles of average diameter about 5 nm, which electron diffraction patterns showed to be magnetite.

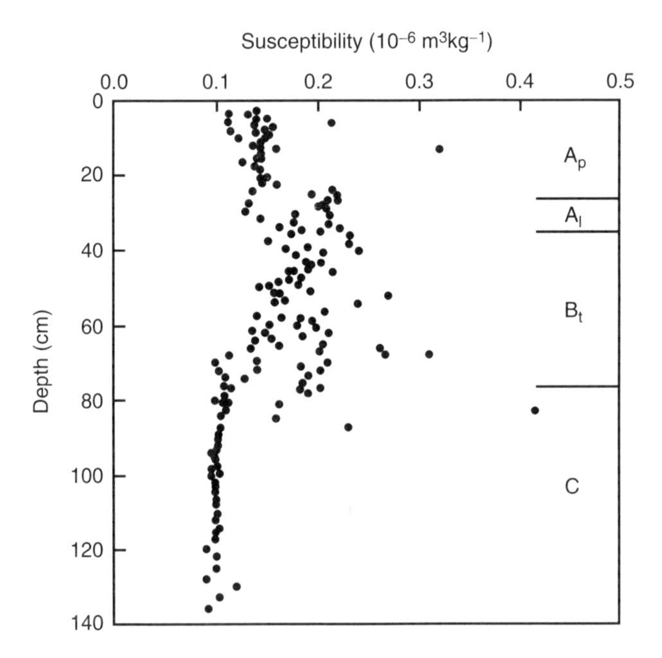

Figure 9.8 Magnetic susceptibility of a soil profile in southern Germany studied by Hanesch and Petersen (1999). © Elsevier Science, with permission of the publishers.

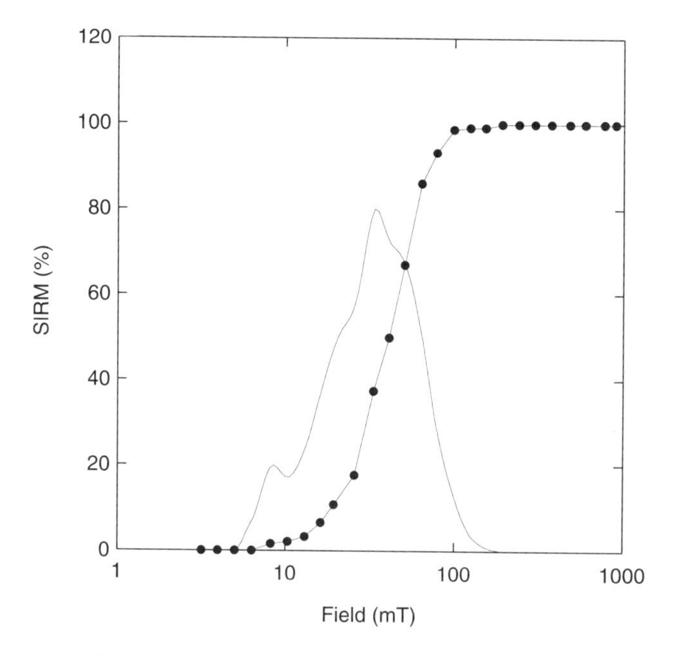

Figure 9.9 Isothermal remanence acquisition curve (and its first derivative) for BIM magnetite produced by the bacterium GS-15. (Redrawn from Lovley *et al.*, 1987.) © Macmillan Magazines Limited. Reprinted with permission.

The work of Lovley *et al.* (1987) mentioned before involved GS-15 bacteria recovered from sediments of the Potomac River, Maryland. When they inoculated the bacteria into culture medium, a highly magnetic black precipitate was formed. Transmission electron microscopy (TEM) showed this to contain aggregates of tiny crystals in the size range 10 to 50 nm, which electron diffraction and X-ray energy-dispersive analysis demonstrated to be magnetite. Progressive acquisition of isothermal remanence (IRM) by the black precipitate resulted in a coercivity spectrum typical of magnetite (Fig. 9.9), reaching 50% of the maximum at 43 mT and saturating at $\sim$100 mT. The ability of this extracellular magnetite to carry a remanence is consistent with the TEM observations, which indicate that at least some of the grains are above the superparamagnetic threshold. This suggests that BIM magnetite — like its BOM cousin — may also be important in paleomagnetism as a significant source of sediment NRM.

9.3.3 Diagnostic Magnetic Tests

The desire to avoid time-consuming electron microscopy and microbiological procedures provides a strong incentive to seek rapid magnetic tests to establish the presence of bacterial magnetite in whole samples or in magnetic extracts. The experimental quantities commonly measured in environmental magnetic work (susceptibility, ARM, IRM, etc.) were described in Chapter 4 and their applications to such issues as mineral identification, granulometry, and domain state were discussed. Sometimes a single experimentally determined quantity (such as the Curie point) is useful on its own, but experience shows that more can often be learned by defining certain combinations (H_{cr}/H_c, for example). Here, we describe certain tests that have been proposed specifically with biogenic magnetic material in mind.

Oldfield (1994) proposes a method for discriminating between fine-grained ferrimagnetic particles of detrital and bacterial origin using the routine room-temperature parameters susceptibility (χ), frequency dependence of susceptibility (χ_{fd}), and susceptibility of anhysteretic remanence (χ_{ARM}). His procedure is to combine the measured parameters into a bilogarithmic scatter plot of the two quotients χ_{ARM}/χ and χ_{ARM}/χ_{fd}. Because both denominators are the same, it is clear that there is some redundancy, but Oldfield retains these definitions in order to facilitate comparison with many other papers in which the two quotients are employed separately. Using a set of samples representing river, reservoir, lake, and marine environments, Oldfield shows that the suggested double-quotient plot successfully defines two fields (labeled A and B in Fig. 9.10), which, on other grounds (e.g., knowledge of likely source materials, comparison with synthetic analogues), can be considered to represent detrital and bacterial particles, respectively. He therefore offers this plot as a template to provide "preliminary discrimination of fine-grained ferrimagnets into dominantly bacterial or detrital assemblages in sediments and soils on the basis of magnetic measurements alone." Because no transmission electron microscopy was undertaken

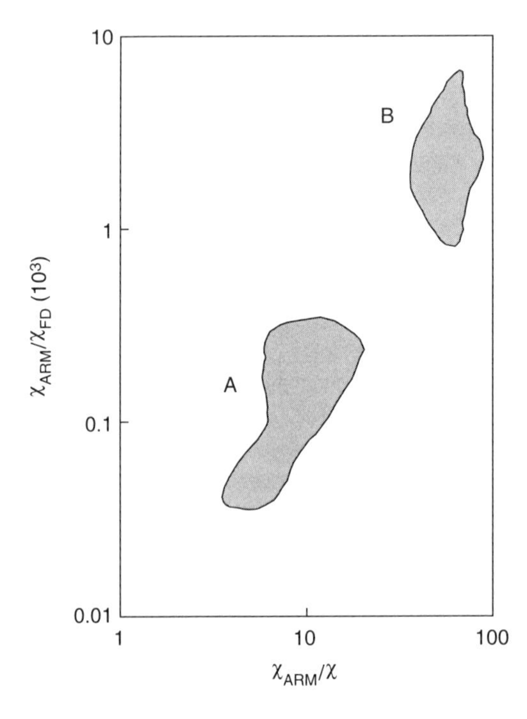

Figure 9.10 Bilogarithmic plot to discriminate between detrital (field A) and bacterial magnetite (field B). (Redrawn from Oldfield, 1994.) © American Geophysical Union. Modified by permission of American Geophysical Union.

to characterize the magnetic grains actually present, this test should be regarded as a working hypothesis and the results treated with appropriate caution. Nevertheless, it does have the merit of being based on routine measurements on complete samples rather than specialized experiments on magnetic separates.

Moskowitz *et al.* (1989) investigated several magnetic properties of carefully cultured samples of both BOM and BIM bacteria (strains MV-1 and GS-15, respectively). For diagnostic purposes, they particularly draw attention to the low-temperature behavior they observed. The samples were cooled from room temperature in zero field, given an isothermal remanence in a field of 2.5 T, and then allowed to warm up to room temperature in zero field. During warming, their IRMs were monitored and a clear difference between BIM and BOM samples became apparent. The BIM material has blocking temperatures as low as 2 K, and the IRM is already reduced to 50% at 40 K (Fig. 9.11). The grain size distribution clearly extends into the superparamagnetic region. Indeed, Moskowitz *et al.* (1989) estimate a mean particle diameter of only 9 nm. The BOM material, on the other hand, decays much more slowly with increasing temperature and still retains more than 60% of the original IRM at room temperature. Furthermore, the BOM strain exhibits a marked magnetite Verwey transition. These differences are entirely due to grain size effects. To be successfully magnetotactic, BOM bacteria necessarily favor stable SD particles. BIM bacteria have no such requirement, but the distribution in this case was found to

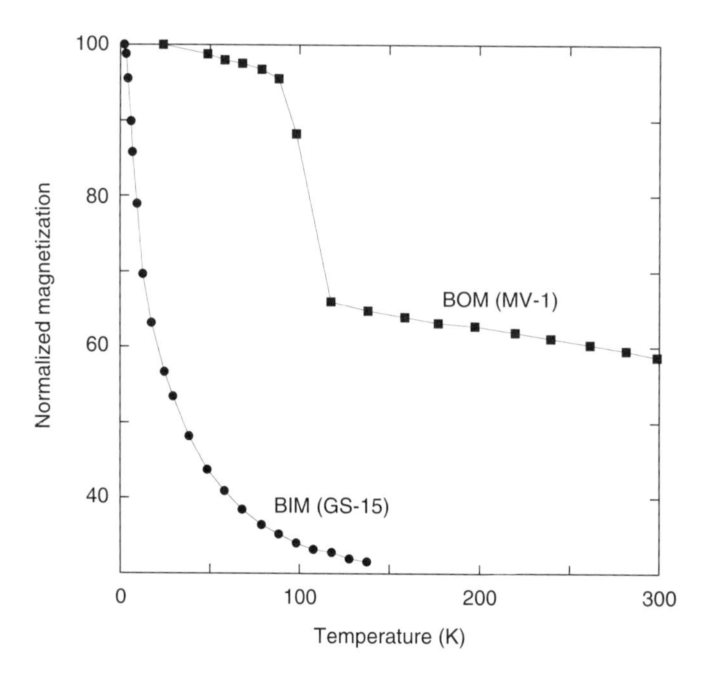

Figure 9.11 Low-temperature behavior of saturation remanence for BIM (strain GS-15) and BOM (strain MV-1) magnetite. (Redrawn from Moskowitz *et al.*, 1989.) © American Geophysical Union. Modified by permission of American Geophysical Union.

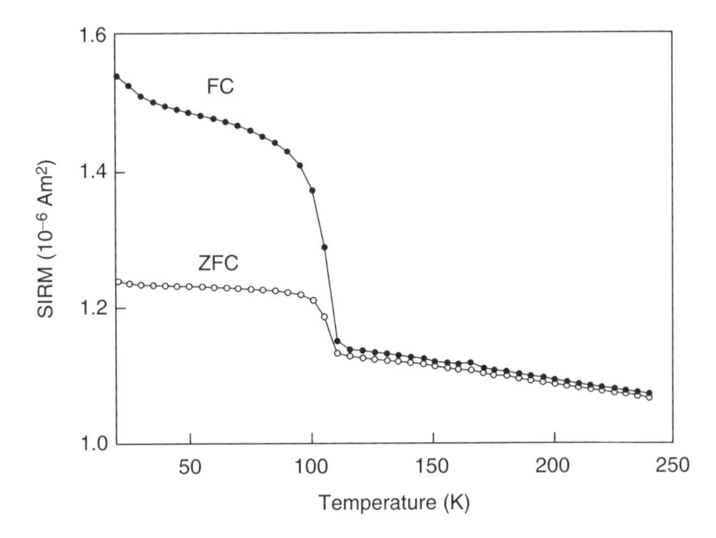

Figure 9.12 Low-temperature data used to establish the delta/delta test. The two curves refer to cooling in a field (FC) or in zero field (ZFC). (Redrawn from Moskowitz *et al.*, 1993.) © Elsevier Science, with permission of the publishers.

extend into the stable single-domain (SSD) range as witnessed by the ability of the material to retain some remanence. However, the amounts are sufficiently small that the Verwey transition is not in evidence.

In a subsequent paper, Moskowitz *et al.* (1993) broaden their analysis to other mineral magnetic properties and conclude that what they call the δ_{FC}/δ_{ZFC} parameter "is potentially the most diagnostic." This is obtained by adding a second low-temperature experiment to the one illustrated in Figure 9.11. Whereas the first experiment involves cooling in zero field, the second involves cooling in the same field (2.5 T) in which the low-temperature IRM is to be given. The ratio employed in their test thus uses field-cooled (FC) and zero field cooled (ZFC) quantities (δ_{FC} and δ_{ZFC}, respectively). These are defined as $\delta = (M_{80} - M_{150})/M_{80}$ (where M refers to the amount of initial SIRM remaining after warming to 80 or 150 K, that is, through the Verwey transition). For whole cells of BOM material, the sudden decrease seen in the ZFC curve is significantly enhanced in the FC curve (compare Figs. 9.11 and 9.12). But this is not the case if the magnetosome crystals are extracted and investigated separately. In other words, the proposed test can pick out intact magnetosome chains. How? Above the Verwey transition the magnetic easy axis in magnetite is <111>, but this switches to <100> below the transition. Moskowitz *et al.* (1993) argue that, in a single magnetosome chain, the equivalence of the various <111> directions is lost: the chain structure produces a uniaxial anisotropy along one particular <111> direction. Below the Verwey transition the FC and ZFC behaviors differ. In zero field each crystal randomly selects any one of the three possible <100> directions, but in the FC case the strong applied field favors the <100> direction

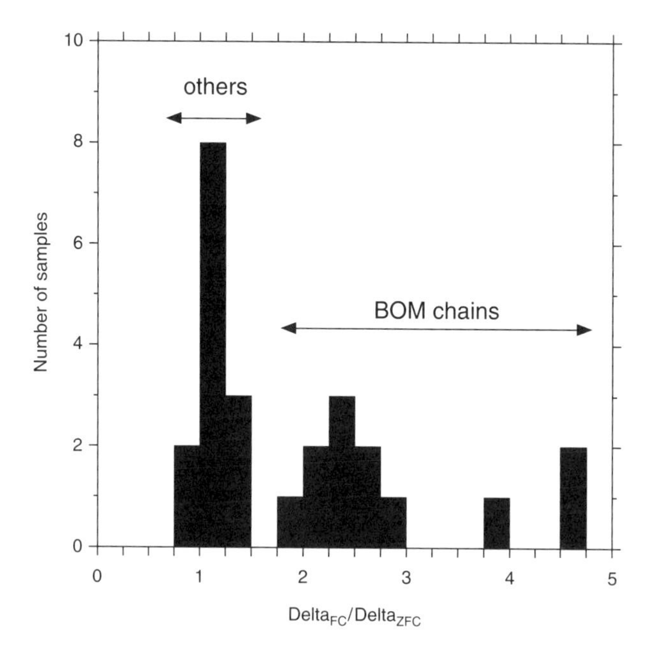

Figure 9.13 Results of the delta/delta test. Intact BOM magnetite chains yield values generally exceeding 2 (see text for explanation). (Compiled from Moskowitz *et al.*, 1993.)

most nearly parallel to itself. This partial alignment leads to a stronger SIRM below the Verwey transition. On subsequent warming through the transition, FC and ZFC curves merge as the direction of magnetization switches back to the original <111> easy axis along the chain. The net result of all this is that $\delta_{FC} > \delta_{ZFC}$. Magnetosome chains are characterized by $\delta_{FC}/\delta_{ZFC} > 2$, whereas other samples have values close to unity (Fig. 9.13). Moskowitz *et al.* (1993) recognize that "real" samples probably contain mixtures of various magnetic components, and they discuss in detail the circumstances under which magnetosome chains may still be identified by their "delta–delta" test.

9.3.4 Bacterial Greigite

As pointed out earlier, certain strains of bacteria manufacture the ferrimagnetic sulfur analogue of magnetite, namely the iron sulfide greigite, Fe_3S_4. Intracellular, BOM-type greigite was first identified by Mann *et al.* (1990a) and Heywood *et al.* (1990) in magnetotactic bacteria collected from brackish sulfide-rich sites (salt-marsh pools) in Massachusetts and California. The magnetosomes observed contain either particles of greigite alone or a mixture of greigite and nonmagnetic pyrite (FeS_2). In a more recent investigation, however, Pósfai *et al.* (1998) found no sign of pyrite. Instead, they reported the presence of tetragonal FeS (mackinawite) as a precursor to greigite (as part of the suggested pathway amorphous Fe sulfide → cubic FeS →

mackinawite → greigite). The morphology of the greigite crystals is either cubo-octahedral or rectangular prismatic. Heywood *et al.* (1990) report that the individual cells they investigated contained an average of 26 of the former (with a mean dimension of 67 nm) or 57 of the latter (with mean dimensions of 69×50 nm, but with axial ratios varying between 1.0 and 2.0). Greigite is only about one quarter as magnetic as magnetite, its spontaneous magnetization (M_s) being about 125 kA/m (Hoffmann, 1992) compared with 480 kA/m for magnetite. Nevertheless, estimates of the magnetic moment of typical cells indicate that they will definitely be magnetotactic, with magnetic energies in the geomagnetic field (~50 μT) exceeding thermal energy at 300 K by more than a factor of 10 (see Box 9.3).

The successful identification of magnetosome (BOM) greigite in living bacteria found in salt-marsh pools prompted Williams (1990) to suggest that Fe/S compounds formed the basis of energy capture in the very early Earth and are "as important as DNA in life's history." It also sparked interest in seeking its "fossil" counterpart in other settings. Within a few years, Stanjek *et al.* (1994) discovered appropriate evidence in a gleyed soil profile near Gerolfingen (10°30′E, 49°03′N) in Bavaria, Germany. Magnetic susceptibility was found to be relatively low (and uniform) for the top ~70 cm of the profile but to peak sharply at ~80 cm depth (Fig. 9.14); total sulfur content shows a very similar profile. Direct measurements

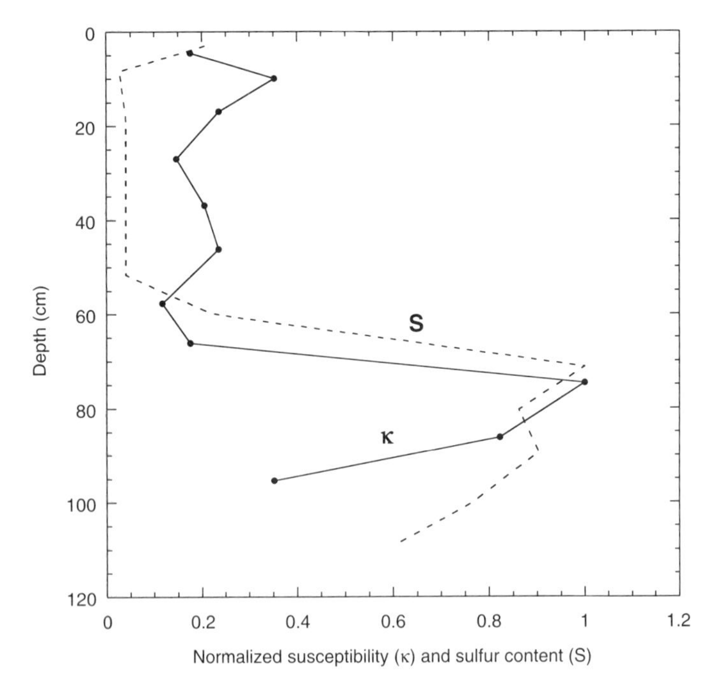

Figure 9.14 Normalized magnetic susceptibility (κ) and sulfur content (S) as a function of depth in a gleyed soil profile in southern Germany. The actual peak values are 486 × 10^{-6} SI and 5.53 mg/g, respectively. (Redrawn from Stanjek *et al.*, 1994.) © Blackwell Publishing, with permission of the publishers.

of Eh and pH indicate that the conditions prevailing at 80 cm depth fall in the greigite stability field. Plant tissues from the high-susceptibility zone could be separated with a hand magnet and were found to contain black irregular patches lining cell walls. X-ray analysis showed that this was greigite, and transmission electron microscopy indicated that this occurred in aggregates with maximum dimensions up to $\sim 1\,\mu$m. On closer inspection, these were often seen to contain bacteria, which themselves contained 30- to 50-nm irregular to elongated greigite crystals. Stanjek *et al.* (1994) also measured the relevant magnetic properties, as did Fassbinder and Stanjek (1994) in a follow-up study. As in the susceptiblity profile, peaks at ~80 cm depth are also seen in the remanence data (NRM, IRM, and ARM). High values of coercivity ($B_c = 35$ mT), coercivity of remanence ($B_{cr} = 57$ mT), ratio of saturation magnetization to saturation remanence ($M_{rs}/M_s = 0.53$), and the Lowrie–Fuller test ($MDF_{ARM} - MDF_{IRM} = +8$ mT) all confirm that the greigite is acting as stable single-domain material. This is supported by the low values observed for χ_{fd} (1.8 to 3.7%, mean = 2.8%), which show that superparamagnetic particles play only a minor role.

Sulfate-reducing bacteria also use BIM processes to respire and thereby produce extracellular iron sulfides that are both magnetic [greigite and pyrrhotite (Fe_7S_8)] and nonmagnetic [pyrite (cubic FeS_2), marcasite (orthorhombic FeS_2), and mackinawite (approximately FeS)]. Unfortunately, few details are currently known. Most studies involve the bacterium *Desulfovibrio desulfuricans*, but Bazylinski and Moskowitz (1997) argue that if excess iron is available, it is likely that all sulfate-reducing species will produce BIM iron sulfides. Because anaerobic bacteria of this kind are phylogenetically and morphologically very diverse, it is possible that BIM greigite is widespread, but whether or not it contributes significantly to the environmental and/or paleomagnetic sedimentary record is still an open question.

Examples of sedimentary greigite are known from several localities including Scotland, Sweden, Czechoslovakia, and Alaska, but there is no clear evidence that it is either BIM or BOM. In the Scottish lake Loch Lomond (5°W, 56°N), Snowball and Thompson (1988) clearly demonstrate the presence of greigite throughout most of a 4.5-m core but particularly in the lowermost 50 cm, where it is the dominant magnetic mineral. However, they argue that it most probably formed inorganically as a result of a Late Glacial intrusion of seawater. In the Swedish example, Snowball (1991) finds a high concentration of greigite in a 4-cm layer in two lakes in the Kårsavagge valley (18°30′E, 68°20′N). On the basis of magnetic measurements, he suggests that the grain size distribution lies in the single-domain and pseudo–single-domain range. As superparamagnetic particles seem to be entirely lacking, it seems unlikely that there is a significant BIM component. The Czech greigite examples (Krs *et al.*, 1990, 1992) occur in Miocene sediments of the coal basins of western Bohemia (12°E, 50°N). The greigite is "confined to sedimentary rocks that contain a fossil micro-organic substance," which provided the reducing environment necessary for its precipitation. This is reminiscent of the German soil example described before (Stanjek *et al.*, 1994), but the Czech greigite occurs as grains 4 to 8 μm in diameter—much larger than typical BIM products (see also Pósfai *et al.*, 2001). Finally, Reynolds *et al.* (1991) find that greigite is widespread in the Upper Cretaceous sediments of the Simpson oil field on the North Slope of Alaska (156°W, 71°N), where, they claim, it is the source of the observed aeromagnetic anomalies (of up to 50 nT). They suggest that the greigite was produced by bacteria using detrital organic matter and/or organic compounds derived from hydrocarbons as food sources (i.e., BIM greigite). Unfortunately, they report no microscopic investigations, so their proposal remains speculative.

9.3.5 Mars Meteorite

An intriguing set of observations concerning the ongoing debate about life on Mars was described by McKay *et al.* (1996). They studied a meteorite (called ALH84001) that was blasted off Mars by an impact event some 16 million years ago. Thereafter, it wandered through space before being captured by the Earth's gravity: it landed in Antarctica 13,000 years ago, where it was found in the Allan Hills in 1984 (hence its name). Fresh fracture surfaces display 50-μm carbonate globules that contain fine-grained magnetite (10 to 100 nm) that bear an uncanny resemblance to magnetosomes found on Earth. Thomas-Keprta *et al.* (2000) have thoroughly investigated several hundred magnetite crystals from ALH84001 and found that they can be classified into three populations: irregular (65%), elongated prisms (28%), and whiskers (7%). On the basis of their morphology, chemistry, and crystallography, the elongated prisms are found to be virtually identical to magnetites produced by the terrestrial magnetotactic bacteria strain MV-1. Thomas-Keprta and her colleagues argue that, collectively, the detailed criteria they used to make the comparison amounts to a "biosignature." The implication is that the meteorite has somehow preserved a magnetofossil record of early Martian life. Nevertheless, the frenzy of scientific research sparked by the original announcement has also produced staunch supporters of an abiotic origin. There is still no agreement, so a skeptical approach is advised (for a balanced assessment, see Treiman, 1999). If they are eventually confirmed by other observations, these BOM magnetites have importance not only for extraterrestrial biology but also for planetary physics. For example, if magnetotactic bacteria once lived on Mars, it presumably had a strong magnetic field? And a liquid conducting core?

9.4 OTHER ORGANISMS

As indicated before, magnetite is found in many different creatures representing many different taxa. In some cases, evidence is growing that this is related to some form of magnetoreception giving the organism the ability to navigate. In other cases, the magnetic properties are not, apparently, relevant. In what follows, we briefly summarize some representative examples.

9.4.1 Molluscs

Lowenstam (1962) first identified magnetite in *chitons*, a type of mollusc belonging to the class *Polyplacophora*. These invertebrates are herbivores that live in intertidal and near-tidal marine environments. They possess a tonguelike organ called the *radula* that they use to scrape microorganisms and algae off the rocky substrate. This feeding mechanism is aided by the presence of teeth on the radula that are hardened by biomineralized magnetite crystals. A great deal is now known about the biological pathways involved in this biomineralization (Webb *et al.*, 1990) and the architecture of the final structure (van der Wal, 1990). The magnetite is formed from a ferrihydrite precursor, which itself is derived from ferritin in the blood. Ferritin consists of a hollow protein shell inside which small iron particles (up to several thousand atoms) can be deposited. It is found throughout the animal world as well as in fungi, bacteria, and plants. In the chiton *Cryptochiton stelleri*, Towe and Lowenstam (1967) found that the magnetite crystallites are typically a few hundred nanometers across, much larger than the bacterial particles discussed previously. This is no surprise because chitons use magnetite solely for its mechanical hardness, not for its intrinsic magnetic properties. Lowenstam also identified the occurrence of another iron biomineral in a different class of Mollusca, namely goethite in the radula of limpets (members of the class *Gasteropoda*). In the species *Patella vulgata*, for example, the mature teeth contain single crystals of acicular goethite with lengths up to $\sim 1\,\mu m$ and widths varying from 20 to 200 nm (Mann *et al.*, 1986). Despite its interest to biologists, it is not clear how important limpet goethite is to sediment magnetism, but Kirschvink and Lowenstam (1979) calculate that chiton magnetite could well be a significant contributor to the natural remanent magnetization of marine sediments [see also later comments by Lowrie and Heller (1982) and Chang and Kirschvink (1989)].

9.4.2 Insects

Bees and butterflies contain magnetic particles, primarily in the abdomen and thorax, respectively. Kuterbach *et al.* (1982) found bands of iron-rich granules in abdominal cells of the honeybee (*Apis mellifera*) but concluded that they consisted of paramagnetic hydrous iron oxides. They point out, however, that if about one third of it were reduced to magnetite, it would suffice to explain the observed magnetic moments. In a case study of the monarch butterfly (*Danaus plexippus*), McFadden and Jones (1985) show that magnetic material (thought to be magnetite on the basis of an observed Verwey transition in low-temperature experiments) is biosynthesized during the life cycle: eggs and larvae are not magnetic, adults are. Furthermore, they find that the monarch butterfly, which migrates thousands of kilometers each year, is substantially more magnetic than other, nonmigratory butterflies. Etheredge *et al.* (1999) have shown that, regardless of the remaining problems concerning the exact nature of the magnetic material and its possible use in a magnetoreceptor, monarch butterflies do, indeed, use magnetic fields to navigate. When released in a laboratory magnetic field created by Helmholtz coils, test individuals flew mostly in one direction; when the field was reversed, they flew in the opposite direction.

9.4.3 Fish and Birds

Magnetite crystals in the 30- to 50-nm size range have been found in tissue samples of several species of fish including yellowfin tuna (*Thunnus albacares*; Walker *et al.*, 1984), chinook salmon (*Oncorhynchus tshawytscha*; Kirschvink *et al.*, 1985), sockeye salmon (*O. nerka*; Mann *et al.*, 1988), and rainbow trout (*O. mykiss*; Walker *et al.*, 1997). In a detailed study of the head of a rainbow trout, Diebel *et al.* (2000) located magnetite particles arranged in a chain about 1 μm long in what they call a *magnetoreceptor cell* located in the olfactory lamellae. They succeeded in directly imaging the magnetic state of single particles by means of magnetic force microscopy (MFM) and were thus able to prove conclusively that they are single domained and have coercivities between 20 and 40 mT, comparable to measurements made on bacterial magnetosomes.

Homing pigeons and migratory birds have long been fascinating for their remarkable navigational powers (Presti, 1985). Holtkamp-Rötzler *et al.* (1997) have demonstrated the presence of superparamagnetic (SP) magnetite within the upper beak of homing pigeons. The crystals are typically 3 nm in size and are arranged in dense clusters, typically 3 μm in size. On the basis of these observations, Shcherbakov and Winklhofer (1999) have proposed a novel magnetoreceptor in which SP particles are dispersed in a liquid surrounded by a biological membrane. Such a cluster of magnetic particles is known as a *ferrofluid*, and the proposed biological arrangement is called a *ferrovesicle*. By magnetizing the ferrofluid, an external magnetic field causes an initially spherical ferrovesicle to become an ellipsoid of revolution whose long axis is parallel to the field. If the vesicle's surface area is constant, its volume decreases, fluid traverses the membrane, and the outside osmotic pressure increases. It is then supposed that this pressure change is sensed by the nerve fibers that Holtkamp-Rötzler *et al.* (1997) observe in close association with the magnetite clusters. Shcherbakov and Winklhofer refer to the whole thing as an *osmotic magnetometer*.

9.4.4 Mammals

The first evidence for mammalian magnetite was obtained from the common Pacific dolphin (*Delphinus delphis*) (Zoeger *et al.*, 1981). Tissue from the heads of four individuals acquired measurable remanences in laboratory fields and alternating field demagnetization showed this to be magnetically soft (median destructive fields $\approx$2 mT). It was assumed to be multidomained magnetite. In one case, a single, large ($\sim$0.5 mm), iron-rich particle was investigated under the scanning electron microscope and was seen to be associated with "a stalk-like" object consisting of what the authors interpret as nerve fibers. Subsequently, several species of whale have been studied magnetically (Bauer *et al.*, 1985), although such large animals are difficult to dissect in sufficiently clean environments. Laboratory-induced remanences indicate the presence of significant amounts of magnetic material not only in the brain but also in other parts of the body. The evidence is mixed as to whether or not this is always magnetite, but a humpback whale (*Megaptera novaeangliae*) sample did yield evidence of single-domain magnetite. In most of the animals studied, large magnetic particles were present but were interpreted as contamination. It is quite possible that this is also the case for the putative dolphin magnetoreceptor.

Human brain tissue also contains magnetite (Kirschvink *et al.*, 1992; Dunn *et al.*, 1995), as do the heart, spleen, and liver. On the basis of laboratory IRMs given to tissue from eight cadavers, Grassi-Schultheiss *et al.* (1997) and Schultheiss-Grassi and Dobson (1999) estimated magnetite concentrations on the order of 1 part in 10^7 (by weight). Values were highest for the heart (mean = 185 ng/g; range 343 to 102), followed by the spleen (75 ng/g; 308 to 14), the liver (68 ng/g, 158 to 34), and the brain (63 ng/g; 164 to 17) (see Fig. 9.15). The function of this material—if, indeed, it has any—is entirely unknown. Nevertheless, it is currently a hot research topic because disruption of normal iron metabolism in the brain is a characteristic of several neurodegenerative disorders such as Alzheimer's disease and Parkinson's disease. Magnetite accumulation apparently upsets the local chemistry and leads to tissue damage and further deterioration. On the other hand, the good news is that such

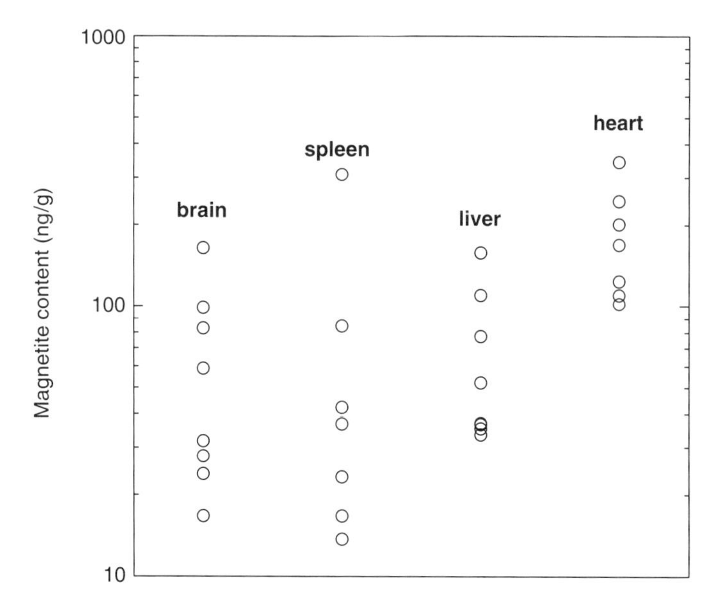

Figure 9.15 Magnetite concentrations in tissue samples from human cadavers deduced from IRM experiments. Data from Grassi-Schultheiss *et al.* (1997) and Schultheiss-Grassi and Dobson (1999). The original authors point out that the magnetic mineral present may be maghemite rather than magnetite. If this is so, the concentrations shown here must be increased by about 25% to allow for the lower saturation magnetization of maghemite [380 kA/m, compared with 480 kA/m for magnetite (Dunlop and Özdemir, 1997)].

magnetic centers may turn out to be detectable by magnetic resonance imaging (MRI) at an earlier stage than current techniques allow and thus lead to more effective treatments (Dobson, 2001). In a similar development, *magnetic drug targeting* is now being tested as a potential clinical technique in which anticancer agents bound to magnetic nanoparticles are injected into the bloodstream and concentrated in the desired area by an external magnetic field (see Alexiou *et al.*, 2000).

10

MAGNETIC MONITORING OF POLLUTION

10.1 INTRODUCTION

Many industrial processes, such as the production of steel and cement, generate airborne magnetic material, but coal-burning power plants are by far the most significant sources. Even with electrofilters working at ~98% efficiency, a single 1000-MW plant emits almost 2 tonnes of fly ash every hour (Konieczynski, 1982). Smokestacks deliver this into the atmosphere and, depending on meteorological conditions, the particles involved may travel hundreds of kilometers before settling back to the surface. Flanders (1994) observes that the amount of magnetic material deposited on trees and buildings varies inversely with distance from its source. As older particles are washed away, new ones are deposited so that a rough equilibrium is achieved. Typically, at a distance of 1 km, this equilibrium amounts to something on the order of 1 μg of magnetic material per square centimeter of surface (Fig. 10.1).

Before being burned, coal is essentially nonmagnetic. The process of combustion causes the pyrite (FeS_2) present — typically a few percent — to dissociate and form pyrrhotite (Fe_7S_8) and sulfur gas. Above about 1350 K, pyrrhotite decomposes into sulfur and iron. Spherical iron particles, typically about 20 μm in diameter, are formed and subsequently oxidize to magnetite (Fe_3O_4). Flanders (1999) calculates that if the magnetite produced by the 830 million tons of coal burned annually by U.S. power stations were spread evenly over the entire country, it would amount to ~1.5 ng/cm^2/day (~5 mg/m^2/year). Naturally, one would not expect absolute geographic uniformity. Flanders confirmed the essential validity of his calculation by measuring the magnetic properties of the material deposited on newly exposed plastic windows located more than 10 km downwind from the nearest coal-fired utility. He reported values ranging from 2 ng/cm^2/day near Philadelphia, Pennsylvania, to 0.2 ng/cm^2/day near Oviedo, Florida.

Once created, mineral particulates may suffer several different fates. At the lower boundary of the atmosphere, they may be deposited on vegetation and buildings, or they may fall directly onto the topsoil. Schiavon and Zhou (1996) have investigated

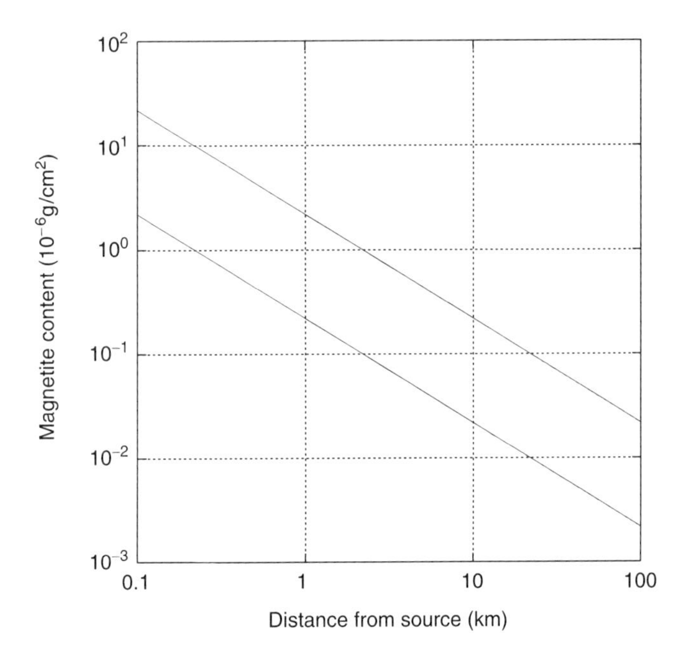

Figure 10.1 Decrease of magnetic deposition with distance from the source. This very generalized representation summarizes the numerous observations collated by Flanders (1994) involving many different sources (e.g., electric power-generating stations) and many different depositional surfaces (e.g., tree bark and leaves). The width of the band reflects variability in source strength and collector efficiency. Note that the observed $1/r$ relationship would be expected from a fixed input spread over a radially increasing sectoral area ($r\,d\theta\,dr$).

the magnetic content of weathering crusts on several historical buildings in England (four cathedrals, a church, and a chapel — all made of limestone). They conclude that coal-burning power plants and domestic fires are the main pollution sources. They also point out that the accumulated content of iron-rich particulate pollutants acts as a catalyst in oxidation reactions leading to limestone decay. An example of fly ash captured by topsoil is provided by the work of Hanesch and Petersen (1999) at a site in southern Germany. Much of their investigation is concerned with pedogenesis and bacterial magnetite (and is described in Chapter 9), but magnetic extracts indicated that industrial magnetic spherules (0.5 to 5 μm in diameter) are abundant in the uppermost 25 cm of the profile, which constitutes the A_p, or ploughed, horizon (see Box 5.3). The ability to capture and retain atmospheric particulates depends very much on the nature of the surface involved. For example, tree trunk bark is some two orders of magnitude more effective than leaves, apparently because of their different roughness. Similar observations were made by Oldfield *et al.* (1979) concerning the trapping of airborne particulates by hummock-forming plants such as are found in peat bogs. Whatever their source — and ultimate fate — magnetic dust particles are certainly ubiquitous. Flanders (1994) reports examples from trees in several countries in Europe, Asia, and America. He also points out that upholstery cloth, household dust balls, and even spider webs are efficient collectors. Using measured values of magnetic moment and coercive force and applying the empirical relationship between particle diameter and coercive force deduced by Heider *et al.* (1987), it appears that dust balls and spider webs may contain up to 10^7 magnetic particles per gram (see Box 10.1).

Fascinating as magnetic spider webs might be, they are essentially a curiosity. The important issues concerning the monitoring of pollution and the concomitant degradation of the environment require a broader overview and sampling strategy. The iron oxide particulates of which we speak are no more than a small fraction of the total dust burden but there is evidence that they can be a health risk, particularly in the smaller grain sizes (Garçon *et al.*, 2000). Their real significance, however, lies in the fact that they provide an excellent tracer that can be exploited to track the total particulate content of the atmosphere and the concentration of associated heavy metals that are potentially hazardous to plants, animals, and humans. We therefore turn our attention to some examples of greater significance, selected to illustrate various environmental components — the air we breathe, the water we drink, and the ground we cultivate.

10.2 SOIL CONTAMINATION

In the decade 1980–1990, more than 10^8 tonnes of coal were burned annually in the power plants of Katowice Province, which constitutes only 2% of Polish territory

Box 10.1 Magnetic Spider Web

Spider webs are effective collectors of atmospheric dust and can apparently possess magnetic moments as high as 10^{-4} Am^2 per gram. Corresponding coercivities of ~ 10 mT are typical. Assuming a simple model consisting of identical spherical particles of magnetite trapped on a nonmagnetic web, we first estimate the particle diameter from the empirical relationship

$$H_c = 17d^{-0.43}$$

(where H_c is in mT and d in μm). This yields $d = 3.4\,\mu$m, with a corresponding volume of 21×10^{-18} m^3. Such a particle has a magnetic moment given by

$$(21 \times 10^{-18}\ \text{m}^3)(480\,\text{kAm}^{-1}) \approx 10^{-11}\ \text{Am}^2$$

The strongly magnetic webs in question therefore have some 10^7 particles per gram. In case this seems like an enormous number, one should realize that the total mass of magnetite involved will amount to no more than ~ 1 mg (the density of magnetite being 5200 kgm^{-3}). The remaining 999 mg consists of the web material itself and other nonmagnetic dust particles, probably mostly the latter. If this is so, then the magnetic fraction accounts for about 0.1% of the dust trapped by the web.

(Fig. 10.2). This highly concentrated industrial activity (16 power plants, 17 steel mills, 56 coal mines) led to the establishment in 1970 of a regular monitoring system to assess atmospheric pollution. At the peak, in 1981, as much as 35 $g/m^2/year$ of iron oxide was deposited over much of Katowice Province. Meanwhile, in the heavily industrialized—but more strictly regulated—Ruhr District of Germany, values 100,000 times smaller were recorded. Subsequent installation of electrofilters in the Polish power stations drastically reduced the dust fall and iron deposition, but the legacy remains in the soils of the district.

Soils of arable land are inevitably disturbed by agricultural activity: a clearer picture emerges if attention is focused on soils in forests that better preserve the natural pedogenic horizons. Strzyszcz *et al.* (1996) and Heller *et al.* (1998b) report data from about 1000 samples collected from 200 soil profiles measured in pits dug in the forests of Upper Silesia (which includes Katowice Province). For local administrative purposes the forests are organized into 39 so-called superforestries. Within each of these, several pits were investigated in order to obtain a general overview of the spatial variability of magnetic susceptibility. A simple pattern emerges that is obviously

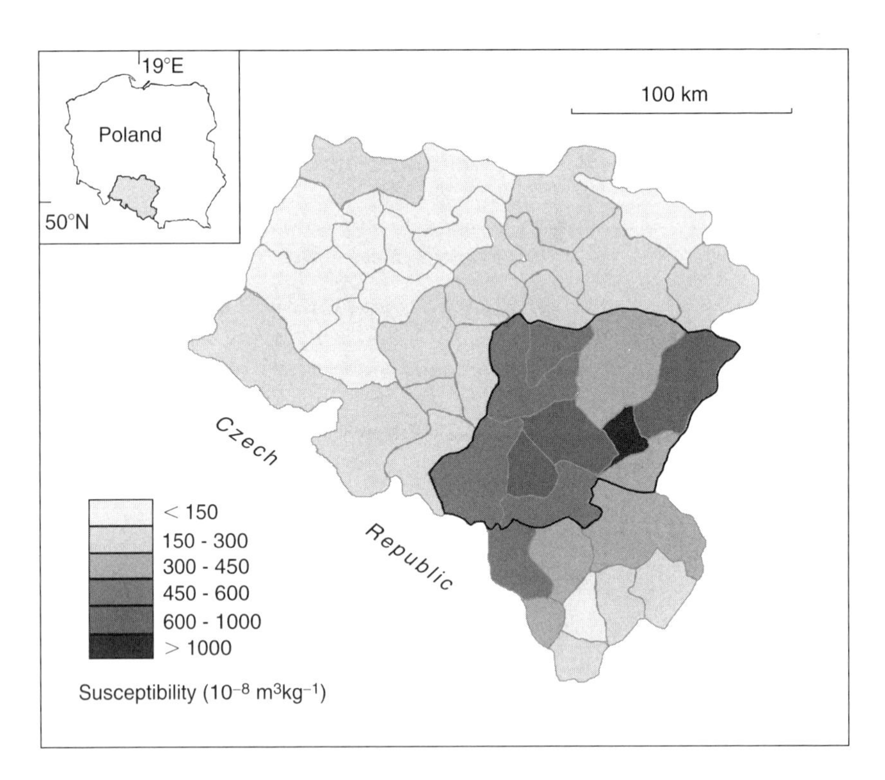

Figure 10.2 Map of Upper Silesia (Poland) showing magnetic susceptibility of soils. Katowice Province is indicated by the bold outline. The other boundaries indicate the individual forest administrative areas for which magnetic susceptibility data are available. (Modified from Heller *et al.*, 1998b.) © American Geophysical Union. Modified by permission of American Geophysical Union.

related to industrial activity (Fig. 10.2). Several superforestry susceptibility values in Katowice Province exceed 600×10^{-8} m^3kg^{-1}, and one of them yields the extraordinarily high value of 1493×10^{-8} m^3kg^{-1}. These values may be compared with the national survey covering the whole of England (1176 data points each representing a 10×10 km square, Dearing *et al.*, 1996). A mean value of 73×10^{-8} m^3kg^{-1} is reported, but some high values are also found; 33 samples have susceptibilities exceeding 500×10^{-8} m^3kg^{-1}, the highest being 1794×10^{-8} m^3kg^{-1}. As in Poland, the high values are closely associated with industrial areas, especially around the cities of London, Liverpool, Manchester, Sheffield, Leeds, and Newcastle-upon-Tyne. By contrast, the highest susceptibility values recorded in the thoroughly investigated paleosols (fossil soils) of the Chinese Loess Plateau are $\sim350\times10^{-8}$ m^3kg^{-1}. Beyond the immediate Katowice area, superforestry susceptibility values drop below 300×10^{-8} m^3kg^{-1} and eventually to $<150\times10^{-8}$ m^3kg^{-1}. But even these values are still considerably higher than typical values measured in unpolluted areas such as the topsoils of the Slowinski National Park near the Baltic Sea ($\sim20\times10^{-8}$ m^3kg^{-1}).

Figure 10.3 shows detailed results from 13 sites distributed along a 140-km transect crossing Katowice Province. It is immediately clear that the magnetic susceptibility of the topsoils is strongly correlated with the measured amounts of dust fall obtained from the monitoring network. Both profiles show two peaks, each of

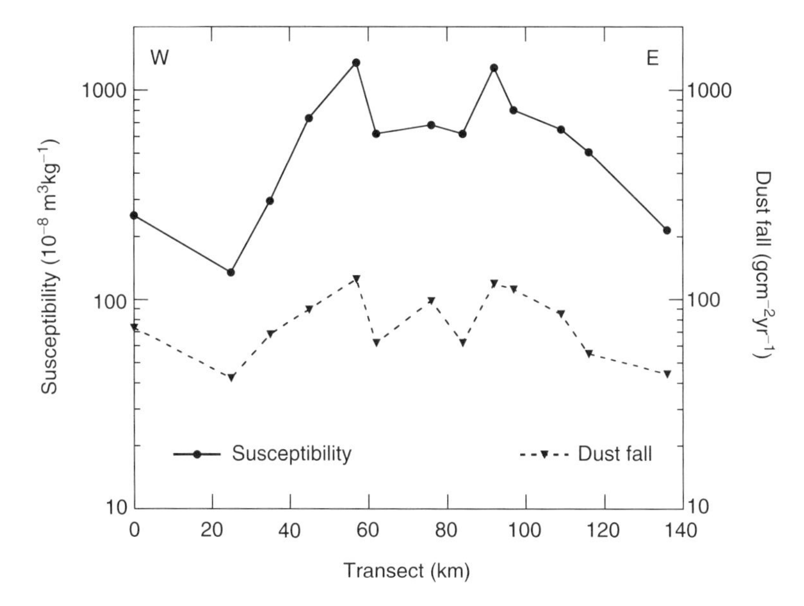

Figure 10.3 Dust fall and magnetic susceptibility along a west–east transect through Katowice Province, Poland (see Fig. 10.2). The dust fall was measured at stations belonging to a permanent pollution-monitoring network. The magnetic susceptibility was determined from a series of topsoil measurements. Both profiles show two peaks, each within a kilometer of a large power station (Laziska near 60 km, and Jaworzno near 100 km). (Modified from Heller *et al.*, 1998b.) © American Geophysical Union, with permission of the publishers.

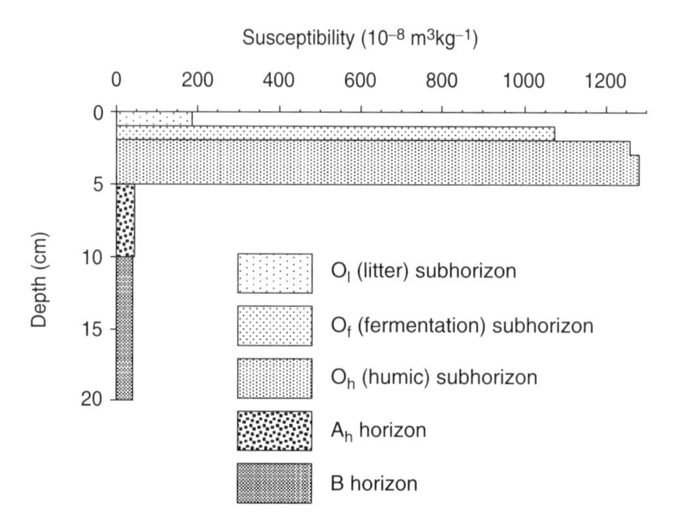

Figure 10.4 Depth profile of magnetic susceptibility in a soil pit in Chrzanow forestry area (the area showing the highest pollution in Fig. 10.2). The magnetic signal peaks in the uppermost 5 cm (i.e., in the organic subhorizons: O_l, litter; O_f, fermentation; and O_h, humic subhorizons). (Modified from Heller *et al.*, 1998b.) © American Geophysical Union. Modified by permission of American Geophysical Union.

which occurs within a kilometer of large power stations: Laziska near 60 km and Jaworzno near 100 km. Soil profiles indicate that the enhanced susceptibility resides in the organic subhorizons constituting the uppermost few centimeters (Fig. 10.4). It is unlikely therefore to be inherited from the underlying bedrock, which, in any case, has measured susceptibilities no more than 10×10^{-8} m^3kg^{-1}. The same general trend is observed in a soil profile at Jaworzno (Fig. 10.5) in which heavy metal concentrations were also determined. These turn out to be strongly correlated with susceptibility. This is perhaps the most important finding of the whole study — it means that measurements of magnetic susceptibility offer a rapid and cheap first step to identify pollution of soils by industrial contaminants.

The anthropogenic origin of these soil contaminants is indicated by several lines of evidence. The strong geographic correlation of industrial activity and elevated soil magnetism is perhaps the most convincing factor. This is supported by susceptibility measurements on 96 samples of fly ash from Upper Silesian power plants, which demonstrate that it is strongly magnetic (median susceptibility $\sim 2000 \times 10^{-8}$ m^3kg^{-1}). Another key point is that the soils studied, despite having high bulk susceptibilities, have low frequency dependence of susceptibility (1–4%), as opposed to values of $\sim 10\%$ typically found in soils that have been magnetically enhanced by normal pedogenic processes (Mullins, 1977). The higher values are due to the presence of superparamagnetic (SP) particles, whereas the Polish soils are magnetically dominated by larger grains typical of those generated during the combustion of coal. Furthermore, diagnostic iron-rich spherules with diameters measuring 2–200 μm have been identified by scanning electron microscopy of samples from the organic layers of the forest soils (Fig. 10.6).

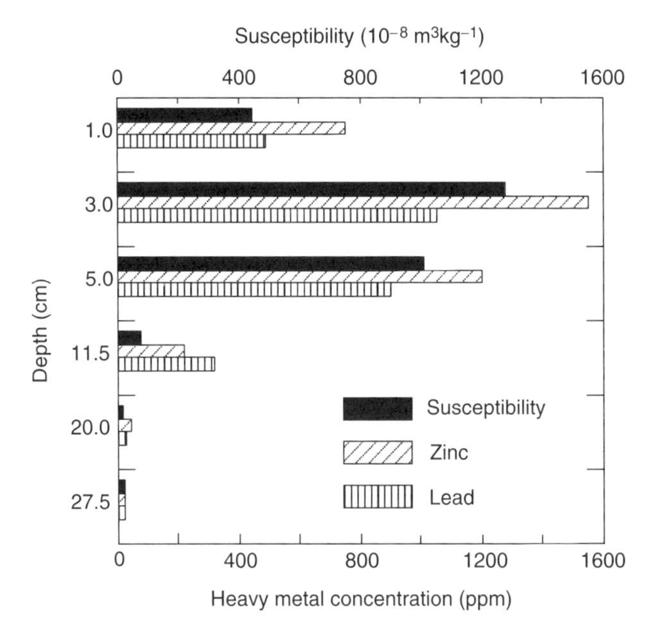

Figure 10.5 Correlation of magnetic susceptibility with lead and zinc content in a soil pit near Jaworzno power station (see Fig. 10.3). (Modified from Heller *et al.*, 1998b.)

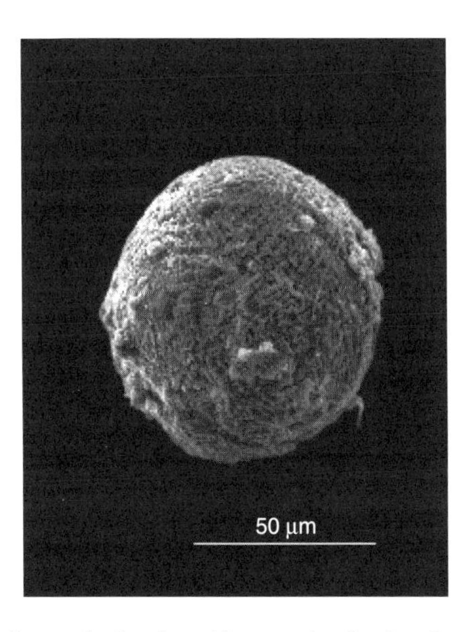

Figure 10.6 Electron micrograph of an iron-rich magnetic spherule collected from the topsoil 800 m from Jaworzno power station (Poland). (Courtesy of Tadeusz Magiera.)

Magnetic monitoring of soil pollution is becoming a widely accepted technique; in addition to the Polish work, investigations have been reported from England (Hay *et al.*, 1997), the Czech Republic (Kapicka *et al.*, 1999), and Estonia (Bityukova *et al.*, 1999). The English investigation involves the same national data set described by Dearing *et al.* (1996) and referred to before, but with the addition of information on heavy metal concentration. It is found that high magnetic susceptibility correlates with elevated levels of copper, lead, zinc, and nickel. Furthermore, useful magnetic criteria are proposed for the identification of soils containing "significant concentrations of pollution particles," namely susceptibility $>380\times10^{-8}$ m^{3}kg^{-1} and frequency dependence of susceptibility $<3\%$.

In the Czech example, an attempt was made to map spatial correlation between susceptibility and heavy metals over a two-dimensional grid covering distances up to $\sim$20 km from the Pocerady coal-burning power plant. The authors admit that the correlation is not convincing. It seems that the whole area — which lies in one of the most heavily industrialized parts of Europe — is affected by the overlapping fallout zones of several major pollution sources that complicate the simple geographic pattern expected from a single, isolated source.

Finally, in the Estonian study, a great deal of effort was put into determining concentration levels of 40 elements in 531 topsoil samples collected in and around the city of Tallinn. A particularly useful suggestion is the introduction of a so-called enrichment index (EI), which combines the observed amounts of the six most important polluting metals. It is obtained by summing the ratios Pb/Pb′, Cu/Cu′, Zn/Zn′, Cr/Cr′, Ni/Ni′, and Mo/Mo′ (where the unprimed symbols represent the measured concentrations and the primed ones represent the corresponding worldwide average concentrations of these elements in noncontaminated soils). Maps of magnetic susceptibility and EI are strikingly similar, and both show strong peaks in the vicinity of metal-working and machine-building factories (Fig. 10.7). The peak values themselves are as high as 11×10^{-3} SI for susceptibility and 30 for EI. In order to mass normalize the susceptibility values, recall that we divide by the density (see Chapter 2). Taking a typical soil density to be 1500 kgm^{-3}, we obtain maxima of $\sim700\times10^{-8}$ m^{3}kg^{-1}, which is well above the Hay *et al.* "pollution threshold" (380×10^{-8} m^{3}kg^{-1}), but less than half the maximum values found in Poland (1493×10^{-8} m^{3}kg^{-1}) and England (1794×10^{-8} m^{3}kg^{-1}). In other words, the Tallinn soils are definitely contaminated but not as severely as soils near coal-burning power plants.

The success of magnetic monitoring is leading to its adoption by various governmental agencies and city administrations. In Austria, for example, Hanesch and Scholger (2002) report soil surveys carried out in the cities of Leoben and Vienna. In the former, high susceptibilities result from centuries of mining and metallurgical activity, whereas in the latter they are correlated with traffic flow (see Section 10.5).

10.3 RIVERS, LAKES, AND HARBORS

Rather than falling on dry land (as in the soil contamination examples discussed before), atmospheric pollutants may, of course, fall directly onto water. This was

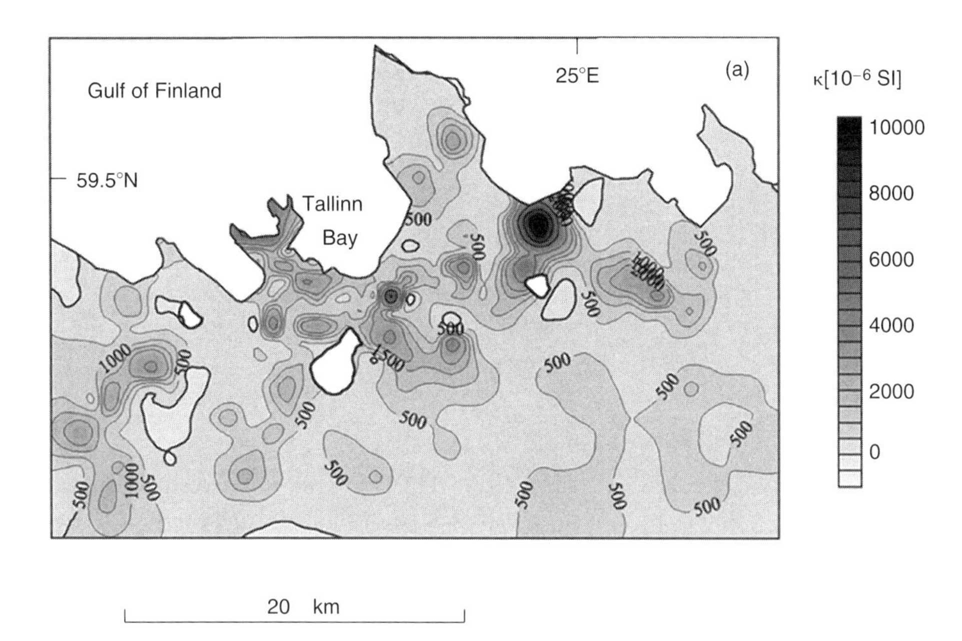

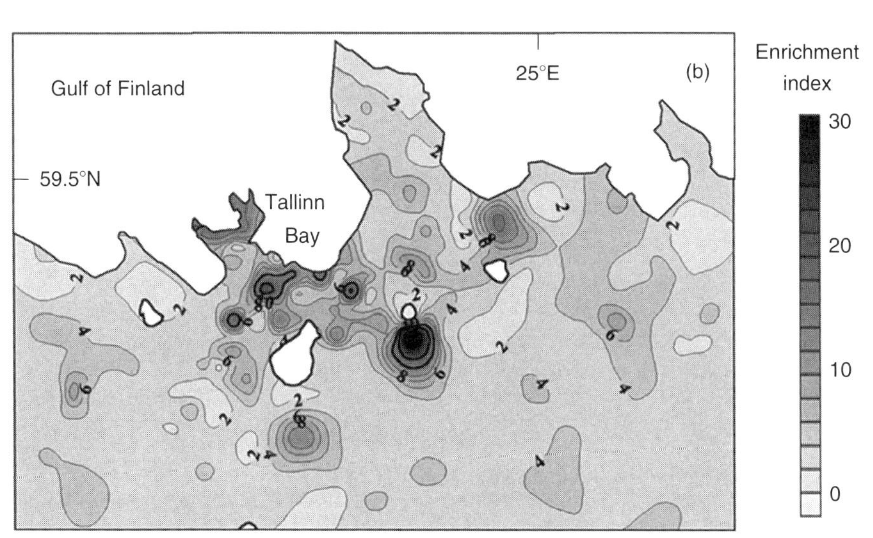

Figure 10.7 Maps of (a) magnetic susceptibility and (b) heavy metal enrichment index (see text for definition) in the area around the city of Tallinn, Estonia. (Redrawn from Bityukova *et al.*, 1999.) © Elsevier Science, with permission of the publishers.

probably the case for Big Moose Lake in the Adirondack Mountains of New York State (Oldfield, 1990), for two lakes in Scotland (Dubh Loch and Loch na Larach) and two in Wales (Llyn Irddyn and Llyn Glas) (Oldfield and Richardson, 1990), and for two lakes in northeastern Pennsylvania (Lake Lacawac and Lake Giles) (Kodama *et al.*, 1997). Or they may initially fall on dry land but then be washed into a river system to be deposited in the river sediments themselves or transported downstream and eventually sequestered in other sedimentary settings, including ultimately the sea. In other cases, there may be no atmospheric path at all, the material involved being flushed directly into a nearby body of water. This happened, for example, with the discharge into Mediterranean coastal waters from the Greek iron and steel works complex studied by Scoullos *et al.* (1979). We have chosen several examples like this to illustrate a variety of hydrological situations in which magnetic monitoring has proved useful, but we focus discussion on three of them. These involve a harbor on the shores of Lake Ontario (Canada), a bay in Lake Geneva (Switzerland), and a river in the province of Styria (Austria).

10.3.1 A Canadian Harbor

Hamilton Harbour lies at the western end of Lake Ontario; it is a triangle-shaped embayment surrounded by extensive urbanization and industrialization that has developed over the last hundred years. Currently, there is a daily discharge of water into the harbor of about 3 million cubic meters, of which about three quarters consist of exchange with Lake Ontario. It is for this reason that Hamilton Harbour has come under scrutiny by the Great Lakes Water Quality Board. It has been shown (Versteeg *et al.*, 1995a) that magnetic susceptibility correlates strongly with the heavy metal content of sediment cores taken from various locations within the harbor.

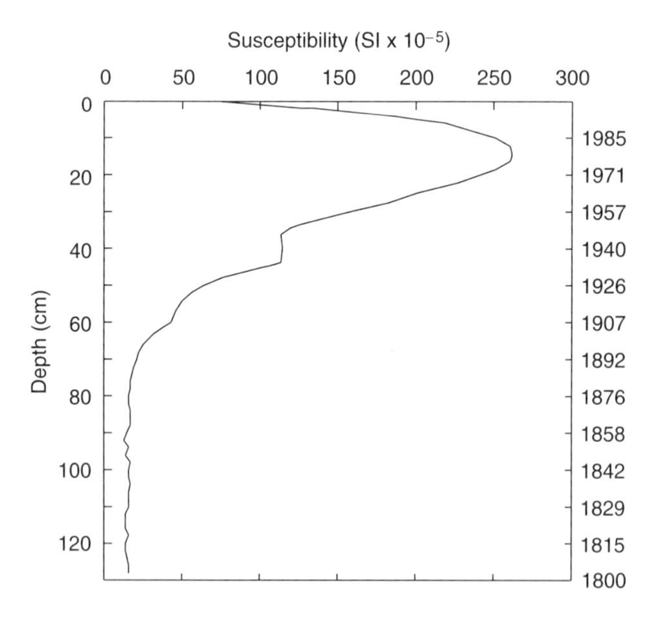

Figure 10.8 Typical magnetic susceptibility profile of a core taken in Hamilton Harbour, Ontario, Canada. Ages were obtained from ^{210}Pb dating. (Redrawn from Versteeg *et al.*, 1995b.) © Geoscience Canada, with permission of the publishers.

Thus, magnetic measurements can be used to track contaminant levels, eliminating the need for prohibitively expensive chemical analyses. A typical susceptibility profile is shown in Fig. 10.8. At the bottom of the core there is about 50 cm of very weakly magnetic sediment representing the natural background. Near the beginning of the 20th century, susceptibility begins to rise rapidly, essentially coincident with the first steel production in the area. Currently, the two largest steel mills in Canada are located on the south shore of the harbor. In recent years, pollution control equipment has been installed, and contaminant levels have fallen. The integrated historical input remains in the habor, of course.

There is a need to find out how much of this contaminated material there is and to map its spatial distribution. To these ends, Versteeg *et al.* (1995b) have studied sediment cores from 40 sites arranged on an approximately rectangular grid throughout the harbor (typical spacing $\sim$500 m). By mapping the spatial variations in the features of profiles like that shown in Fig. 10.8, they were able to produce the desired map (Fig. 10.9) and thereby to estimate the total volume of contaminated sediments to be about 10^7 m^3. The cores themselves were collected in plastic tubes and all the measurements were done on a pass-through susceptibility meter. This not only is a very rapid technique but also means that the core tubes need never be opened. The sediments thus remain undisturbed and are available for further study by other methods. In a subsequent paper concerning Hamilton Harbour, Mayer *et al.* (1996)

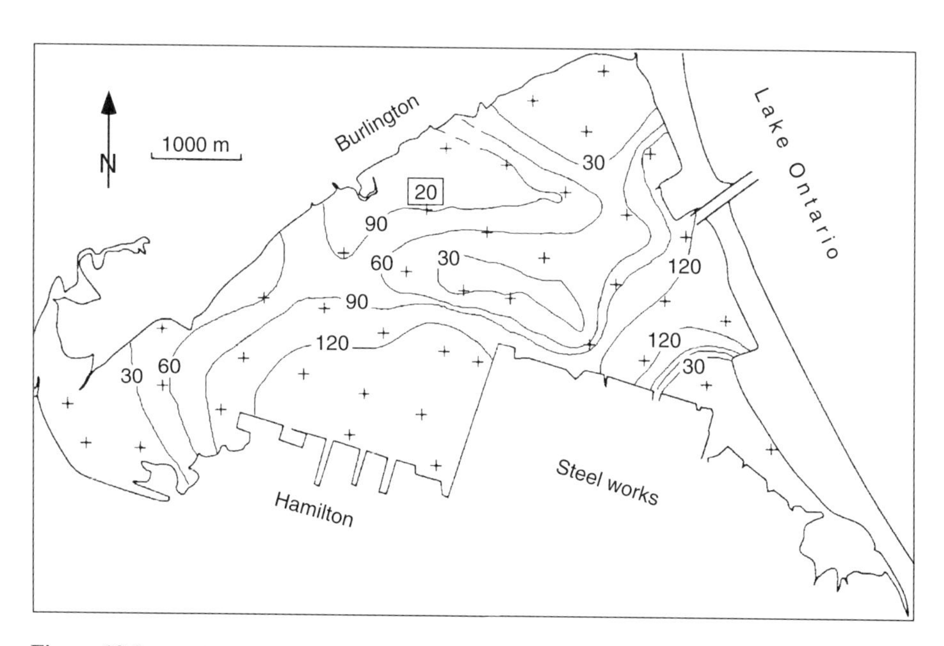

Figure 10.9 Hamilton Harbour, Ontario, Canada, showing contours of sediment thickness (in cm) deduced from profiles like that of Fig. 10.8. Sampling sites are indicated by crosses. Site 20 is the location from which the profile shown in Fig. 10.8 was obtained. The cities of Hamilton and Burlington and the major steel works (the largest in Canada) are indicated. (Redrawn from Versteeg *et al.*, 1995b.) © Geoscience Canada, with permission of the publishers.

succeeded in determining the magnetic properties of the contaminants actually suspended in the water. A total of 53 water samples were collected for this purpose, and a high degree of correlation between susceptibility and heavy metal content was found. For zinc, iron, lead, and cadmium the correlation is significant at the 99.9% probability level, for copper it is significant at the 99% level, and for nickel it is significant at the 95% level. Manganese, the only other element tested for, shows no correlation.

Two studies very similar to those carried out in Hamilton Harbour are described by Georgeaud *et al.* (1997) and Chan *et al.* (1998). In the former, clear correlations are reported between magnetic parameters (susceptibility and saturation isothermal remanent magnetization) and heavy metal concentrations (Zn, Cd, and Cr) in sediments from the Etang de Berre, a coastal lake close to the industrial area of Marseille, southern France. In the latter, magnetic susceptibility was compared with Pb, Cu, Cr, Zn, and Ni content in sediments cored from Hong Kong Harbour; all correlation coefficients, except that for Ni, were significant at the 95% level.

10.3.2 A Swiss Lake

The Bay of Vidy is a small embayment on the north shore of Lake Geneva that lies within the urban area of the city of Lausanne. Since 1971, sewage has been subjected to dephosphatization treatment before being discharged into the bay. This treatment involves addition of iron chloride ($FeCl_3$, a paramagnetic salt) to the sewage and has resulted in high iron concentrations in the sediments near the discharge outlet (Fig. 10.10). Pass-through scans of cores recovered from the zone of iron enrichment indicate a sharp rise in the magnetic susceptibility of the recent sediments, the date of which (deduced from two ^{137}Cs peaks interpreted as the signatures of the 1963–1964 nuclear weapons testing and the 1986 Chernobyl accident) links it with the

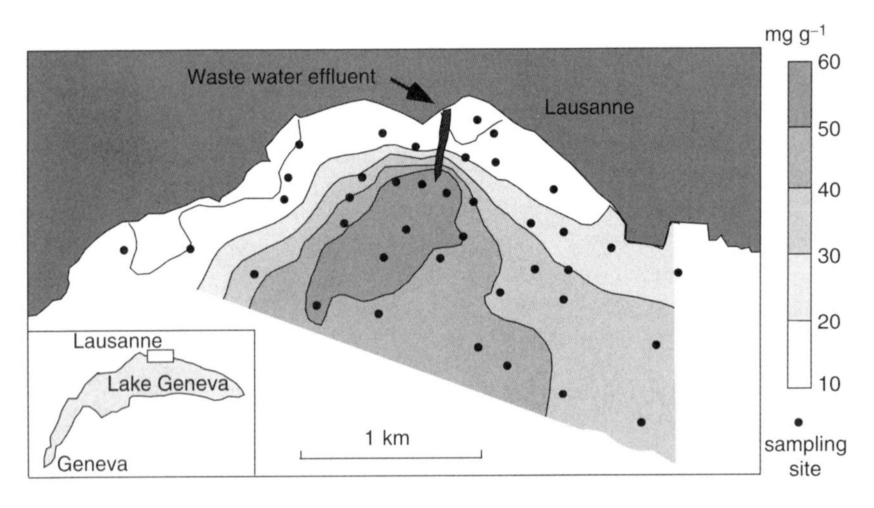

Figure 10.10 Bay of Vidy, Lausanne, Switzerland, showing contours of iron enrichment in the sediments in the vicinity of treated sewage discharge. (Redrawn from Gibbs-Eggar *et al.*, 1999.) © Elsevier Science, with permission of the publishers.

start of dephosphatization treatment (Gibbs-Eggar *et al.*, 1999). Samples removed from the opened core tubes confirm the post-1971 changes. Mass susceptibilities are one to two orders of magnitude stronger in the recent sediments, typically 500×10^{-5} m^3kg^{-1} compared with $5\text{–}50\times10^{-5}$ m^3kg^{-1} in the pretreatment sediments. Frequency dependence of susceptibility is also higher (8–10%), indicating the presence of ultrafine magnetic grains near the SSD/SP threshold (see Chapter 2). These were identified in transmission electron micrographs, and by susceptibility versus temperature measurements, as being magnetite produced from paramagnetic iron by *Geobacter*-type bacteria (see Chapter 9).

The Bay of Vidy investigation differs from most other magnetolimnological studies in that it is focused on the effects of urban water treatment and direct discharge into the body of water in question. There are many more studies concerning situations in which pollutants have reached the lake via an atmospheric route, as in the British and U.S. lakes mentioned earlier. A particularly interesting case concerns a small lake (surface area 2.5 km^2) in England called Crummock Water (McLean, 1991). The magnetic susceptibility profile of a 6-m core shows a sharp rise over the uppermost 50 cm interpreted as increased pollution due to the industrial revolution. There is no shortage of culprits: Liverpool, Manchester, Newcastle, and Glasgow are all within 150 km of the site. The case is really proved, however, by the extraction of magnetic spherules (like those produced in the combustion of coal) from the core. At a depth of 37 cm (~AD 1730) McLean reports 46 spherules/gram of sediment. This rises to 82 in ~AD 1860 before jumping dramatically to 968 in the first decade of the 20th century when coal-fired power-generating stations became common. As with the susceptibility profile, the spherule concentration continues to increase right up to the top of the core with counts of 1562 in ~1940 and 1847 in ~1970.

10.3.3 An Austrian River

The River Mur in the province of Styria, Austria, is a tributary of the Drava, which is itself a tributary of the Danube. With its own tributary, the River Mürz, the Mur drains urbanized industrial regions with numerous metal-producing and metalworking facilities. The most important aquifers in Styria are Holocene and Pleistocene river gravels, which are now under erosive attack by the rivers themselves due to increased velocities resulting from an extensive river regulation program during the late 19th century. There is thus concern that any pollutants carried by the rivers may infiltrate into the water supply. Scholger (1998) has investigated this potential hazard magnetically by means of some 500 sediment samples covering a 190-km stretch of the river between Judenburg and Spielfeld. He finds that the magnetic susceptibility is determined by the presence of *iron scale* that results from high-temperature processes such as forging and rolling. This takes the form of small flakes of metal (Fig. 10.11) that manage to survive the settling tanks designed to trap them. As well as iron, the scale often contains chromium, nickel, and copper (which are used as alloying elements) or lead and zinc (probably originating from steel production and processing). The ability of magnetic susceptibility to monitor quantitatively the scale content

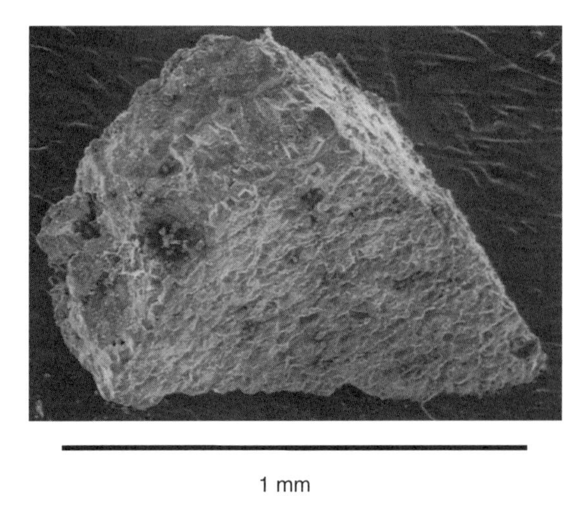

Figure 10.11 Electron micrograph of a millimeter-sized flake of *iron scale* recovered from an industrial sedimentation tank. (Courtesy of Robert Scholger.) © Geophysical Press, with permission of the publishers.

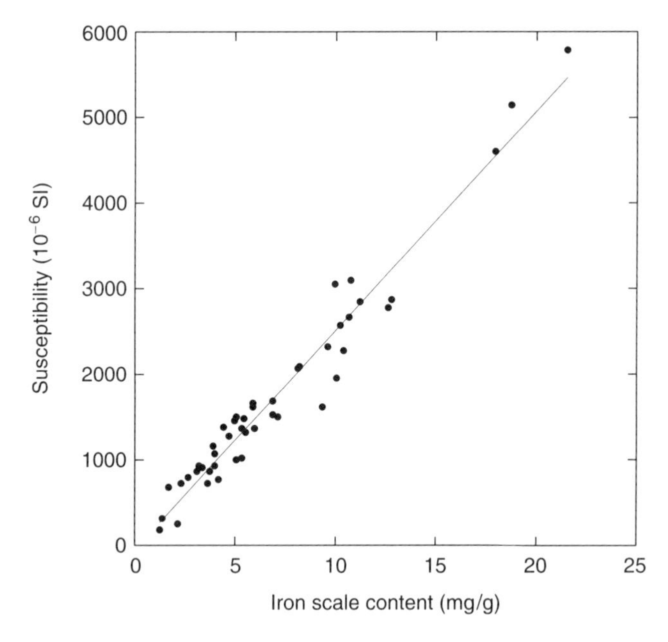

Figure 10.12 Correlation of *iron scale* content with magnetic susceptibility for 46 sediment samples from the River Mur, Austria. (Redrawn from Scholger, 1998.) © Geophysical Press, with permission of the author and the publishers.

is clearly demonstrated by Fig. 10.12. Given the strong correlation observed ($R = 0.99$), it is apparent that a very rapid measurement of susceptibility suffices to monitor effectively the total scale fraction—and thus the total heavy metal contamination — present in a sediment sample. In terms of magnetic variations along the course of the river system, Scholger particularly points out a strong susceptibility peak at the point where the River Mur ("which drains an important industrial zone") enters, bringing with it its pollutants (sharp increases in measured concentrations of Pb, Zn, Cr, Ni, and Cu are observed at the confluence). Upstream, along the Mur, a very strong, localized Ni peak associated with quarrying in bedrock serpentine is also reflected in the magnetic susceptibility measurements.

A similar study in the Czech Republic is reported by Petrovsky *et al.* (2000). Soil samples collected along the left bank of the River Litavka show an abrupt increase in magnetic susceptibility immediately downstream from a lead smelter in the town of Pribram (near Prague). Upstream from the smelter, susceptibility is typically $\sim 1 \times 10^{-5}$ SI, but this rises to $\sim 9 \times 10^{-5}$ SI at the smelter before settling down to a steady value of $\sim 5 \times 10^{-5}$ SI for the entire downstream distance investigated, some 15 km.

10.4 ATMOSPHERIC CONTAMINANTS

Respirable mineral particles pose a serious health risk (Guthrie, 1995). Air quality is therefore of great concern to everyone, and monitoring programs are now routine in many countries. Hitherto, however, there has been relatively limited application of magnetic methods to material collected *directly* from the atmosphere (as opposed to studies of material already deposited, such as in the numerous soil examples already discussed). The earliest studies — carried out in the 1980s — are summarized by Oldfield *et al.* (1985b). These already indicated that airborne dusts from different sources could be distinguished on the basis of their magnetic properties. Morris *et al.* (1995) have succeeded in identifying the magnetic signature of respirable airborne particulate matter (usually abbreviated as PM) collected in an urban environment. They studied filters deployed between May 1990 and June 1991 at an air-monitoring station in downtown Hamilton, Ontario, a few kilometers from the two largest steel mills in Canada. Each filter sampled 1630 m^3 of air in any given 24-hour period. Magnetic susceptibility was determined by simply folding each filter and placing it in the sensor cavity of a commercial susceptibility meter. Scanning electron microscopic examination indicated that magnetic susceptibility varied according to the abundance of iron-rich spherules of the kind that result from the combustion of coal (see Sections 10.1 and 10.2). Also adsorbed on the filters were various organic compounds (polycyclic aromatic hydrocarbons, PAHs) that pose health risks because of the damage they cause to DNA. This *mutagenicity* was quantified by extracting the organic compounds from the filters and submitting them to standard bioassays. A strong correlation ($R = 0.89$) was observed between magnetic susceptibility and mutagenicity (Fig. 10.13). It appears that magnetic monitoring provides an inexpensive and rapid procedure for selecting appropriate samples for further analytical chemistry and bioassay tests. Such tests are needed to determine the health risk posed by the particulate content of the air we breathe.

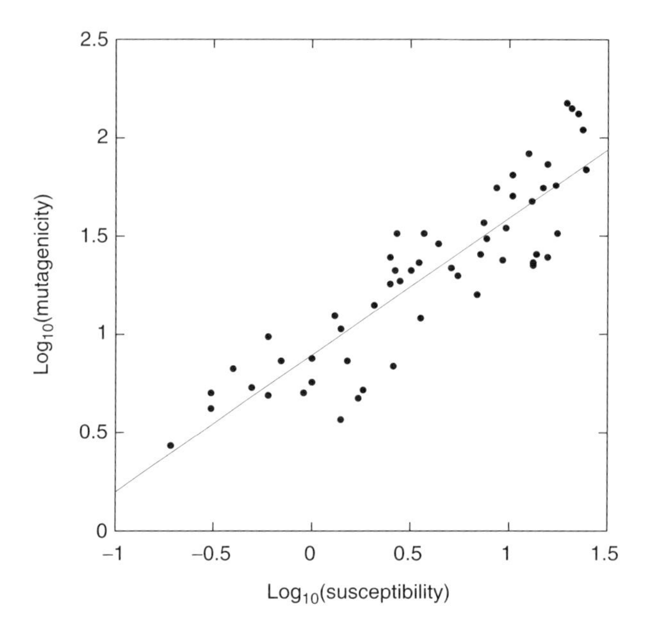

Figure 10.13 Mutagenicity of filters taken from an air monitoring station in Hamilton, Ontario, plotted against their magnetic susceptibility. (Redrawn from Morris *et al.*, 1995.) © Elsevier Science, with permission of the publishers.

Work of this kind has also been carried out in Shanghai, China, where daily atmospheric dust samples were collected at 11 sites in and around the city for seven consecutive days in November 1998 (Shu *et al.*, 2001). Three sites lying within a kilometer of the Baoshan iron and steel manufacturing complex had the highest susceptibility values (743 to 1521×10^{-8} m^3kg^{-1} averaged over the whole week). The other sites, which are located 6 to 10 km from the complex, yield values between 299 and 524×10^{-8} m^3kg^{-1}. As one would expect, the dust trapped in any given sampler on any given day depends on meteorological conditions, particularly wind speed and direction. A good example is provided by site 10, which lies ~6 km southwest of the Baoshan complex. On a day when the wind was from the north–northeast, the frequency-dependent susceptibility of the airborne particulates was found to be 5%, but when the wind was from the south, this rose to 13%. This change reflects the increased relative input of superparamagnetic particles in windblown soil when the wind is coming from the direction away from the industrial area.

Experiments have been conducted to see whether atmospheric monitoring of this kind can be extended by using deposition on common natural surfaces, thereby avoiding the limitations and expense of using artificial filters. It appears that pine needles and tree leaves are suitable collectors. Both are readily obtainable in most urban and industrial situations of interest, and both can usually be removed (without fear of prosecution) and studied directly in the laboratory. This avoids possible collection inefficiencies involved in wiping the surface as in the procedure used by

Flanders (1994, 1999; see earlier). In the industrial area of Leipzig-Halle (Germany), Schädlich *et al.* (1995) found that pine needles on trees (*Pinus sylvestris*) in the path of power station fly ash emissions were more magnetic than their uncontaminated counterparts. The maximum value measured was 9.6×10^{-8} m^3 kg^{-1}, which corresponds to a fly ash coating of slightly more than 1% of the total needle mass according to a calibration derived by adding known amounts of fly ash to "clean" needles. The most thorough study of tree leaf collectors (Matzka and Maher, 1999) concerns pollution related to road traffic, to which we now turn.

10.5 ROADSIDE POLLUTION

Vehicular traffic is a significant source of pollution, but relatively little research has been done in terms of magnetic monitoring. Prior to the widespread introduction of unleaded fuel, it was established that there was a tendency for lead-based contaminants to be associated with magnetic minerals. This facilitated various studies involving the extraction of contaminants for detailed chemical and microscopic investigations by allowing the use of routine magnetic separation techniques. Apparently, the iron-rich magnetic material does not come from the fuel itself. It results from rusting of the bodywork, wear of the moving parts, and ablation from the interior of the exhaust system. These factors are still important even though the amount of lead present is now drastically reduced.

A magnetic investigation of pollution in an urban highway environment in London, England, was carried out by Beckwith *et al.* (1990). For the road center, road gutter, and sidewalk, they obtain mean susceptibility values of 5.2×10^{-6}, 2.4×10^{-6}, and 1.8×10^{-6} m^3kg^{-1}, respectively. They also report similar patterns of Cu, Fe, Pb, and Zn concentrations but do not give details. Possible contributions from direct atmospheric fallout were checked by sampling the roofs of nearby buildings, for which a much lower mean susceptibility of only 0.7×10^{-6} m^3kg^{-1} was obtained. Their conclusion is that these data imply that the dominant source is "most probably associated with motor vehicles." A similar conclusion is reached by Matzka and Maher (1999), who investigated the use of tree leaves as pollution samplers (see earlier). They collected leaves from roadside trees in the city of Norwich (England), which is situated in a largely agricultural area with no heavy industry. Sampling was restricted to a single species of birch tree (*Betula pendula*) in order to avoid possible species-dependent effects. Several leaves were collected from the outer canopy of each tree at a height of 1.5 to 2 m and were given a laboratory IRM in a field of 300 mT. All values were normalized to the area of the leaf, which was obtained by digital scanning. Leaves from rural settings were found to be 10 times less magnetic than those collected near busy urban roads. The connection with vehicular traffic was demonstrated particularly clearly by a detailed study of a single tree in the city center (see Box 10.2). Tree leaf pollution monitoring based on magnetic measurements is now being carried out in several European cities (for summaries, see http://www.geo.uu.nl/~magnet/ or http://www.ig.cas.cz/magprox/).

Box 10.2 Magnetic Tree

A survey of a single birch tree 5 m from a busy street in the city center of Norwich (England) was carried out by Matzka and Maher (1999). Groups of six leaves were collected at 30° intervals around the outer canopy 1.5–2 m above the ground. In the laboratory, isothermal remanent magnetizations (IRMs) were given to the leaves in a field of 300 mT and normalized to the leaf's area (obtained by digital scanning). Because the magnetic moment is in Am^2, the normalized results are simply in amperes. The accompanying graph shows the results as a function of angular position around the tree (θ, measured from true north).

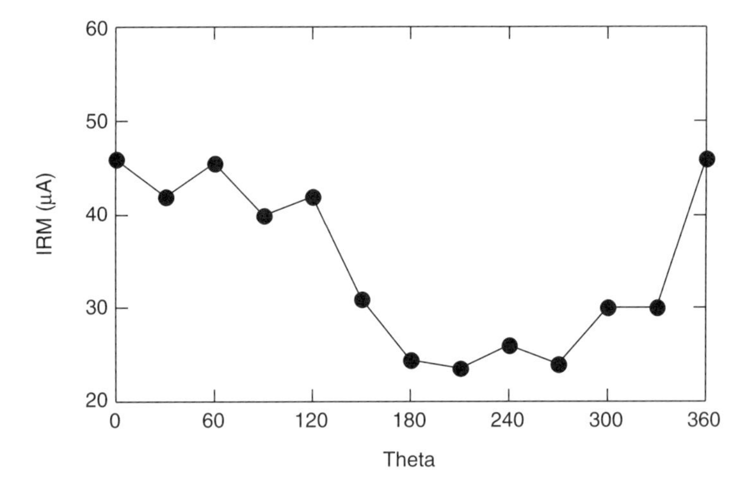

The side of the tree facing the street runs from 305° to 125° and is consistently more magnetic (mean = 41 μA, compared with 27 μA for the side away from the street). This strongly implies that the magnetic signal is tracking traffic pollution and that the neighborhood can be somewhat protected by vegetation. Whether this is the good news or the bad news depends on whether you are the neighbor or the vegetation!

Another study (Hoffmann *et al.*, 1999) illustrates the magnetic pattern associated with a busy highway in Germany. Near the road margin, significantly elevated susceptibility readings are observed in the soil, but these fall rapidly as one moves away from the road. From an average peak value of $\sim 1.5 \times 10^{-3}$ SI, the readings drop by 50% within 2 m and are indistiguishable from the background beyond 5 m. It seems that the magnetic flux emanating from road traffic is easily monitored but is rather localized. Of course, this result concerns only the large particles that deposit quickly. There may well be finer aerosols that travel farther, but as far as magnetic studies go, these remain as topics for future research. Eventually, most of the material that deposits close to the road will be washed into the local drainage system and probably end up in a nearby river. A good example is provided by the River Bonde as it traverses the small town of Etrépagny, France (Brilhante *et al.*, 1989). The town drainage system delivers the integrated urban runoff to the river and increases the observed susceptibility of the river sediments by a factor of 8.5 immediately downstream from the town compared with the upstream "background" value where the river runs through agricultural land. Furthermore, a very similar pattern of Pb and Zn concentrations is observed.

Xie *et al.* (2000) studied 97 samples of street dust collected within ~1.5 km of the city center of Liverpool in northwest England. They were particularly concerned with the organic content, especially toxic components such as polycyclic aromatic hydrocarbons (PAHs; see preceding discussion concerning air quality monitoring in the city of Hamilton, Canada). Some of the magnetic properties (low-frequency susceptibility, frequency-dependent susceptibility, susceptibility of anhysteretic remanent magnetization, and high-field susceptibility) of the Liverpool street dust samples correlate positively with organic content estimated by the loss-on-ignition method and thus provide a simple, rapid, and nondestructive proxy for environmentally significant organic material in street dust.

In the city of Munich (southern Germany), Muxworthy *et al.* (2002) have investigated airborne dust related to traffic flow in a busy downtown street. They deployed plastic sheets and trays that collected PM dust directly from the atmosphere. Using Mössbauer spectroscopy and a variety of magnetic measurements, they concluded that the dominant magnetic minerals present were maghemite (60–70 %) and metallic iron (30–40 %). It appears that the maghemite is emitted from automobiles whereas the iron comes from street trams.

10.6 PNEUMOMAGNETISM

Inhaled particulate matter often contains a magnetic fraction that can—under the right circumstances—be detected by magnetometers situated outside the body. Already in the very first paper on this topic, Cohen (1973) pointed out that the magnetic signals involved are often strong enough to be detected by a simple fluxgate magnetometer without the need for an expensive installation consisting of a SQUID magnetometer in a shielded room. Indeed, Junttila *et al.* (1985) describe a fluxgate gradiometer arrangement that they installed in a mobile healthcare unit. Cohen's original *magnetopneumography* (MPG) technique was to map the subject's torso on a 5×5 cm grid after magnetizing the dust particles in the lungs with an external 50-mT field. Among other things, he was able to demonstrate that significant amounts of magnetizable dust had accumulated in the lungs of an arc welder and an asbestos mine worker. Cohen also proposed that voluntary inhalation of magnetite dust provides a way of investigating how the lungs clear themselves of respirable airborne dust, thereby avoiding the use of radioactive tracers. In a subsequent study, Cohen *et al.* (1979) followed up this suggestion to compare lung clearance in smokers and nonsmokers. After 11 months, they found that smokers still retained 50% of the magnetite dust inhaled at the start of the test, whereas the nonsmokers retained only 10%.

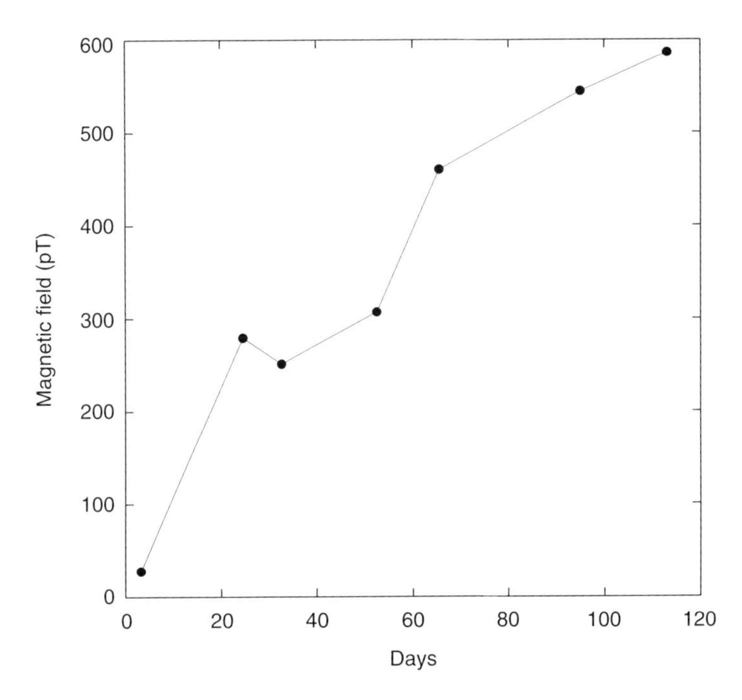

Figure 10.14 Magnetic field measured a few centimeters from a trainee welder's back as a function of days since the start of training. (Compiled from Forsman and Högstedt, 1989.)

This development led to several uses of magnetometry as a medical tracer, which, although important, lie beyond the scope of this book [for summaries, see Valberg and Zaner (1989) and Kalliomäki (1998)]. Instead, we briefly describe two examples of pollution in the workplace. Forsman and Högstedt (1989) monitored the retention of magnetizable lung dust in eight trainees in welding school who had no previous exposure to arc welding fumes. The gradual buildup of dust in the lungs was monitored by the magnetic field measured a few centimeters from the subject's back. In the most extreme case, this increased steadily to a maximum of $\sim$600 pT after 114 days (Fig. 10.14). The other example concerns long-term effects and involves the magnetic properties of *postmortem* lung tissue samples from a number of asbestos workers (Rassi *et al.*, 1989). The procedure was to magnetize the tissue samples and then measure their remanence with a SQUID magnetometer. This was converted to the mass of magnetite present by calibrating against the remanence measured for known quantities of magnetite embedded in polyurethene foam. The natural magnetite content of asbestos-bearing rocks varies from deposit to deposit but is fairly constant within each mining area. This permits the extent of dust exposure to be gauged from worker to worker in any given mine. For example, at Wittenoom (Western Australia) miners had magnetite contents up to 200 μg/g but millhands had values up to 800 μg/g. This reflects the gradual enrichment in asbestos (and therefore in magnetite) that takes place along the production line.

11

ARCHEOLOGICAL AND EARLY HOMINID ENVIRONMENTS

11.1 INTRODUCTION

Magnetic investigations have long been a source of productive interaction between archeology and geophysics. More than a century ago, Giuseppe Folgheraiter (1856–1913)—the father of archeomagnetism—demonstrated that ancient ceramic vases retain a record of the Earth's magnetic field as it was at the time they were made (Folgheraiter, 1899). The great Hungarian geophysicist Baron Loránd Eötvös (1848–1919) immediately pursued this idea and made a number of relevant measurements (Mikola, 1900). Shortly thereafter, Pierre David investigated the remanent magnetization of paving slabs used in the construction of the Temple of Mercury by the Romans in the first century BC at the summit of the Puy-de-Dôme in the Auvergne district of France (David, 1904).

All of these early examples involve thermoremanent magnetization (TRM). In the case of the ceramics, this results from the anthropogenic firing process. The paving slabs, on the other hand, consist of igneous rocks that carry a natural TRM dating from the time they were originally formed. Although they were critical in establishing a tradition of magnetic studies of archeological features, these early achievements—and a host of others that followed—are mostly of interest to geomagnetists wishing to determine the past history of the Earth's magnetic field, as discussed in Chapter 6 (for a summary, see Gallet *et al.*, 2002). By contrast, the application of magnetic studies to what can be thought of as the environmental aspects of archeology is a much more recent development. It essentially grew out of the work of Eugène Le Borgne (1913–1978) in the 1950s concerning the magnetic susceptibility of soils. At first, Le Borgne was interested in this topic because of its possible relevance—essentially as noise—to the investigation of magnetic anomalies arising from the geological bedrock. But he was also cognizant of its potential impact elsewhere and eventually published a summary of his results specifically aimed at an archeological audience (Le Borgne, 1965).

11.2 ARCHEOLOGICAL SOILS

During an investigation of magnetic anomalies in central Brittany (northern France), Le Borgne (1950, 1951) observed that the uppermost few centimeters of soils in the area have a much higher magnetic susceptibility than the underlying bedrock. Intrigued by the possible significance of this observation, he set out to study several hundred soil samples from around the world. He soon concluded that the magnetic enhancement is almost universal and is largely independent of bedrock lithology (Le Borgne, 1955). A few years later, a comprehensive survey of the major soil types in the United States and Panama, involving 250 sites, produced a similar result (Cook and Carts, 1962). Nor did the latter authors find any correlation between susceptibility and soil color determined by matching to the standard Munsell color chart.

Initially, Le Borgne (1955) attributed the observed topsoil magnetism to the so-called fermentation process (see Chapter 5), requiring alternating moist and dry conditions, but later the effect of fire was considered to be important (Le Borgne, 1960). In both mechanisms, the ultimate outcome is the production of strongly magnetic maghemite (γ-Fe_2O_3) from weakly magnetic hematite (α-Fe_2O_3). When the soil is moist, or when the overlying vegetation is being burned, anaerobic conditions prevail and hematite is reduced to magnetite (Fe_3O_4). During the subsequent drying, or cooling, aerobic conditions are reestablished allowing reoxidation to maghemite. Le Borgne himself experimentally established the validity of these suggestions.

Tite and Mullins (1971) pursued this topic by investigating soils from a variety of archeological sites in Britain. They tested the thermal enhancement of magnetic susceptibility of 22 samples from 14 sites by subjecting them to what they refer to as the "nitrogen-then-air" procedure, as Le Borgne had done. It involved heating the samples to 550°C in a nitrogen atmosphere. This temperature was then maintained for 1 hour, the first 40 minutes of which were in nitrogen. The final 20 minutes and the whole of the cooling process were carried out in air. This experimental setup was intended to mimic ancient agricultural practice. During the first stage, the nitrogen atmosphere excludes air and a reducing atmosphere is produced by the combustion of organic matter in the soil. Once the air is introduced, oxidizing conditions are established. Tite and Mullins found that all their samples were magnetically enhanced by this procedure. The largest increase amounted to a factor of 61. Overall, in 13 cases the susceptibility increased by a factor of more than 10, in a further 8 cases the increase exceeded a factor of 4, and the remaining sample was enhanced by a factor of 1.4.

Peters and Thompson (1998b) report a magnetic enhancement factor of over 200 from a Norse archeological settlement on the island of Papa Westray, Scotland (2.9°W, 59.4°N). The site is located on glacial deposits (till) on which natural soils have developed that have, in turn, been further modified by human activity. The underlying till has magnetic susceptibility values as low as $0.1\times10^{-6}\,m^3/kg$ but this is typically increased to $\sim2\times10^{-6}\,m^3/kg$ by natural pedogenesis. Thereafter, burning causes a further order of magnitude increase, to a maximum observed value of $22\times10^{-6}\,m^3/kg$. Linford and Canti (2001) have investigated this effect by conducting carefully monitored experimental fires on sandy and clayey substrates, backed up with laboratory heating of fresh substrate samples. They also find susceptibility enhancement by factors of 100 or more. It is useful to place these values in context by comparing them with other observations—for example, a survey of unheated soils from 54 northern hemisphere sites yielded a maximum susceptibility of $6\times10^{-6}\,m^3/kg$ and an average of value of about $1\times10^{-6}\,m^3/kg$ (Maher and Thompson, 1995).

Based on their careful analysis of hysteresis loops of the Norse material, Peters and Thompson (1998b) argue that the increase observed on Papa Westray is due to the production, by burning, of superparamagnetic (~10 nm in diameter) grains of either maghemite (γ-Fe_2O_3) or magnetite (Fe_3O_4). In a survey of heated soils from 60 sites spread throughout Bulgaria, Jordanova *et al.* (2001) find enhanced susceptibility values up to $10\times10^{-6}\,m^3/kg$ with an average frequency dependence (see Chapters 2 and 3) of almost 8%. From the corresponding susceptibility versus temperature data, they conclude that the mineral responsible is magnetite (or titanomagnetite with a low Ti content), much of which is superparamagnetic.

An interesting example of the archeological application of magnetic enhancement by burning is described by Marshall (1998). Magnetic susceptibility maps were made at six levels as the excavation of an Early Bronze Age burial mound (round barrow) progressed. Each level was 10 cm below the previous one, with the final survey being on the old land surface. A Bartington instrument (see Chapter 4) equipped with a 20-cm-diameter field probe was used to survey $100\,m^2$ at 25-cm spacing. The results reveal a prominent circular pattern of high susceptibility caused by a ring pyre ~4 m in diameter. This large size "suggests an emphasis on spectacle rather than merely providing a means for basic, efficient incineration of a corpse" (Marshall, 1998, p. 162). The site (in Gloucestershire, southwest England) seems to have had considerable significance as a place for funerary rituals because many satellite pyres were revealed by a broader magnetic susceptibility survey.

Magnetic enhancement by heating is clearly an established fact, but the mineralogical details are still debated. Fassbinder and Stanjek (1993) claim that hematite is not necessarily present in soils, and even if it were, it is unlikely that typical fires (natural or anthropogenic) would achieve soil temperatures able to reduce it to magnetite. Instead, they list four alternative maghemite-forming processes: (1) oxidation of magnetite inherited from the parent material, (2) dehydration of lepidocrocite (γ-FeOOH), (3) dehydration of goethite (α-FeOOH), and (4) oxidation of siderite ($FeCO_3$). Process (1) simply results from natural weathering at ambient temperatures, but the others involve modest heating ($<300°C$). One of them—the dehydration of goethite—has been verified for soil from an English archeological site using a combination of Mössbauer spectroscopy and magnetic susceptibility (Longworth and Tite, 1977). Regardless of the exact mineralogical details, therefore, the fact remains that fire can, indeed, promote magnetic enhancement in archeological and other settings.

Fassbinder and Stanjek (1993) propose that magnetic enhancement can also take place in the absence of fires and without any lithogenic magnetite inherited from the parent material. Their own magnetic surveys of archeological sites in Germany detect

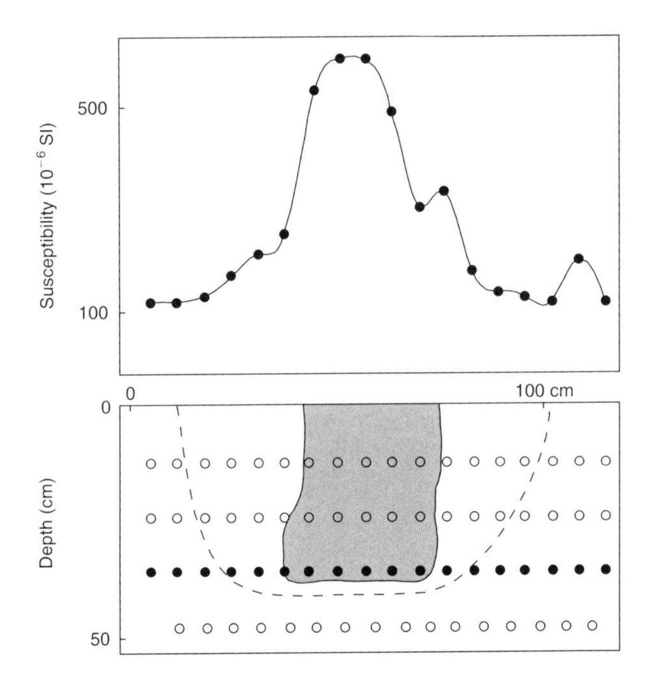

Figure 11.1 Excavation of a single posthole forming part of a palisade at a Neolithic site near Vilsbiburg, Bavaria (southern Germany). The outline of the postpit is indicated by the dashed line, and the remains of the wooden post itself are represented by shading. Circles indicate where samples were removed for magnetic analysis. The curve above is a cubic spline fit to the susceptibility values of the row of samples indicated by the filled circles in the excavation. (Redrawn from Fassbinder and Stanjek, 1993.) © Polish Academy of Sciences, with permission of the publishers.

magnetic signals from buried wooden palisades that, upon excavation, prove to be unburned. A detailed investigation of a single posthole on a neolithic site near Vilsbiburg (Bavaria) yields firm evidence of increased magnetic susceptibility (Fig. 11.1). Thermal measurements on magnetic extracts from the remains of the post clearly reveal the Curie point and Verwey transitions diagnostic of magnetite, and electron microscopy indicates the presence of ultrafine magnetite grains with a mean crystallite size of about 40 nm. Following the findings of Fassbinder *et al.* (1990), they argue that this magnetite is bacterial in origin. At other localities, however, Maher and Taylor (1988, see also Maher, 1990) favor an inorganic fermentation origin (see earlier) for the ultrafine-grained magnetite they observe in two British soils. Certainly, both options are viable, and there is no reason to suppose that they cannot both be valid: individual cases will be dominated by one or the other as environmental conditions dictate. As Maher (1998) emphasizes in her review article, the interpretation of soil magnetism requires site-by-site assessment — there is no golden rule. This is very much the case, for example, in the discovery that some ancient human burials are closely associated with strong magnetic susceptibility enhancement (Linford, 2002). An excellent example is shown in Fig. 11.2, but the actual source of the signal awaits further magnetomineralogical investigations.

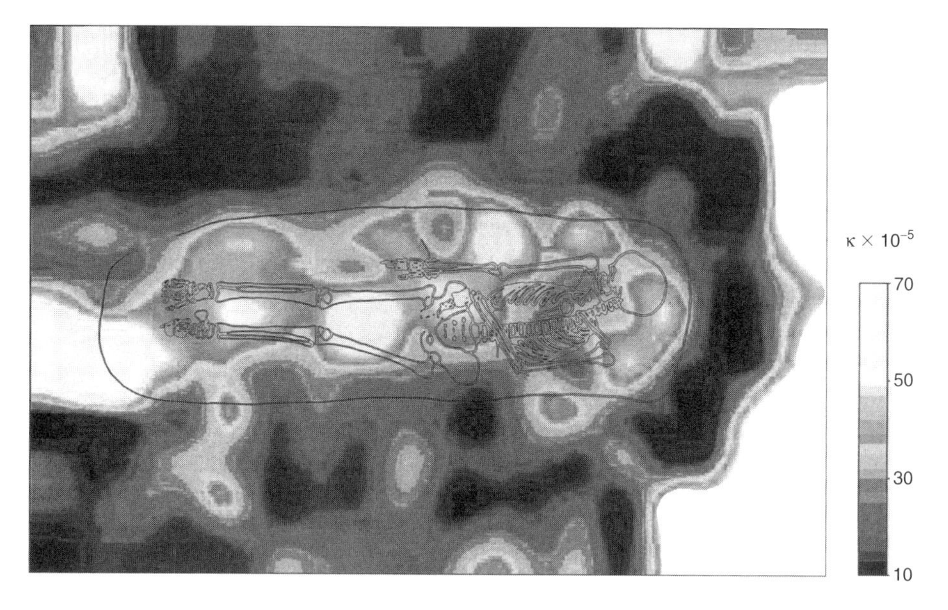

Figure 11.2 Magnetic susceptibility values recorded over an Anglo-Saxon grave at Lakenheath, England (0.6°E, 52.4°N). The subsequently excavated skeletal remains are indicated. The color scale represents SI susceptibility (κ) values ranging from 10×10^{-5} (black) to 70×10^{-5} (white). The area shown measures 1.4×2.1 m. The data were collected under the auspices of *English Heritage* using a Bartington MS2 meter (see Chapter 4). (Figure kindly provided by Neil Linford.) See color plate.

So far, our discussion has been centered on magnetic susceptibility. This is to be expected because historically by far the majority of investigators have considered it to be the important parameter. However, there is a growing trend to employ a wider variety of magnetic parameters, such as hysteresis properties—as in the study by Peters and Thompson (1998b). A summary of the various soil magnetic parameters and their use in archeology is given by Dalan and Banerjee (1998), who also provide a very thorough bibliography of relevant publications. In their own study of the Cahokia Mounds State Historic Site in southwestern Illinois, these authors found a combination of susceptibility and anhysteretic remanent magnetization to be particularly useful, as we shall see in the discussion in Section 11.4.

11.3 ARCHEOLOGICAL MAGNETIC PROSPECTION SURVEYS

In his early work, Le Borgne (1950, 1951) already drew attention to the magnetic properties of soil in connection with magnetic mapping of the underlying bedrock. He was worried about how the signal from the soil might mask, or confuse, the bedrock signal that was his real interest. The same problem arises in the archeological context (Graham and Scollar, 1976), but variations in soil magnetism may also provide archeological signals rather than troublesome magnetic noise. The archeologist and the geophysicist must cooperate in an effort to decide what is noise and what is signal. Many publications discuss the location of buried archeological remains by geophysical methods, including magnetometry (e.g., Aitken, 1974; Gibson, 1986; Clark, 1990; Becker, 1999; Garrison, 2001; http://www.cast.uark.edu/nadag/). Nor are such magnetic surveys restricted to land, as is dramatically demonstrated by the exploration — using submersible magnetometers — of the drowned part of ancient Alexandria on the Mediterranean shore of Egypt (http://www.franckgoddio.org). As we have seen, one prominent way in which archeological remains acquire magnetic signals is by means of thermoremanent magnetization (TRM) resulting from heating in antiquity. This is particularly significant for pottery kilns, which — if not buried too deeply — can sometimes produce anomalous fields up to 500 nT, although 100 to 200 nT is more typical. Here, we focus attention on the aspects relevant to soil magnetism. This means that we will be concerned mostly with magnetic susceptibility, the resulting signals arising from the induced (not the remanent) magnetization (to recap this distinction, see Chapter 2). Consequently, there is a great deal of common ground between this section and the previous one. We adopt a division based essentially on the objectives of the original authors whose work is described.

An excellent starting point is offered by a state-of-the-art investigation of a Scythian settlement in Siberia (Becker and Fassbinder, 1999). This survey (which was completed in only 3 days) revealed the plan of a fortified settlement consisting of more than a hundred pit-houses (*Grubenhäuser*) dug into loess (Fig. 11.3). The source of the susceptibility contrast between the houses and their surroundings is not completely understood, but the working hypothesis is that these shallow pits have

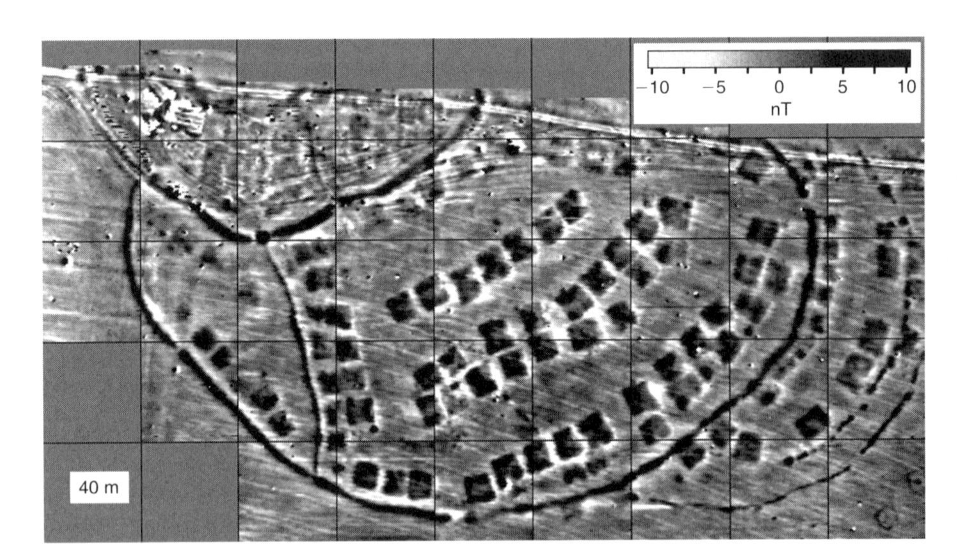

Figure 11.3 Magnetic signature of eighth to seventh century BC pit-houses at a Scythian settlement in Siberia. The houses are relatively uniform in size (8×10 m) and are arranged in streets. The complex includes a series of ditches and palisades. The survey was carried out with a Scintrex cesium magnetometer (SM4G-Special in duo-sensor configuration). The survey squares measure 40 m on a side. The gray scale spans a range of 20 nT. (From Becker and Fassbinder, 1999, with permission of the authors.)

been filled by recent chernozemic topsoil that has higher magnetic susceptibility than the surrounding loess. Maximum magnetic anomalies are about 10 nT (= 10^{-8} tesla), which represents a variation of only 1 part in 6000 of the ambient geomagnetic field. This is a small signal, but the instrument sensitivity is yet another thousand times smaller, being about 10 pT (=10^{-11} tesla). Indeed, magnetic surveying has now reached a level such that individual postholes can often be resolved. A simple calculation (see Box 11.1) shows that this is quantitatively reasonable.

As pointed out before, the posthole magnetism is thought to be an example of bacterial magnetite. But several authors have argued that the alternative mechanism of enhancing magnetic susceptibility by fire is important in the creation of detectable magnetic anomalies (Marmet *et al.*, 1999). These reports appeal to the mineralogical transformations effected in the soil as a result of heating, as discussed in Section 11.2. To investigate the conditions under which these transformations take place, Canti and Linford (2000) deployed arrays of thermocouples around five experimental fires to monitor temperatures in the air and in the soil. At a depth of only 1 cm, the overall maximum temperature recorded was 570°C (close to the Curie point of magnetite), but in four out of the five cases the maximum was markedly lower (433–456°C). At 4 cm depth, maximum temperatures lie in the range 199 to 318°C. In addition to their own experimental data, Canti and Linford summarize the earlier literature on this topic (much of which has an agricultural and/or forestry context rather than an archeological one). They find that maximum subsurface temperatures exceed 400°C in only 6 cases out of a total of 71. Firing of pottery involves temperatures on the order of 1000°C, but all the evidence suggests that, in general, hearths, ovens, campfires, and so forth heat their immediate surroundings to only a few hundred degrees. The effect of this is twofold. The likely enhancement of magnetic susceptibility will lead to a stronger induced magnetization and thereby a larger overhead anomaly. But as the material cools down it will inevitably acquire a TRM (or, if the Curie point was not reached, a *partial* TRM) that may well overwhelm any increase of induced magnetization resulting from thermally enhanced susceptibility. In general, therefore, the magnetic anomaly over material that has been heated and cooled *in situ* will be due, in large part, to thermoremanent magnetization. If, on the other hand, the material is moved after cooling (for example, the dumping of burned domestic garbage into a pit), thermally enhanced susceptibility will be important for the overhead anomaly.

A somewhat different explanation of magnetic anomalies over burned archeological features is put forward by McClean and Kean (1993). They report several magnetic experiments related to fire pits and hearths and find that the residual ash itself, rather than the substrate, is strongly magnetic. In one case (at Ellicottville, New York), the ash layer had a magnetic susceptibility 22 times greater than that of a nearby control soil sample. This is similar to the findings of Tite and Mullins (1971) described in the previous section, but with the important distinction that McLean and Kean are dealing directly with the ash produced from the wood that was used to make the fire. Actually, it is worth noting that although the reported enhancement factor is only 22, compared with the maximum value of >200 obtained by Peters and Thompson (1998b), the Ellicottville ash susceptibility is almost three times higher than that of the burned soil studied by Peters and Thompson — 17×10^{-6} compared

Box 11.1 Magnetic Posthole

Consider a buried sphere whose top just touches the surface as shown in the accompanying diagram. Assume that the ambient geomagnetic field is vertical (i.e., the sphere is vertically magnetized). The magnetic anomaly caused by such a model attains its maximum value directly over the center of the sphere and is given by

$$F = (\mu_0 2M)/(4\pi d^3)$$

where μ_0 is the permeability of free space and the sphere's magnetic moment (M) is

$$M = \Delta k H v$$

H being the ambient field ($= B/\mu_0$), Δk the susceptibility contrast between the sphere and its surroundings, and v the sphere's volume. Taking the values given in the diagram, we obtain

$$F = 0.5\,\text{nT}$$

In practice, Fassbinder and Irlinger (1994) find that single posts produce anomalies about half this size. This is not unexpected because our calculation is no more than a crude estimate. Nonverticality of the ambient geomagnetic field will reduce the magnitude of the anomaly and change its shape. Furthermore, reference to Figs. 2 and 3 of Fassbinder and Stanjek (1993) indicates that the susceptibility contrast we used is the maximum value observed; the average value throughout the entire volume of the post is probably somewhat smaller. Nevertheless, the predicted anomaly is still well above the 10-pT noise level of modern cesium magnetometers.

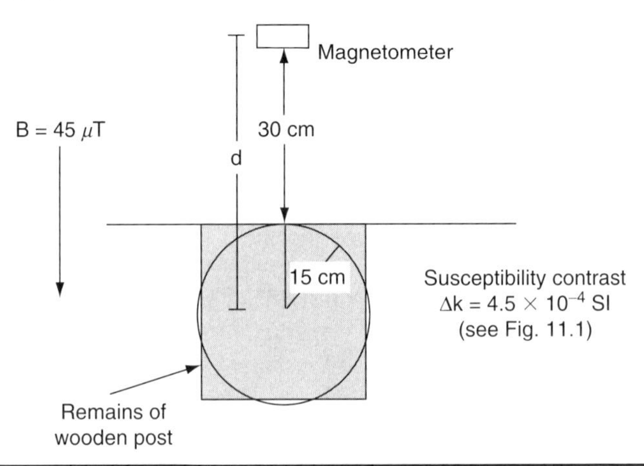

with 6×10^{-6} m^3/kg. McLean and Kean suggest — but do not prove — that the source of this very strong magnetism is magnetite derived from phytoferritin, a biomineralized iron–protein complex (Hyde *et al.*, 1963). Although the main purpose of the investigations undertaken by McLean and Kean was to assess the role of fire in giving rise to magnetic anomalies over relevant archeological features, it is interesting to note that their work also provides a vivid example of the interconnectedness of the subject matter of this book. First the archeologist appeals to the geophysicist to locate buried cultural features, then the geophysicist turns to the mineralogist to discover the source of the magnetic signal, and finally the mineralogist seeks the assistance of the plant physiologist.

In summary, one has to admit that the origin of magnetic anomalies in archeological settings is likely to be complex. For convenience, the main possibilities are summarized as follows:

- Enhanced susceptibility by (1) burning or (2) fermentation
- Bacterial magnetite from magnetotactic bacteria
- Residual, magnetically enhanced ash
- Thermoremanent magnetization of *in situ* material

11.4 ECONOMY, INDUSTRY, AND ART

Regardless of the specific mechanisms responsible for magnetic enhancement, it is instructive to consider the illuminating case study of the Cahokia Mounds Site in southwestern Illinois (Dalan and Banerjee, 1998). It illustrates how a variety of magnetic techniques provide "a rapid, cost-effective, and minimally destructive means of understanding prehistoric landscapes and landscape change." The mounds are a vivid testimony to the Cahokians' engineering skill in earth moving. The largest feature (Monks Mound) measures 291×236 m at its base and rises to a height of about 30 m. By using magnetic properties (particularly susceptibility and anhysteretic remanence) to map the site, Dalan and Banerjee revealed that large areas from which material had been removed to construct the mounds (borrow pits) were reclaimed to provide a level surface for the so-called Grand Plaza, a broad open space covering some 175,000 m^2. Evidence from ceramics indicates that this landscape-modifying activity was initiated late in the Emergent Mississippian period (AD 800–1000) and that the site was probably abandoned about AD 1400. The authors admit that prior to the magnetic work, their understanding of this area had been in "in error." The magnetic results were crucial in providing a means of characterizing the various materials present and in deciphering the "cultural processes involved in molding the Cahokia landscape." In particular, a very large borrow pit ($\sim$40,000 m^3) — probably used for Monks Mound — seems to have become a communal receptacle for trash (a midden), which was systematically mixed with unmodified soil to reclaim the large flat area necessary for the Grand Plaza. The midden material has a saturation magnetization $\sim$12 times greater than that of the natural soil, whereas the landfill mixture is ~4 times more magnetic. Dalan and Banerjee point out that the reclamation project therefore mixed 3 parts soil to 1 part midden.

A very similar example (Jing and Rapp, 1998), using the same magnetic parameters, has been reported from archeological sites in the Shangqiu area of China (115°E, 34°N). This is the homeland of the Shang civilization (1750–1100 BC), the first literate civilization in East Asia. In this case, the sediments could be characterized magnetically into two main groups that owe their different magnetic properties to changes in the drainage system over the last 2000 years. Prior to the 12th century, the area was part of the Hauai River drainage system, but between the 12th and 19th centuries, the Yellow River flowed southward through the area. In the first case, the sediments brought into the area came from a weakly magnetic source (mean susceptibility $= 10\times10^{-8}\,m^3/kg$), but for 700 years thereafter the Yellow River brought in more magnetic material (mean susceptibility $= 45\times10^{-8}\,m^3/kg$). The marked difference between these two sources allowed Jing and Rapp to identify the sedimentological context of various anthropogenic features.

An innovative application of magnetism to archeological problems has been reported by Church *et al.* (2001). As in the work of McLean and Kean (1993), this also involves fire ash but looks at the diagnostic value of variations in mineral magnetic properties from hearth to hearth rather than considering possible anomalies facilitating magnetic location. It is found that the magnetic properties measured (susceptibility and its frequency dependence, isothermal and anhysteretic remanence) can be combined to provide a fingerprint of the type of fuel used in domestic fires at sites on the Western and Northern Isles of Scotland. Sites on the Isle of Lewis (7°W, 58°N), for example, indicate that well-humidified peat was the dominant fuel source for thousands of years. This implies a stable system of managing the peat banks involving issues of ownership and organization that is obviously of interest to archeologists concerned with the way of life and general economy in ancient settlements (e.g., Ceron-Carrasco *et al.*, 2001).

Similar examples exist in which the intrinsic magnetic properties of particular archeological artifacts are used to identify where they came from (their provenance). Of special interest in this context is prehistoric obsidian—a volcanic glass possessing very desirable conchoidal fracture making it ideal for arrowheads, blades, scrapers, and the like. Determination of the chemical composition (particularly the trace elements) has been successfully employed to associate certain obsidian artifacts with specific volcanic outcrops, but such tests are expensive, time consuming, and destructive. On the other hand, McDougall *et al.* (1983) found that rapid, cheap, and nondestructive fingerprints could be obtained magnetically. Specifically, they demonstrated that obsidians from known Mediterranean, central European, and near Eastern sources define distinct, restricted, fields on plots of susceptibility versus saturation magnetization. The implication is that slight compositional differences between obsidians from different sources are reflected in their bulk magnetic properties. Given that only a few sources seem to have been available in antiquity, the possibility of magnetic matching exists and, because obsidian was a significant prehistoric trade item, the possibility of determining patterns of cultural contact arises. On the other hand, similar studies of obsidians from the American Southwest

(Church and Caraveo, 1996) and Mexico (Borradaile *et al.*, 1998) appear to be less promising.

An investigation by Williams-Thorpe *et al.* (1996) concerns the provenance of granite columns used in ancient buildings in Rome (including the Pantheon, the Temple of Venus, and the Baths of Caracalla), particularly those made of the *granito del foro* (so named because of its abundant use in the Roman Forum). Magnetic susceptibility measurements confirm other observations that indicate that this material comes from Mons Claudianus in the Eastern Desert of Egypt. Furthermore, a contour map of 1119 magnetic susceptibility values covering the entire 9-km^2 area of quarrying activity there allows some columns to be sourced to a single quarry, of which there are 130. It appears that quarrying did not evolve in any systematic spatial pattern; rather, several parts of the quarry field were opened up within the first century AD, and they all continued in use during the second and third centuries.

We have already considered the importance of fired archeological features in extending the geomagnetist's knowledge of the secular variation. Studies in Italy have extended this kind of research to an entirely different (and somewhat surprising) magnetic recorder, namely mural paintings (Chiari and Lanza, 1997). Careful removal of thin layers of pigment (using adhesive tape) shows that the direction of the ambient magnetic field in which the murals were painted can be recovered. Controlled laboratory experiments demonstrate that the iron oxide particles (hematite) present in the pigment are aligned by the magnetic field during the drying process, in essentially the same way as the depositional remanence (DRM) mechanism illustrated in Box 5.1. Zanella *et al.* (2000) report a number of results from murals in Pompeii. Of particular interest are those in the *Thermae Stabianae* (Stabian Baths), which are known to have been painted just a few years before the AD 79 eruption of Vesuvius and which yield an archeomagnetic direction indistiguishable from that obtained by Evans and Mareschal (1989) from a nearby pottery kiln—the one featured in Fig. 6.5, in fact.

11.5 SPELEOMAGNETISM

In many places throughout the world, caves provided convenient ready-made housing for our distant ancestors and many important examples have been thoroughly investigated by archeologists. Geophysicists, on the other hand, have paid relatively little attention to cave deposits. Nevertheless, the sediments on the floor of a cave as well as the ubiquitous speleothems (the collective term for stalagmites, stalactites, and flowstones) have been exploited as geomagnetic recorders (Latham and Ford, 1993; Perkins and Maher, 1993). Consider, for example, the evolution of a stalagmite—as it grows in girth, each successive layer records the ambient magnetic field of the day, the actual recording being done by trace amounts of magnetic minerals in the calcium carbonate that makes up the deposit. The outcome is rather like a set of magnetic tree rings. If the resulting secular variation patterns can be successfully matched to reference master curves, then records of this kind have obvious potential in terms of chronological control like the other examples discussed in Chapter 6.

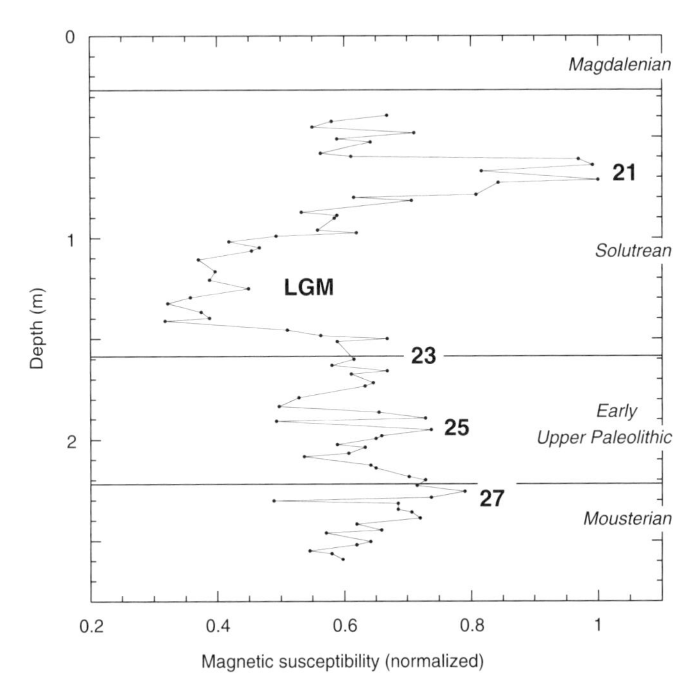

Figure 11.4 Magnetic susceptibility profile of sediments in Caldeirão Cave, Portugal. The various cultures — identified by stone tools found in the cave — are indicated as are the MSEC zones SE-21 through 27. LGM = last glacial maximum. (Modified from Ellwood *et al.* 1998, 2001.) © John Wiley & Sons Limited. Reproduced with permission.

In terms of environmental magnetism, cave sediments are still largely *terra incognita*. One notable exception is the work of Ellwood *et al.* (1998) in the Caldeirão Cave near the town of Tomar in Portugal (8.4°W, 39.5°N). A few meters of Middle and Upper Paleolithic sediments have accumulated in this cave and these have yielded evidence of several successive cultures, including Mousterian, Solutrean, and Magdalenian. Samples collected over a 2-m stratigraphic section yielded a smooth pattern of magnetic susceptibility variations (Fig. 11.4) that was interpreted as a paleoclimate signal, high values corresponding to warm intervals, low to cold. Ellwood and his coauthors argue that the climate signal is controlled by pedogenesis outside the cave followed by wind and/or water transport and preservation of the resulting material in the protected environment within the cave. When warmer conditions exist, the soil-forming processes are enhanced and more magnetic material is produced (see Chapters 5 and 7). Using the cultural remains (Paleolithic stone tools) and ^{14}C data, they attribute the strong minimum between 1.0 and 1.4 m depth to the last glacial maximum (LGM). This is an excellent demonstration of how enviromagnetic information ties together anthropology and climatology. Another convincing example is provided by the detailed work of Sroubek *et al.* (2001) on some 6 m of sediments ($\sim$700 samples) in Kulna Cave in the Moravian karst country of the Czech Republic.

They also find a strong correlation between magnetic susceptibility and climatic conditions and again appeal to the pedogenic production of magnetic minerals (magnetite and/or maghemite) during warm interglacial periods. In particular, for the time interval covered by the sediments (110–15 kyr BP), they observe a close match between their susceptibility profile and the record of sea surface temperature deduced by Ruddiman (1987) from core K708-1 in the North Atlantic (24°W, 50°N).

This kind of cave research is being expanded. Ellwood *et al.* (2001) have summarized their magnetic investigations of cave sediments, which now include examples from Albania and Spain in addition to the Portuguese data described previously. Their success in identifying cycles of climatic change and in correlating these from cave to cave leads them to introduce a specific name for their procedure, namely the *magnetosusceptibility event and cyclostratigraphy* (MSEC) method. This they use to propose a system of numbering of warm and cold intervals in a manner reminiscent of the well-known marine oxygen isotope (MOI) stages (see Chapter 6). But two important differences must be noted. First, they do not claim that their system is entirely global. For this reason they label their intervals SE-1, SE-2, and so on, the SE standing for southern Europe, with odd numbers corresponding to warm intervals. Second, the timescale associated with the cave sediments is very much shorter than that of the oxygen isotope stages: MSEC zones span a few centuries, whereas MOI stages typically last for 10 to 50 kyr. To relate the MSEC zones to previously established schemes, it is convenient to consider two intervals of particular interest — the Younger Dryas and the LGM. These correspond to zones SE-12 and SE-22, respectively. However, it should be remembered that the whole MSEC system is still rather speculative. More data are urgently needed to establish (or reject) its validity.

11.6 HOMINID EVOLUTION

The reversal polarity sequence that was so important in providing the chronology now universally accepted for the deposition of the vast deposits of eolian sediments (loess) in China also provides important control for the age of an early hominid cranium (*Lantian man*) found in the vicinity of Xian. Shaw *et al.* (1991) report that the cranium was found in a reversely magnetized stratum just below the Cobb Mountain excursion. On the revised GPTS (see Chapter 6), this implies an age close to 1.2 million years, making this the oldest reliably dated Chinese hominid site. This procedure relies on the remanent magnetization preserved in the sediments, but magnetic susceptibility — via its tracking of climatic changes — has also been used to date cultural remains found in loess/paleosol sequences. At Karamaidan, for example, correlating the susceptility profile with the oxygen isotope stages enabled Shackleton *et al.* (1995; see also Chapter 8) to establish the correct ages for the pebble tools found there. Material previously thought to be about 100,000 years old is now seen to be more than three times older. The new chronology is in much better agreement with data from elsewhere and clears up what had been a confused interpretation of the Paleolithic archeology of central Asia.

Magnetic susceptibility time series have also been exploited in the context of hominid evolution. In particular, deMenocal and Bloemendal (1995) investigate Plio-Pleistocene climatic change in subtropical Africa by magnetic monitoring of the paleoenvironment in which major evolutionary events took place. Their procedure is to use magnetic susceptibility as a proxy for the terrigenous material transported by monsoon winds from the African continent and now found in cores recovered offshore in the Arabian Sea [see also Bloemendal and deMenocal (1989) and discussion in Chapter 7]. The validity of this proxy is based on the excellent correlation observed between susceptibility and terrigenous content determined independently. This makes it possible to obtain a continuous record reaching back to 7.3 million years, with 1500-year resolution. The enormous advantage offered by the magnetic measurements can be appreciated by simply realizing that the time series needs at least 5000 data points to obtain the resolution claimed. The time-consuming extraction process necessary to verify independently the terrigenous content makes this option very unattractive. It is unlikely that such remarkable records would have ever been obtained without the speed and ease offered by magnetic susceptibility measurements.

The high-resolution records obtained by deMenocal and Bloemendal (1995) indicate that the pattern of variability of terrigenous content was not constant throughout the last 7 million years. In the earliest part of the record (prior to $\sim$2.8 million years ago), the Milankovitch precession cycle ($\sim$20 kyr) is dominant (see Fig. 7.23). Thereafter, the obliquity signal ($\sim$40 kyr) becomes more prominent. Still later, at $\sim$1.0 million years ago, the eccentricity cycle ($\sim$100 kyr) starts to play a bigger role. These changes reflect important climatic shifts that are also captured in high-latitude marine records (Shackleton *et al.*, 1984; Ruddiman *et al.*, 1989). Furthermore, numerical general circulation model (GCM) experiments demonstrate the sensitivity of climatic conditions in subtropical East Africa to the size and elevation of the Fennoscandian ice sheet (deMenocal and Rind, 1993).

deMenocal and Bloemendal (1995) go on to consider the significance of these climatic changes to human evolution. They point out that the major shift at $\sim$2.8 Myr coincides with the time at which the ancestral lineage *Australopithecus afarensis* gave rise to later australopithecines, on the one hand, and the line from which our own genus (*Homo*) arose, on the other. Furthermore, the earliest major geographic expansion of our direct ancestor (*Homo erectus*) took place $\sim$1 Myr ago when the *Homo* lineage first radiated out of Africa and occupied sites in Europe and western Asia. It was also at this time that the entire australopithecine lineage became extinct.

Although speculative, these suggestions are not unreasonable — the influence of climate on biological (including hominid) evolution is widely appreciated. It is the use of magnetic monitoring by deMenocal and Bloemendal (1995) that is important for our present purposes. As they conclude, "it was a change in mode of subtropical climatic *variability* [their italics] rather than a wholesale, stepwise change in climate that prompted evolutionary responses." This was discovered only because extended, high-resolution, magnetic records became available for spectral analysis.

12

OUR PLANETARY MAGNETIC ENVIRONMENT

12.1 INTRODUCTION

Throughout this book we have been concerned with the application of magnetic methods to monitor our natural and cultural environments, but we should not lose sight of the planetary setting in which the various effects involved take place. The geomagnetic field is a key factor in such diverse topics as magnetostratigraphy, sea floor spreading, bacterial navigation, and anomalies over buried archeological structures. Moreover, it is fundamental for the creation of the magnetosphere, which is the most important entity controlling the near-Earth space environment. It, too, plays a key role in many phenomena, being intimately connected with the northern and southern lights (*aurora borealis* and *aurora australis*), radio communication disruptions, power outages, and satellite failures. In what follows, a brief outline of the main features of the Earth's magnetic environment is given so that the preceding chapters can be placed in their proper framework.

The fact that the Earth has a magnetic field of its own means that everything around us is penetrated by magnetic lines of force — including the page you are currently reading and, indeed, your whole body. Every schoolchild knows that the shape of this field outwardly resembles that of a simple bar magnet, a fact that was first clearly enunciated by William Gilbert (1544–1603) in his seminal work *De Magnete* published in 1600. As he put it, "magnum magnes ipse est globus terrestris" (the terrestrial globe itself is a great magnet). Wilson (2000) provides a fascinating analysis of Gilbert's tome, which is often regarded as the first modern scientific textbook because of its reliance on repeatable experimental procedures. He points out that Gilbert probably focused his attention on magnetism because the lodestones (naturally occurring lumps of magnetite) with which he worked exhibited powerful effects that were easily observed. These effects, of course, arise from the magnetization acquired by the lodestones in the geomagnetic field. This prompts Wilson to ask, "How would science have progressed if we ourselves had evolved on Mercury, Venus or Mars? Those planets have very weak or zero magnetic fields, so that if lodestone exists on those three planets, it would quite possibly have acquired no significant magnetization during its formation. What then?"

12.2 THE GEOMAGNETIC FIELD

The field surrounding a bar magnet is mathematically equivalent to that produced by a uniformly magnetized sphere. Such a field is *dipolar*, possessing two poles that are conventionally called *north* and *south*. The Earth, however, is a little more complicated. First of all, there is the matter of nomenclature: the pole in the Arctic is a south magnetic pole (this is why the north end of a compass needle points toward it — recall that like poles repel, opposite poles attract). Second, the hypothetical bar magnet is only an approximation to the observed geomagnetic field. This is not surprising when one considers the actual origin of the field.

The Earth is certainly not the permanently magnetized sphere imagined by William Gilbert. Most of its volume is far too hot to sustain such a condition. Common ferromagnetic materials have Curie points (see Chapter 2) of only a few hundred degrees, and such temperatures are reached a few tens of kilometers below the surface. This is proportionally no more than the shell of an egg. The earth's shell, or *crust*, does contain magnetic material, but even the highest observed values of crustal magnetic remanence fail by several orders of magnitude to account for the strength of the geomagnetic field. To find the real source, we must look deeper into the planet by appealing to electrical currents flowing in the outer core — a zone of highly conducting molten iron starting 2900 km below the surface. Fluid motions in this region sustain a *geodynamo*. The mathematical treatment necessary to deal with the geodynamo is rather difficult, and only recently has computing power risen to an adequate level to allow realistic models to be investigated (Glatzmaier and Roberts, 1995).

Superficially, the pattern of the field at the Earth's surface resembles a meteorological map. There are broad features that are fairly stable over the long term (trade winds, for example), but there are very significant small-scale, rapid fluctuations (such as the pressure highs and lows of the daily weather map). The vagaries of the fluid motions in the core produce a wide spectrum of temporal changes, some of which were discussed in Chapter 6 in connection with dating methods. The outcome is that the geomagnetic field not only fluctuates in time but also is not exactly central, nor is it aligned along the spin axis, and it is not even dipolar! For some purposes, the so-called GAD (geocentric axial dipole model) is adequate (see Box 12.1), but a more sophisticated procedure is generally needed. The universally accepted treatment is based on *spherical harmonics*, a technique invented by the great German mathematician C. F. Gauss (1777–1855). The spherical harmonic components can be added together to provide a mathematical representation of the total field (in a manner similar to adding the fundamental tone and overtones of a violin string to get a complete representation of the overall sound it produces). In the case of the Earth's magnetism, the spherical harmonics constitute what is called the International

Box 12.1 GAD

The characteristics of a *g*eocentric *a*xial *d*ipole are readily derived from the expression for the potential of a dipole [$V = (\mu_0/4\pi)(m \cos \theta/r^2)$, where m is the dipole moment, μ_0 is the permeability of free space, r is the distance from the Earth's center, and θ is the colatitude measured from the geographic north pole]. At any point on, or above, the Earth's surface, the radial and tangential components are $Z = (\mu_0/4\pi)(2m \cos \theta/r^3)$ and $H = (\mu_0/4\pi)(m \sin \theta/r^3)$, respectively. On the equator, $Z = 0$ and $H = H_{max} = (\mu_0/4\pi)(m/r^3)$. On the magnetic axis (i.e., at the poles), $Z = Z_{max} = (\mu_0/4\pi)(2m/r^3)$ and $H = 0$. Thus $Z_{max} = 2H_{max}$. The field strength (F) is given by $F = (Z^2 + H^2)^{1/2} = (\mu_0/4\pi)(m/r^3)(1 + 3 \cos^2 \theta)^{1/2}$.

The present dipole moment of the Earth is $\sim 8 \times 10^{22}$ Am^2, so that on the surface Z_{max} turns out to be $\sim 60\,\mu$T. Finally, the inclination (I) is given by $\tan(I) = Z/H = 2 \cot \theta$. The accompanying figure shows pole-to-pole profiles of I and F.

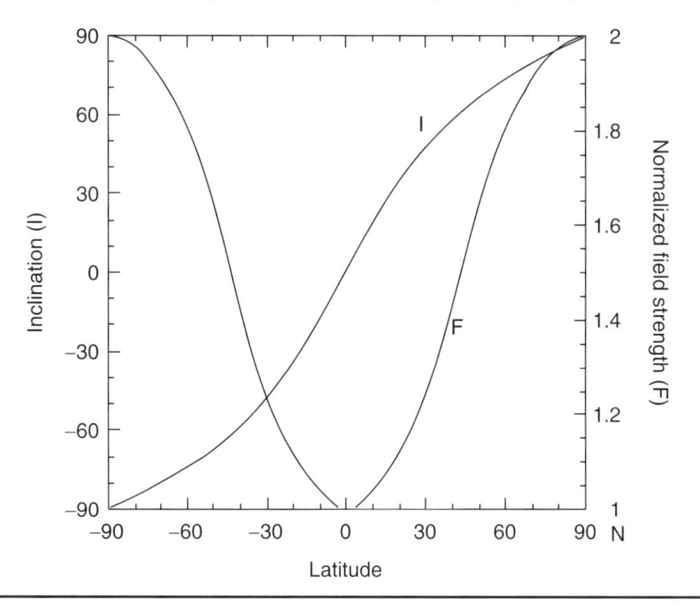

Geomagnetic Reference Field (IGRF) (check out http://www.ndgc.noaa.gov). Lowrie (1997) gives an excellent introduction.

The departure of the IGRF from a GAD field can be appreciated by inspecting Fig. 12.1, which shows profiles of inclination (I) and field strength (F) along the zero (Greenwich) meridian from the equator to the geographic north pole. Because the best-fitting dipole is tilted by about 11° toward northern Canada, the IGRF inclinations along this meridional profile are lower than the GAD model, particularly in tropical latitudes. Similar discrepancies arise in the total field strength, but now the differences are greatest in high latitudes. The full spatial pattern of the IGRF is illustrated in Fig. 12.2. The corresponding map for a GAD field would consist of a set

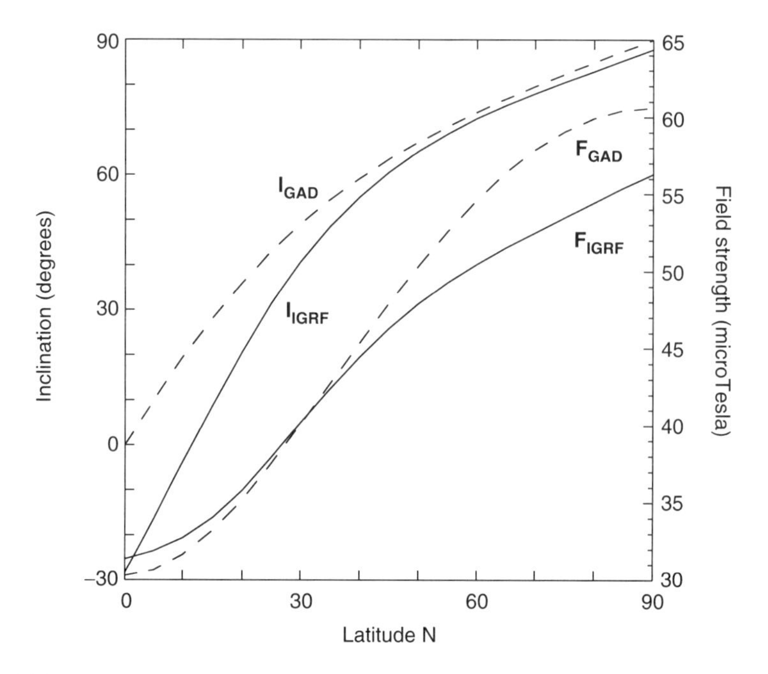

Figure 12.1 Latitudinal dependence of inclination (*I*, left ordinate) and field strength (*F*, right ordinate) for the International Geomagnetic Reference Field (IGRF2000) and a geocentric axial dipole (GAD). The profiles shown here run along zero longitude (the Greenwich meridian) from the equator to the north geographic pole.

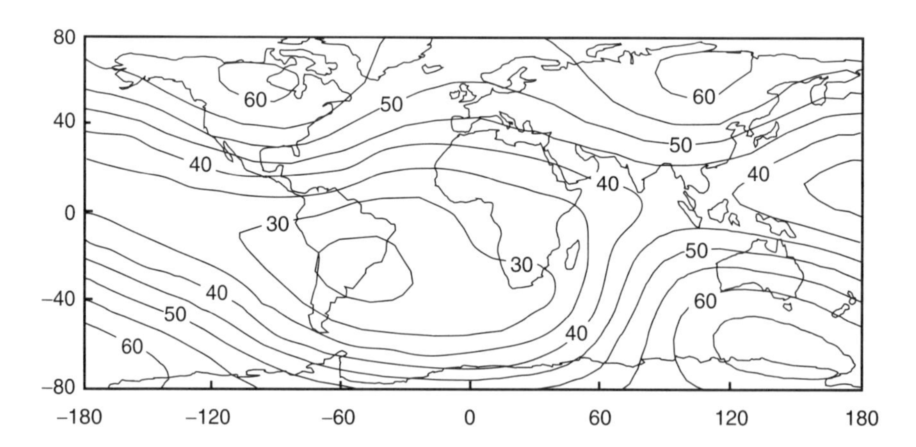

Figure 12.2 Isodynamic contours (lines of equal field strength, in μT) for IGRF2000. For the GAD model, the contours would consist of a set of horizontal lines ranging from ~30 μT at the equator to ~60 μT at the geographic poles. [Similar maps for inclination (isoclinics) and declination (isogonics) are often useful. For the GAD model, the isoclinics would also be a set of horizontal lines ranging from +90° (vertically down) at the north geographic pole to –90° (vertically up) at the south geographic pole and equal to zero (horizontal) along the entire equator. The isogonic map would be blank because GAD declination is zero everywhere.]

of horizontal lines representing increasing values from $\sim 30\,\mu T$ at the equator to $\sim 60\,\mu T$ at latitudes $+90°$ and $-90°$. We see that, rather than having a maximum at each pole and a linear minimum along the equator, the actual field has three maxima (in Canada, in Siberia, and south of Australia) at latitudes of about $\pm 60°$ and a single minimum over southern Brazil (~28°S). Maps of this kind, when projected down to the Earth's core, show that the magnetic field is patchy; some areas possess flux concentrations, others are relatively barren. Bloxham and Jackson (1992) show that much of the geomagnetic secular variation can be understood by tracking the changes that have occurred to these flux patches over the last 300 years. Some of them are essentially stationary (but may grow or decay); others definitely move about. Using data from archeomagnetic artifacts, lava flows, and lake sediments, this evolutionary pattern has now been successfully extended back to 1000 BC (Constable *et al.*, 2000).

One very important observation is that the overall integrated effect is that the best-fitting GAD has decayed by almost 10% in the last century and a half. Archeomagnetic data (see Chapters 6 and 11) show that this trend has been in effect for the last two millennia, during which time the Earth's dipole moment has decreased from $\sim 11 \times 10^{22}$ Am^2 to its present value of $\sim 8 \times 10^{22}$ Am^2 (McElhinny and Senanayake, 1982). Is the next polarity reversal coming? A summary (Hulot *et al.*, 2002; see also Olson, 2002) comparing the results obtained from the Magsat satellite (which operated in 1979–1980) with those currently being gathered by the Oersted satellite suggests that this is a strong possibility.

Fluctuations in the strength of the geomagnetic field (whether or not they are associated with full polarity reversals) have been of considerable use in providing chronometric control in some sedimentary sequences (see Chapter 6). They also play important roles in controlling the rates at which ^{14}C (see, e.g., Laj *et al.*, 1996) and ^{10}Be (see, e.g., Robinson *et al.*, 1995) are produced in the atmosphere—as discussed in Chapters 6 and 7, respectively. The effect on the carbon clock, for example, can be readily appreciated by reference to Fig. 12.3, which shows that, at certain times in the past, the rate of ^{14}C production was much higher than at present. According to Laj *et al.* (1996), the ultimate effect on radiocarbon dates is that measured ages have to be increased by up to 3500 years during the 20- to 40-kyr interval.

12.3 THE MAGNETOSPHERE

The Earth is by no means unusual in possessing a magnetic field; the magnetic moments of Saturn and Jupiter, for example, are 550 and 19,000 times greater than the Earth's, respectively. The sun also has a strong magnetic field, as do many other stars. In fact, the entire solar system is bathed in an *interplanetary magnetic field* (IMF) created by the flow of charged particles constantly being emitted by the sun. This flux constitutes the solar wind, a low-density gas of ionized particles (*plasma*)—mostly protons, electrons, and helium nuclei. In our part of the solar system, the IMF currently has a strength of ~6 nT. At an early stage in the evolution of the solar system, some theories appeal to the presence of a much stronger IMF that played a central role in transferring angular momentum from the sun to the planets by a

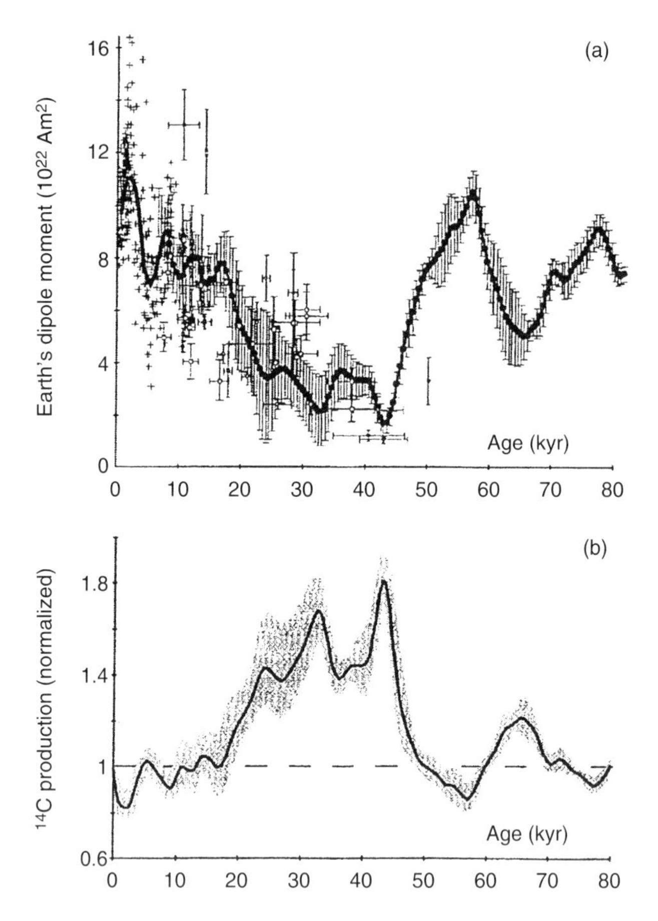

Figure 12.3 (a) Variations in the Earth's dipole moment deduced from oceanic sediments (smooth curve) and volcanic lava flows (individual points) (see also Fig. 6.10). (b) Changes in the production of ^{14}C atoms expected as a result of the geomagnetic history indicated in (a). (Modified from Laj *et al.*, 1996.) © American Geophysical Union. Modified by permission of American Geophysical Union.

mechanism known as *magnetic braking*. Some process of this kind is necessary in order to explain the curious distributions of mass and angular momentum in the solar system: more than 99% of the mass but only 2% of the angular momentum resides in the sun. This situation would not arise in a straightforward development of a cooling, contracting solar nebula. The former existence of an enhanced IMF has been sought by investigating the remanent magnetization of meteorites, and some support for the concept has been forthcoming. In particular, paleofields as high as 300 μT have been reported from the Allende meteorite that formed more than 4.5×10^9 years ago [see Dunlop and Özdemir (1997) for a summary of extraterrestrial paleomagnetism].

Regardless of the actual strength and history of the IMF, the Earth presents an obstacle to the solar wind, which is obliged to flow around it like water past a rock in a stream (Fig. 12.4). The speed of the solar wind — typically a few hundred km/s — is

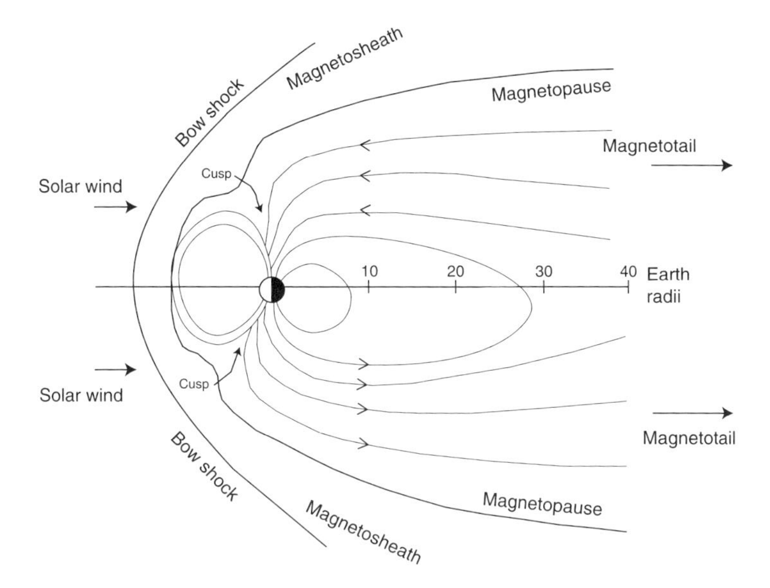

Figure 12.4 Sketch of the magnetosphere indicating the main features discussed in the text.

such that a *bow shock* (similar to the sonic boom effect of a supersonic aircraft) is set up in front of the Earth (at ~10 Earth radii along the sun–Earth line). Inside the shock front, the field is compressed, but downstream from the Earth it is stretched out into the so-called *magnetotail*. These distortions confine the geomagnetic field to the region inside the magnetosphere (the boundary of which is termed the *magnetopause*) and establish high-latitude *cusps*. The ionized particles of the solar wind are slowed down as they traverse the bow shock and are then deflected around the magnetosphere, but the cusps provide a means by which some particles leak through. Thus, the geomagnetic field provides an umbrella to protect us from the solar wind—perhaps parasol would be a better word!

The particles that find their way into the magnetosphere are involved in a variety of important phenomena. Some are trapped by the magnetic field in the so-called *Van Allen belts*, two doughnut-shaped regions encircling the Earth, one (containing mostly protons) about 2000 km above the Earth's surface, the other (containing mostly electrons) about 25,000 km above the Earth's surface. Other particles eventually make their way into the upper atmosphere, where they bombard the constituent atoms and molecules and thereby cause the aurorae (for example, the green light with a wavelength of 5577 Å, commonly seen in auroral displays, results from electrons interacting with oxygen). In addition to their striking visual displays, aurorae have significant magnetic effects due to their associated electric currents that flow in the *ionosphere*. This is the part of the atmosphere where the constituent atoms and molecules have been ionized by solar ultraviolet radiation. Auroral effects can lead to magnetic fields at the Earth's surface as large as 1 μT. Even without aurorae, fluctuations in the sun's activity can lead to *magnetic storms* during which irregular

Box 12.2 Spacecraft, Astronauts, and the SAA

The low values of the Earth's field associated with the region centered over the coast of southern Brazil (see Fig. 12.2) allow charged particles of the Van Allen belts to dip to lower altitudes, resulting in increased radiation flux to spacecraft in low orbits (100 to 1000 km). On June 15, 2001, for example, NASA's Moderate Resolution Imaging Spectroradiometer (MODIS) was knocked out of service as it crossed over South America at latitude 20°S (Heirtzler *et al.*, 2002). Collective experience shows that satellite problems commonly occur over South America and the South Atlantic. This region is regarded as a sort of "Bermuda triangle" that experts refer to as the South Atlantic anomaly (SAA). Scientific instruments are not the only victims. Radiation dosimeters on the Shuttle and Mir spacecrafts indicate that astronauts are also subject to peak radiation exposure while traversing the SAA. With extended missions on the International Space Station now being planned, the Earth's magnetic environment clearly needs close attention.

variations of up to 1 μT are again observed at the Earth's surface. Although the magnitude of these effects is small compared with surface IGRF values (typically 40 μT), they are sufficient to disrupt radio transmissions and have been known to generate such large surges in power lines as to cause major blackouts. At higher altitudes, equipment on satellites is also vulnerable at such times, as are orbiting astronauts and even the people on board high-flying aircraft. The threat is real enough (see Box 12.2) that major efforts are now under way to establish means of forecasting what has come to be called *space weather* (http://www.sel.noaa.gov).

Appendix

MAGNETIC UNITS IN THE SI AND CGS SYSTEMS

Appendix 1. Magnetic Units in the SI and cgs Systems

Parameter	SI	cgs	Conversion
Magnetic field	A/m	Oersted (Oe)	$1\ A/m = 4\pi \times 10^{-3}$ Oe
Magnetic induction	Tesla (T)	Gauss (G)	$1\ T = 10^4$ G
Magnetization	A/m	emu/cm^3	$1\ A/m = 10^{-3}\ emu/cm^3$
Mass magnetization	Am^2/kg	emu/g	$1\ Am^2/kg = 1\ emu/g$
Magnetic moment	Am^2	emu	$1\ Am^2 = 10^3$ emu
Susceptibility	None	None	$\kappa_{SI} = 4\pi \times \kappa_{cgs}$
Mass susceptibility	m^3/kg	$emuOe^{-1}g^{-1}$	$1\ m^3/kg = 4\pi \times 10^{-3}\ emuOe^{-1}g^{-1}$
Permeability, μ_0	$4\pi \times 10^{-7}$ Vs/Am	1	

SI, Système Internationale; cgs, centimeter-gram-second; emu, electromagnetic units.

GLOSSARY

This fairly extensive glossary is offered because environmental magnetism is a relatively new subject that has attracted the attention of a wide variety of specialists—from astronomers to zoologists—who are not always familiar with each other's jargon. To avoid introducing further arcane terminology, we have tried to use plain language (which may annoy some experts). To avoid compiling a complete dictionary, we have tried to be brief (which may annoy others). For the rest of you, we hope it proves acceptable—and useful.

anhysteretic remanent magnetization (ARM): a magnetization imparted in the laboratory by means of an initially strong alternating magnetic field that is gradually reduced to zero in the presence of a weak, but constant, magnetic field

anisotropy of susceptibility: the variation of magnetic susceptibility with direction in a sample

antiferromagnetism: a form of ferromagnetism in which the crystalline material involved contains two sublattices that are oppositely and equally magnetized leading to zero net magnetism

ARM susceptibility: anhysteretic remanent magnetization per unit bias field, usually given in mass-normalized form $Am^2kg^{-1}/Am^{-1} = m^3kg^{-1}$

assimilatory: a term used to describe the action of certain bacteria that take up external material and incorporate it into their cells, specifically the acceptance of iron ions by magnetotactic bacteria to fabricate magnetite and/or greigite

asthenosphere: the soft part of the Earth underlying the lithosphere—it deforms plastically during the slow convection that drives plate tectonics

authigenesis: changes that occur *in situ* during and shortly after deposition of a sediment, e.g., secondary overgrowths on mineral grains

biogeocoenosis: coupled functioning of life and earth

biologically controlled mineralization (BCM): same as BOM

biologically induced mineralization (BIM): mineralization produced by bacteria extracellularly

biologically organized mineralization (BOM): mineralization produced by bacteria intracellularly

biomimetics: engineering technology that attempts to fabricate materials copied from nature

blocking temperature: the temperature during cooling at which the relaxtion time of TRM becomes sufficiently long that the remanence can be considered "frozen in"

Bohr magneton: a fundamental unit of magnetism equal to 9.27×10^{-24} Am^2

boundary organized biomineralization (BOB): same as BOM

chemical remanent magnetization (CRM): remanent magnetism acquired as a result of chemical or crystallization processes

chron: a major time interval dominated by either reversed or normal geomagnetic polarity

circle of confidence (α_{95}): the 95% confidence region within which the true orientation of a Fisherian mean vector lies

coercive force (H_c): the "backward" magnetic field necessary to reduce to zero a "forward" saturation magnetization

coercivity of remanence (H_{cr}): the "backward" magnetic field that must be applied (and then removed) in order to obtain zero remanence after saturation in the "forward" direction

colatitude: angular distance from the point in question to the Earth's spin axis, i.e., the complement of the latitude

Coriolis force: a "sideways" force experienced by an object moving in a rotating framework. It is particularly important for the dynamics of the atmosphere and oceans and is the reason for the existence of cyclones and anticyclones [named after the French engineer–mathematician Gustave-Gaspard Coriolis (1792–1843)]

Curie point: the temperature at which ferromagnetism is lost, named after Pierre Curie (1859–1906)

Dansgaard–Oeschger cycles: long-term cooling cycles that terminate abruptly, represented by clusters of millennium-length oxygen isotope fluctuations in the Greenland ice cores but also seen in the abundance of cold-water forams in North Atlantic sediments

Dansgaard–Oeschger events: oxygen isotope fluctuations in the Greenland ice cores that represent millennium-length intervals of cold air temperatures

declination: the angle in the horizontal plane between magnetic and true north

demagnetizing factor: a numerical factor that describes the extent to which a magnetized object demagnetizes itself by means of the (backward) internal magnetic field (for a sphere it equals $1/3$)

dendrochronology: a dating method based on counting tree rings

depositional remanent magnetization (DRM): remanent magnetism acquired as a result of sediment deposition (often referred to as *detrital remanent magnetization*)

diagenesis: the processes (e.g., compaction, lithification) that affect a sediment while it is close to the Earth's surface (i.e., at low temperature and pressure), as opposed to metamorphic changes brought about by deeper burial and/or tectonism

diamagnetism: magnetism — in all materials — arising from the orbital motion of electrons

diatom: a microscopic plant with a siliceous skeleton

dissimilatory: a term used to describe the action of certain bacteria that affect their immediate environment in such a way as to produce certain chemicals outside their own cells, specifically the use of iron ions to create magnetite and/or greigite

eccentricity: departure from circularity — specifically referring to the Earth's orbit

eolian: pertaining to the wind

epoch: an older term for intervals dominated by a single geomagnetic polarity, now superseded by the term *chron*

eutrophy: a term used to describe high organic productivity in lakes

event: an older term for short geomagnetic polarity intervals within the longer epochs, now superseded by the term *subchron*

exchange coupling: the interaction between neighboring atoms (due to overlap of electron orbitals) that gives rise to ferromagnetism

exsolution: a crystallographic process by which two (or more) materials separate, usually leading to an intergrown pattern of two (or more) distinct minerals

fermentation: microbial iron reduction during metabolic respiration

ferrihydrite: a common oxyhydroxide ($5Fe_2O_3 \cdot 9H_2O$), also known as limonite, that often undergoes chemical alteration to produce strongly magnetic magnetite and/or hematite

ferrimagnetism: a form of ferromagnetism in which the crystalline material involved contains two sublattices that are oppositely but unequally magnetized

ferritin: a biosubstance consisting of a hollow protein shell inside which small iron particles (up to several thousand atoms) can be deposited

ferromagnetism: a form of magnetism characterized by strong interaction between atoms in the crystal lattice

Fisherian statistics: a mathematical treatment of unit vectors distributed in three-dimensional space, developed by Ronald Aylmer Fisher (1890–1962) (Fisher, 1953)

fluviatile: pertaining to rivers

fluvioglacial: pertaining to glacial rivers
flux density: an alternative term for magnetic induction
foram: abbreviated term for a unicellular animal with a calcite skeleton belonging to the foraminifera
foreland basin: a crustal depression formed in front of a mountain range as a result of orogenic compression
GAD: geocentric axial dipole, a useful approximation (for some purposes) to the Earth's magnetic field
general circulation model (GCM): a computer model designed to simulate the circulation of the atmosphere and oceans
geomagnetic excursions: brief, "larger than normal" fluctuations in the direction and/or strength of the local magnetic field, currently poorly understood
geomagnetic polarity timescale (GPTS): the reference sequence of normal and reversed geomagnetic polarity intervals
glacial flour (rock flour): finely comminuted lithic material resulting from the grinding action of glaciers and ice sheets
gleying (also gleization): a process that takes place in soils when they become waterlogged, leading to oxygen starvation and the production of ferrous minerals
goethite: an important iron oxyhydroxide (α-FeOOH); it is hexagonal and (almost) antiferromagnetic but possesses a weak ferromagnetism thought to be due to unbalanced numbers of atomic moments in its crystal lattice
greigite: Fe_3S_4, the sulfide analogue of magnetite
hectare (ha): a measure of area equal to 100×100 m($= 2.47$ acres)
hematite: a common hexagonal iron oxide (α-Fe_2O_3)
hemipelagic: associated with oceanic settings that are close enough to the continents to be influenced by them — hemipelagic sediments (usually muds) thus contain significant input from land (terrigenous material)
Heinrich events: times at which large amounts of lithic material were delivered by icebergs from the Laurentide ice sheet into the North Atlantic
high-field susceptibility (χ_{hifi}): the slope of the magnetization curve beyond the closure of the hysteresis loop
Holocene: the latest period of geological time, usually referring to postglacial time (approximately the last 10,000 years)
hysteresis loop: the magnetization curve obtained by cycling a sample through a sequence of magnetic fields between large positive and negative extrema. Much of the early work in this area was carried out in the 19th century by James Alfred Ewing (1855–1935)
IGRF: International Geomagnetic Reference Field
illite: a common clay mineral
ilmenite: a common naturally occurring iron titanium oxide ($FeTiO_3$)
IMF: the interplanetary magnetic field created by the flow of charged particles in the solar wind
inclination: the angle that the local geomagnetic field vector makes with the horizontal, conventionally taken as positive downward
induced magnetization: the magnetization acquired by a sample while it is held in an external magnetic field
initial susceptibility: the magnetic susceptibility measured in a low field (χ_{lf})
interglacial: a major warm period between two glacials, characterized by the retreat of glaciers and ice sheets
interstadial: a pause, or minor retreat, during a period of ice advance (shorter in time and smaller in amplitude than an interglacial)
intrinsic susceptibility: the "real" susceptibility of a substance measured by avoiding internal demagnetization (by using a Rowland ring, for example)
inversion temperature: the temperature at which maghemite coverts to hematite
isothermal remanent magnetization (IRM): the magnetization acquired by a sample after exposure to (and subsequent removal from) a preset magnetizing field, all at a fixed temperature (usually, but not necessarily, room temperature)
lacustrine: pertaining to lakes
Larmor precession: the "wobbling" of an electron's orbit in the presence of a magnetic field (named after Joseph Larmor, 1857–1942)

laterite: residual material (e.g., hydrated iron oxides, bauxite) formed by weathering in the tropics, especially where clear wet and dry seasons exist

lepidocrocite: a common oxyhydroxide (γ-FeOOH) that often undergoes chemical alteration to produce strongly magnetic magnetite or hematite

limnology: the study of lakes

limonite: a common oxyhydroxide ($5Fe_2O_3 \cdot 9H_2O$), also known as ferrihydrite, that often undergoes chemical alteration to produce strongly magnetic magnetite or hematite

lithosphere: the stiff outer skin of the Earth involved in plate tectonics

Little Ice Age: a period of widespread cooling that took place between AD ~1510 and ~1850

lodestone: an obsolete term for a lump of naturally occurring magnetite

loess: a windblown dust deposit

low-field susceptibility: same as initial susceptibility (χ_{lf})

maar: a lake occupying the central depression of a volcano

mackinawite: an iron sulfide (approximately FeS) commonly found in aqueous environments such as lakes

maghemite: a common cubic iron oxide (γ-Fe_2O_3)

magnetic domain: a region within a crystal wherein all the atomic magnetic moments are parallel

magnetic hysteresis: the phenomenon in which a magnetized material does not return to zero magnetization when the external magnetic field is removed

magnetic induction: often called the B field, it is measured in tesla and (*in vacuo*) is given by the product of the magnetizing (H) field and μ_0 (the permeability constant)

magnetic moment: the property of a magnetic object that determines its tendency to twist into alignment with a magnetic field (like the needle of a compass), it is given by the product of the object's magnetization and its volume, Mv (in Am^2)

magnetic remanence: permanent magnetization retained after removal of any magnetizing field

magnetic susceptibility (κ): a dimensionless ratio given by the magnetization per unit field (M/H)

magnetite: the common cubic iron oxide Fe_3O_4

magnetizing field: often called the H field, measured in A/m

magnetofossil: the remains of magnetosomes preserved in geological sediments

magnetopneumography (MPG): medical technique used to monitor contamination in the lungs by means of external magnetic sensors (SQUIDs)

magnetoreception: a general term referring to the detection of external magnetic fields by certain organisms (e.g., homing pigeons)

magnetosome: a structure found in magnetotactic bacteria, usually consisting of a chain of tiny magnetic particles held together by a "sausagelike" membrane

magnetosphere: a region of space around the Earth to which the geomagnetic field is confined by the effects of the solar wind

magnetostratigraphy: a broad term referring to the use of any magnetic parameter in the context of stratigraphic problems, key examples being geomagnetic polarity reversals and magnetic susceptibility variations

magnetotaxis: the response (both passive and active) of certain living bacteria to an ambient magnetic field

magnetotropism: the tendency of some plant roots to be deflected when growing in a (usually very strong) magnetic field

malacology: the study of molluscs

marcasite: orthorhombic FeS_2

marine magnetic anomalies: highs and lows in the magnetic field measured over the oceans caused by alternating bands of normally and reversely magnetized crust (the so-called ocean stripes) resulting from sea floor spreading

marine oxygen isotope (MOI) stages: a standard sequence of geological intervals defined by fluctuations in the isotopic composition of seawater

mass susceptibility (χ): the volume susceptibility divided by the density (ρ), $\chi = \kappa/\rho$, it has units of reciprocal density (m^3/kg)

Maunder minimum: the interval between AD 1645 and 1715 when there were virtually no sunspots (named after E. W. Maunder, 1851–1928)

Medieval Warm Period: a widely documented period of generally warmer climates in many parts of the world, lasting—in some cases—from AD 400 to 1200 (e.g., in the 10th century, corn was grown in Norway beyond 69°N)
micromagnetism: a generalized approach to calculating the internal structure of magnetic material without presupposing any specific domain pattern
micron: one millionth of a meter
midden: a garbage dump on an archeological site
Milankovitch cycles: specific (quasi-)periodic fluctuations in the Earth's orbital parameters (eccentricity, obliquity, and precession), named after Milutin Milankovitch (1879–1958)
Morin transition: a change that takes place in hematite at about −15°C whereby the weak ferromagnetism is lost
Mössbauer effect: a phenomenon arising from the interaction of gamma-rays with atomic nuclei, particularly useful for determining energy levels in iron compounds (named after its discoverer, Rudolph Ludwig Mössbauer, born 1929, Nobel Prize in Physics in 1961)
MSEC: magnetosusceptibility event and cyclostratigraphy—a proposed scheme for identifying (and numbering) climatic changes based on magnetic susceptibility profiles, particularly in cave deposits
multidomain (MD): a state in which a magnetic material is divided into several regions (domains); the magnetization is uniform in each domain but differs in direction from domain to domain
mu-metal (μ-metal): a high-permeability iron–nickel alloy commonly used for magnetic shielding
mutagen: an agent that increases mutation
natural remanent magnetization (NRM): a neutral term for the remanence of a natural sample as first measured in the laboratory; it implies nothing about its origin, which could be TRM, CRM, DRM, etc.
obliquity: the tilt of the Earth's spin axis
obsidian: a black, entirely glassy volcanic rock exhibiting a characteristic style of fracture (conchoidal) that produces very sharp cutting edges
oligotrophy: a term used to describe low organic productivity in lakes
opal: a hydrated amorphous variety of silica (SiO_2) used by many simple organisms for the construction of their skeletons
orbital forcing: a term used to express the idea that climate changes are influenced by variations in the Earth's motion
orogenesis: the geological process of mountain building
oxygen isotope stratigraphy: the application of the marine oxygen isotope stages to stratigraphic problems
PAH: polycyclic aromatic hydrocarbon
paleointensity: a term used in paleomagnetism to denote an experimentally determined value of the strength of the geomagnetic field in the past
paleosol: fossil soil
palynology: the study of fossil plant spores
paramagnetism: a form of magnetism arising from the spin of electrons
parasitic magnetism: an alternative term for the weak ferromagnetism arising from spin canting in antiferromagnetics
pedogenesis: the combined physical, chemical, and biological processes leading to soil production
pedosphere: the discontinuous skin of soil over the surface of the continents
pelagic: associated with the deep oceans—pelagic sediments are very fine-grained oozes and clays containing a high fraction of organic remains
pentlandite: $(Fe, Ni)_9S_8$, the world's most important source of nickel, found in abundance in the mining district of Sudbury, Canada
permeability constant (μ_0): a fundamental quantity that relates the B and H fields ($\mu_0 = B/H$); its value is $4\pi \times 10^{-7}$ Vs/Am
phytoferritin: a botanical version of ferritin
Pleistocene: the time interval dominated by the geologically recent ice ages (approximately the last 2 million years)
pneumomagnetism: the study of lung contamination by means of external magnetic sensors
podzol: a soil typical of areas with a cool temperate humid climate

precession of the equinoxes: the gradual change in the date of the equinoxes caused by the slow "wobble" of the Earth's spin axis

pseudo–single domain (PSD): magnetic structure, and behavior, intermediate between single domain (SD) and multidomain (MD) in which particles contain more than one domain but exhibit many of the properties typical of SD particles

pyrite: a common, paramagnetic, iron sulfide (FeS_2) (fool's gold)

pyrrhotite: a common iron sulfide that crystallizes in several forms, the most common being Fe_7S_8, which is monoclinic and ferrimagnetic, and Fe_9S_{10}, which is hexagonal and antiferromagnetic

radula: a tonguelike rasp used by molluscs to effect feeding

relative permeability (μ_r): the B/H ratio of a material (it is dimensionless)

relaxation time: the time required for a quantity to fall to $1/e$ ($\sim$37%) of its initial value

remanence (M_r): the permanent magnetization that remains when the magnetizing field is removed

Rowland ring: a doughnut-shaped sample of magnetic material used to avoid internal demagnetization by simply not having any ends

saturation isothermal remanent magnetization (SIRM): the remanence acquired by a sample after it has been exposed to a saturating magnetic field at a fixed temperature, usually room temperature

saturation magnetization (M_s): the maximum magnetization of a sample measured while it is still in the magnetizing field

saturation remanence (M_{rs}): the magnetization that remains after a saturating magnetic field is removed

secular variation: slow changes in the direction and strength of the Earth's magnetic field

self-reversal: a general term referring to several possible processes that lead to a sample acquiring a remanent magnetization in the opposite direction to the applied magnetic field

sequence slotting: a mathematical technique for combining two time series of observed data

Shannon index: a statistical measure of biological diversity

siderite: iron carbonate ($FeCO_3$), a paramagnetic mineral common in carbonate sediments

single domain (SD): a type of magnetic structure in which a particle is uniformly magnetized, i.e., all the atomic moments are aligned parallel

smectite: a common clay mineral

speleothems: a collective term for stalagmites, stalactites, and flowstones

spin canting: the slight departure from exact antiparallelism of atomic magnetic moments that gives rise to the weak ferromagnetism of hematite

spinel: a term referring to a specific crystal structure, based on that of $MgAl_2O_2$

spontaneous magnetization: the characteristic magnetism that arises in ferromagnetic materials by virtue of their internal properties, without the need for an external magnetizing field

SQUID: a superconducting quantum interference device, particularly useful for measuring very small magnetic fields

superparamagnetism: ferromagnetism of very small particles that have relaxation times on the laboratory timescale

tephra: a collective term for all the volcanic fragments ejected through the vent (e.g., ash, pumice)

terrigenous: sedimentary material derived from the land

Tesla (T): the SI unit for magnetic induction (sometimes called flux density), it is measured in Vs/m^2 (named after Nikola Tesla, 1856–1943)

thermohaline circulation: circulation of deeper ocean waters driven by density variations arising from differences in temperature and salinity

thermoremanent magnetization (TRM): remanent magnetism acquired as a result of cooling from an elevated temperature

till: a geological term for the material deposited by glaciers

titanohematites: hematites (α-Fe_2O_3) in which a variable number of iron atoms are substituted by titanium

titanomagnetites: magnetites (Fe_3O_4) in which a variable number of iron atoms are substituted by titanium

tonne (t): an international unit of mass equal to 1000 kg (1 Mg), sometimes called a metric ton

troilite: an iron sulfide (FeS) that is common in meteorites and lunar samples but does not occur on Earth

ulvöspinel: a cubic iron–titanium oxide (Fe_2TiO_4)

Van Allen belts: two doughnut-shaped regions encircling the Earth that contain trapped charged particles

varve: an annually layered sediment in which each year's deposit consists of coarse material deposited during the summer and finer material deposited during the winter (usually found in glacial meltwater lakes)

Verwey transition: a crystallographic change in magnetite (Fe_3O_4) that occurs at about $-150°C$ and involves a redistribution of the iron cations such that the previously cubic framework is slightly distorted to monoclinic symmetry; named after its discoverer, E. J. W. Verwey (1905–1981)

vivianite: hydrated ferrous phosphate ($Fe_3P_2O_8 \cdot 8H_2O$)

volume susceptibility (κ): the magnetization (M) acquired per unit field (H) ($\kappa = M/H$)—in SI units, it is dimensionless

VSM: vibrating sample magnetometer

Wheatstone bridge: a classic electrical circuit for determining the precise value of an unknown resistance, named after its inventor, Charles Wheatstone (1802–1875)

Younger Dryas: a climatically cold period lasting approximately a millennium some 11 to 12 kyr ago, named after the Arctic flower *Dryas octopetala*, which flourished at the time

REFERENCES

Aitken, M. J. "Physics and Archaeology," 2nd Ed. Oxford: Clarendon Press, 1974.

Alexiou, C., W. Arnold, R. J. Klein, F. G. Parak, P. Hulin, C. Bergemann, W. Erhardt, S. Wagenpfeil, and A. S. Lübbe. Locoregional cancer treatment with magnetic drug targeting. *Cancer Research* **60**, 6641–6648, 2000.

Anderson, N. J., and B. Rippey. Diagenesis of magnetic minerals in the recent sediments of a eutrophic lake. *Limnology and Oceanography* **33**, 1476–1492, 1988.

Antervo, A., R. Hari, T. Katila, T. Ryhänen, and M. Seppänen. Magnetic field pattern produced by eye blinking. *In* "Biomagnetism: Applications and Theory" (H. Weinberg, G. Stroink, and T. Katila, eds.), pp. 373–377. New York: Pergamon Press, 1985.

Babanin, V. F., V. I. Trukhin, L. O. Karpachevskiy, A. V. Ivanov, and V. V. Morosov. "Soil Magnetism." Moscow-Yaroslavl: Yaroslavl State University Press, 1995.

Banerjee, S. K., S. P. Lund, and S. Levi. Geomagnetic record in Minnesota lake sediments — Absence of the Gothenburg and Erieau excursions. *Geology* **7**, 588–591, 1979.

Barthès, V., J. P. Pozzi, P. Vibert-Charbonnel, J. Thibal, and M. A. Mélières. High-resolution chronostratigraphy from downhole susceptibility logging tuned by palaeoclimatic orbital frequencies. *Earth and Planetary Science Letters* **165**, 97–116, 1999.

Barton, C. E., and M. W. McElhinny. A 10,000 yr geomagnetic secular variation record from three Australian maars. *Geophysical Journal of the Royal Astronomical Society* **67**, 465–485, 1981.

Basnakova, G., and L. E. Macaskie. Microbially enhanced chemisorption of nickel into biologically synthesized hydrogen uranyl phosphate: A novel system for the removal and recovery of metals from aqueous solutions. *Biotechnology and Bioengineering* **54**, 319–328, 1997.

Bassinot, F. C., L. D. Labeyrie, E. Vincent, X. Quidelleur, N. J. Shackleton, and Y. Lancelot. The astronomical theory of climate and the age of the Brunhes–Matuyama magnetic reversal. *Earth and Planetary Science Letters* **126**, 91–108, 1994.

Bauer, G. B., M. Fuller, A. Perry, J. R. Dunn, and J. Zoeger. Magnetoreception and biomineralization of magnetite in cetaceans. *In* "Magnetite Biomineralization and Magnetoreception in Organisms. A New Biomagnetism" (J. L. Kirschvink, D. S. Jones, and B. J. McFadden, eds.), pp. 489–507. New York: Plenum Press, 1985.

Baule, G. M., and R. McFee. Detection of the magnetic field of the heart. *American Heart Journal* **66**, 95–96, 1963.

Bazylinski, D. A. Anaerobic production of single-domain magnetite by the marine, magnetotactic bacterium, strain MV-1. *In* "Iron Biominerals" (R. B. Frankel, and R. P. Blakemore, eds.), pp. 69–77. New York: Plenum Press, 1991.

Bazylinski, D. A., and B. M. Moskowitz. Microbial biomineralization of magnetic iron minerals: microbiology, magnetism and environmental significance. *In* "Geomicrobiology: Interactions between

Microbes and Minerals" (J. F. Banfield, and K. H. Nealson, eds.), *Reviews in Mineralogy* **35**, 181–223, 1997.
Bazylinski, D. A., R. B. Frankel, and H. W. Jannasch. Anaerobic magnetite production by a marine, magnetotactic bacterium. *Nature* **334**, 518–519, 1988.
Becker, H. Duo-and quadro-sensor configuration for high-speed/high resolution magnetic prospecting with cesium magnetometer. *In* "3rd International Conference on Archaeological Prospection" (J. W. E. Fassbinder, and W. E. Irlinger, eds.), pp. 100–105, 1999.
Becker, H., and J. W. E. Fassbinder. Magnetometry of a Scythian settlement in Siberia near Cicah in the Baraba Steppe 1999. *In* "3rd International Conference on Archaeological Prospection" (J. W. E. Fassbinder, and W. E. Irlinger, eds.), pp. 168–172, 1999.
Beckwith, P. R., J. B. Ellis, and D. M. Revitt. Applications of magnetic measurements to sediment tracing in urban highway environments. *The Science of the Total Environment* **93**, 449–463, 1990.
Beer, J., C.-D. Shen, F. Heller, T.-S. Liu, G. Bonani, B. Dittrich, M. Suter, and P. W. Kubik. ^{10}Be and magnetic susceptibility in Chinese loess. *Geophysical Research Letters* **20**, 57–60, 1993.
Begét, J. E., and D. B. Hawkins. Influence of orbital parameters on Pleistocene loess deposition in central Alaska. *Nature* **337**, 151–153, 1989.
Begét, J. E., D. B. Stone, and D. B. Hawkins. Paleoclimatic forcing of magnetic susceptibility variations in Alaskan loess during the Late Quaternary. *Geology* **18**, 40–43, 1990.
Begét, J. E., M. Keskinen, and K. Severin. Mineral particles from Asia found in volcanic loess on the island of Hawaii. *Sedimentary Geology* **84**, 189–197, 1993.
Bergametti, G. Atmospheric cycle of desert dust. "Encyclopedia of Earth System Science," pp. 171–182. San Diego: Academic Press, 1992.
Berger, A. Milankovitch theory and climate. *Reviews of Geophysics* **26**, 624–657, 1988.
Berger, A., and M. F. Loutre. Insolation values for the climate of the last 10 million years. *Quaternary Science Reviews* **10**, 297–318, 1991.
Berger, A., H. Gallée, and M. F. Loutre. The earth's future climate at the astronomical timescale. *In* "Future Climate Change and Radioactive Waste Disposal" (C. M. Goodess, and J. P. Palutikof, eds.), pp. 148–165. Norwich: Climate Research Centre, University of East Anglia, 1991.
Berger, G. W., B. J. Pillans, and Z. Palmer. Dating loess up to 800 ka by thermo-luminescence. *Geology* **20**, 402–406, 1992.
Berggren, W. A., and D. R. Prothero. Eocene–Oligocene climatic and biotic evolution: An overview. *In* "Eocene–Oligocene Climatic Evolution" (D. R. Prothero, and W. A. Berggren, eds.), pp. 1–28, Princeton, NJ: Princeton University Press, 1992.
Bethke, C. M. Roles of sediment compaction, tectonic compression and topographic relief in driving deep groundwater migration. *Geological Society of America, Program and Abstracts* **18**, 540, 1986.
Bingham Müller, T. B. Microbially mediated iron mineral transformations: A case study in Lake Greifen, Switzerland. Ph.D. thesis, ETH Zürich, 1996.
Biscaye, P. E., F. E. Grousset, M. Revel, S. Van der Gaast, G. A. Zielinski, A. Vaars, and G. Kukla. Asian provenance of glacial dust (stage 2) in the Greenland Ice Sheet Project 2 Ice Core, Summit, Greenland. *Journal of Geophysical Research* **102**, 26,765–26,781, 1997.
Bityukova, L., R. Scholger, and M. Birke. Magnetic susceptibility as indicator of environmental pollution of soils in Tallinn. *Physics and Chemistry of the Earth* **24**, 829–835, 1999.
Blakemore, R. P. Magnetotactic bacteria. *Science* **190**, 377–379, 1975.
Blakemore, R. P. Magnetotactic bacteria. *Annual Review of Microbiology* **36**, 217–238, 1982.
Blakemore, R. P., and R. B. Frankel. Magnetic navigation in bacteria. *Scientific American* 245:6, 58–65, 1981.
Blakemore, R. P., D. Maratea, and R. S. Wolfe. Isolation and pure culture of a freshwater magnetic spirillum in chemically defined medium. *Journal of Bacteriology* **140**, 720–729, 1979.
Blakemore, R. P., K. A. Short, D. A. Bazylinski, C. Rosenblatt, and R. B. Frankel. Microaerobic conditions are required for magnetite formation within *Aquaspirillum magnetotacticum*. *Geomicrobiology Journal* **4**, 53–71, 1985.
Bleil, U., and T. von Dobeneck. Geomagnetic events and relative paleointensity records — Clues to high-resolution paleomagnetic chronostratigraphies of Late Quaternary marine sediments? *In* "Use of Proxies

in Paleoceanography: Examples from the South Atlantic" (G. Fischer, and G. Wefer, eds.), pp. 635–65. Berlin: Springer-Verlag, 1999.

Bleil, U., and T. von Dobeneck. Late Quaternary terrigenous sedimentation in the western equatorial Atlantic: South American versus African influences documented by magnetic mineral analysis, in press, 2002.

Bloemendal, J., and P. B. deMenocal. Evidence for a change in the periodicity of tropical climate cycles at 2.4 Myr from whole-core magnetic susceptibility measurements. *Nature* **342**, 897–900, 1989.

Bloemendal, J., F. Oldfield, and R. Thompson. Magnetic measurements used to assess sediment influx at Llyn Goddionduon. *Nature* **280**, 50–53, 1979.

Bloemendal, J., J. W. King, F. R. Hall, and S.-J. Doh. Rock magnetism of Late Neogene and Pleistocene deep-sea sediments: Relationship to sediment source, diagenetic processes and sediment lithology. *Journal of Geophysical Research* **97**, 4361–4375, 1992.

Bloxham, J., and A. Jackson. Time-dependent mapping of the magnetic field at the core–mantle boundary. *Journal of Geophysical Research* **97**, 19,537–19,563, 1992.

Bogue, S. W., S. Gromme, and J. W. Hillhouse. Paleomagnetism, magnetic anisotropy and mid-Cretaceous paleolatitude of the Duke Island (Alaska) ultramafic complex. *Tectonics* **14**, 1133–1152, 1995.

Bond, G., W. S. Broecker, S. Johnson, J. McManus, L. Labeyrie, J. Jouzel, and G. Bonani. Correlations between climate records from North Atlantic sediments and Greenland ice. *Nature* **365**, 143–147, 1993.

Bonhommet, N., and J. Babkine. Sur la présence d'aimentations inversées dans la Chaîne des Puys. *Comptes Rendus de l'Academie des Sciences* **264**, 92–94, 1967.

Bormann, F. H., and G. Likens. The watershed–ecosystem concept and studies of nutrient cycles. *In* "The Ecosystem Concept in Natural Resource Management" (G. M. van Dyne, ed.), pp. 49–76. New York: Academic Press, 1969.

Borradaile, G. J., J. D. Stewart, and W. A. Ross. Characterizing stone tools by rock-magnetic methods. *Geoarchaeology* **13**, 73–91, 1998.

Boudreau, B. P. The mathematics of early diagenesis: From worms to waves. *Reviews of Geophysics* **38**, 389–416, 2000.

Bradley, R. S. "Paleoclimatology," 2nd Ed. San Diego: Academic Press, 1999.

Brady, H. B. Report of the Scientific Results of the Voyage of H.M.S. *Challenger*, vol. IX (Zoology). London, 1884.

Brennan, E. W., and W. L. Lindsay. Reduction and oxidation effect on the solubility and transformation of iron oxides. *Soil Science Society of America Journal* **62**, 930–937, 1998.

Brilhante, O., L. Daly, and P. Trabuc. Application du magnétisme à la détection des pollutions causées par les métaux lourds dans l'environnement. *Comptes rendus de l'Academie des Sciences* **309**, Série II, 2005–2012, 1989.

Broecker, W. S. The great ocean conveyor. *Oceanography* **4**, 79–89, 1991.

Broecker, W. S., and M. L. Bender. Age determinations on marine strandlines. *In* "Calibration of Hominid Evolution" (W. W. Bishop, and J. A. Miller, eds.), pp. 19–38. Edinburgh: Scottish Academic Press, 1972.

Broecker, W., G. Bond, M. Klas, E. Clark, and J. McManus. Origin of the northern Atlantic's Heinrich events. *Climate Dynamics* **6**, 265–273, 1992.

Bronger, A., and Th. Heinkele. Micromorphology and genesis of paleosols in the Luochuan loess section, China: Pedostratigraphic and environmental implications. *Geoderma* **45**, 123–143, 1989.

Bullard, E. C. Reversals of the Earth's magnetic field. *Philosophical Transactions of the Royal Society A* **263**, 481–524, 1968.

Burov, B., D. K. Nurgaliev, and P. G. Jasonov. "Paleomagnetic Analysis." Kazan: Kazan University Press, 1986 [in Russian].

Burton, B. P. Interplay of chemical and magnetic ordering. *In* "Oxide Minerals: Petrologic and Magnetic Significance" (D. H. Lindsley, ed.), pp. 303–321. Mineralogical Society of America, 1991.

Butler, R. F. "Paleomagnetism: Magnetic Domains to Geologic Terranes." Boston: Blackwell Scientific Publications, 1992. http://www.geo.arizona.edu/Paleomag/book/

Butler, R. F., and S. K. Banerjee. Theoretical single domain grain size range in magnetite and titanomagnetite. *Journal of Geophysical Research* **80**, 4049–4058, 1975.

Cande, S. C., and D. V. Kent. Revised calibration of the geomagnetic polarity timescale for the Late Cretaceous and Cenozoic. *Journal of Geophysical Research* **100**, 6093–6095, 1995.

Canfield, D. E. Reactive iron in marine sediments. *Geochimica et Cosmochimica Acta* **53**, 619–632, 1989.

Canfield, D. E., and R. A. Berner. Dissolution and pyritization of magnetite in anoxic marine sediments. *Geochimica et Cosmochimica Acta* **51**, 645–659, 1987.

Canti, M. G., and N. Linford. The effects of fire on archaeological soils and sediments: Temperature and colour relationships. *Proceedings of the Prehistoric Society* **66**, 385–395, 2000.

Carter-Stiglitz, B., B. Moskowitz, and M. Jackson. Unmixing magnetic assemblages and the magnetic behavior of bimodal mixtures. *Journal of Geophysical Research* **106**, 26,397–26,411, 2001.

Ceron-Carrasco, R., M. J. Church, and J. Thoms. Towards an economic landscape of the Bhaltos Peninsula, Lewis during Mid to Late Iron Age. *In* "Tall Stories; Broch Studies Past, Present and Future" (V. Turner, ed.), Oxford: Oxbow, in press, 2001.

Champion, D. E., M. A. Lanphere, and M. A. Kuntz. Evidence for a new geomagnetic reversal from lava flows in Idaho: Discussion of short polarity reversals in the Brunhes and late Matuyama polarity chrons. *Journal of Geophysical Research* **93**, 11,667–11,680, 1988.

Chan, L. S., C. H. Yeung, W. W.-S. Yim, and O. L. Or. Correlation between magnetic susceptibility and distribution of heavy metals in contaminated sea-floor sediments of Hong Kong harbour. *Environmental Geology* **36**, 77–86, 1998.

Chang, S.-B. R., and J. L. Kirschvink. Magnetofossils, the magnetization of sediments and the evolution of magnetite biomineralization. *Annual Review of Earth and Planetary Sciences* **17**, 169–195, 1989.

Chang, S.-B. R., J. F. Stolz, and J. L. Kirschvink. Biogenic magnetite as a primary remanence carrier in limestone deposits. *Physics of the Earth and Planetary Interiors* **46**, 289–303, 1987.

Chang, S.-B. R., J. F. Stolz, J. L. Kirschvink, and S. M. Awramik. Biogenic magnetite in stromatolites. II. Occurrence in ancient sedimentary environments. *Precambrian Research* **43**, 305–315, 1989.

Chen, F. H., J. Bloemendal, J. M. Wang, J. J. Li, and F. Oldfield. High-resolution multi-proxy climate records from Chinese loess: Evidence for rapid climatic changes over the last 75 kyr. *Palaeogeography, Palaeoclimatology, Palaeoecology* **130**, 323–335, 1997.

Chiari, G., and R. Lanza. Pictorial remanent magnetization as an indicator of secular variation of the Earth's magnetic field. *Physics of the Earth and Planetary Interiors* **101**, 79–84, 1997.

Chlachula, J., M. E. Evans, and N. W. Rutter. A magnetic investigation of a Late Quaternary loess/palaeosol record in Siberia. *Geophysical Journal International* **132**, 128–132, 1998.

Chu, K. A preliminary study on the climatic fluctuations during the last 5000 years in China. *Scientia Sinica* **16**, 226–256, 1973.

Church, M. J., C. Peters, and C. M. Batt. Sourcing fire ash on archaeological sites on the western and Northen Isles of Scotland, using mineral magnetism. *Geoarchaeology*, submitted, 2001.

Church, T., and C. Caraveo. The magnetic susceptibility of southwestern obsidian: An exploratory study. *North American Archaeologist* **17**, 271–285, 1996.

Cioppa, M. T., I. S. Al-Aasm, D. T. A. Symons, M. T. Lewchuk, and K. P. Gillen. Correlating paleomagnetic, geochemical and petrographic evidence to date diagenetic and fluid flow events in the Mississippian Turner Valley Formation, Moose Field, Alberta, Canada. *Sedimentary Geology* **131**, 109–129, 2000.

Cioppa, M. T., J. S. Lonnee, D. T. A. Symons, I. S. Al-Aasm, and K. P. Gillen. Facies and lithological controls on paleomagnetism: An example from the Rainbow South field, Alberta, Canada. *Bulletin of Canadian Petroleum Geology* **49**, 393–407, 2001.

Cisowski, S. Interacting vs. non-interacting single-domain behavior in natural and synthetic samples. *Physics of the Earth and Planetary Interiors* **26**, 77–83, 1981.

Clark, A. J. C. "Seeing Beneath the Soil." London: Batsford, 1990.

Clement, B. M., D. V. Kent, and N. D. Opdyke. A synthesis of magnetostratigraphic results from Plio-Pleistocene sediments cored using the hydraulic piston corer. *Paleoceanography* **11**, 299–308, 1996.

CLIMAP Project Members. The last interglacial ocean. *Quaternary Research* **21**, 123–224, 1984.

Coe, R. S., L. Hongre, and G. A. Glatzmaier. An examination of simulated geomagnetic reversals from a palaeomagnetic perspective. *Philosophical Transactions of the Royal Society* **358**, 1141–1170, 2000.

Cohen, D. Ferromagnetic contamination in the lungs and other organs of the human body. *Science* **180**, 745–748, 1973.
Cohen, D., E. A. Edelsack, and J. E. Zimmerman. Magnetocardiograms taken inside a shielded room with a superconducting point-contact magnetometer. *Applied Physics Letters* **16**, 278–280, 1970.
Cohen, D., S. F. Arai and J. D. Brain. Smoking impairs long-term dust clearance from the lung. *Science* **204**, 514–517, 1979.
Collinson, D. W. "Methods in Rock Magnetism and Palaeomagnetism: Techniques and Instrumentation." New York: Chapman & Hall, 1983.
Constable, C. G., C. L. Johnson, and S. P. Lund. Global geomagnetic field models for the past 3000 years: Transient or permanent flux lobes? *Philosophical Transactions of the Royal Society* **358**, 991–1008, 2000.
Cook, J. C., and S. L. Carts, Jr. Magnetic effects and properties of typical topsoils. *Journal of Geophysical Research* **67**, 815–828, 1962.
Cox, A., R. R. Doell, and G. B. Dalrymple. Geomagnetic polarity epochs and Pleistocene geochronology. *Nature* **198**, 1049–1051, 1963.
Creer, K. M., and N. Thouveny. The Euromaars Project. *Quaternary Science Reviews* **15** (2/3), 99–100, 1996.
Creer, K. M., R. Thompson, L. Molyneux, and F. J. H. Mackereth. Geomagnetic secular variation recorded in the stable remanence of recent sediments. *Earth and Planetary Science Letters* **14**, 115–127, 1972.
Creer, K. M., T. E. Hogg, P. W. Readman, and C. Reynaud. Palaeomagnetic secular variation curves extending back to 13,400 years B.P. recorded by sediments deposited in Lac de Joux, Switzerland. *Journal of Geophysics* **48**, 139–147, 1980.
Creer, K. M., P. W. Readman, and S. Papamarinopoulos. Geomagnetic secular variations in Greece through the last 6000 years obtained from lake sediment studies. *Geophysical Journal of the Royal Astronomical Society* **66**, 193–219, 1981.
Creer, K. M., D. A. Valencio, A. M. Sinito, P. Tucholka, and J. F. A. Vilas. Geomagnetic secular variations 0–14,000 yr. B.P. as recorded by lake sediments from Argentina. *Geophysical Journal of the Royal Astronomical Society* **74**, 199–221, 1983.
Cronin, T. M. "Principles of Paleoclimatology." New York: Columbia University Press, 1999.
Dalan, R. A., and S. K. Banerjee. Solving archaeological problems using techniques of soil magnetism. *Geoarchaeology* **13**, 3–36, 1998.
Dalrymple, G. B., and M. A. Lanphere. "Potassium–Argon Dating: Principles, Techniques and Applications to Geochronology." San Francisco: W. H. Freeman, 1969.
Daly, L., and M. Le Goff. An updated and homogeneous world secular variation database. 1. Smoothing of the archeomagnetic results. *Physics of the Earth and Planetary Interiors* **93**, 159–190, 1996.
Dankers, P. H. M. Relationship between median destructive field and coercive forces for dispersed natural magnetite, titanomagnetite and hematite. *Geophysical Journal of the Royal Astronomical Society* **64**, 447–461, 1981.
Dansgaard, W., S. J. Johnsen, H. B. Clausen, D. Dahl-Jensen, N. Gundestrup, C. U. Hammer, and H. Oeschger. North Atlantic climatic oscillations revealed by deep Greenland ice cores. *In* "Climate Processes and Climate Sensitivity" (J. E. Hansen, and T. Takahasi, eds.), *Geophysical Monographs* **29**, 288–298, 1984.
Dansgaard, W., S. J. Johnsen, H. B. Clausen, D. Dahl-Johnsen, N. Gunderstrup, C. U. Hammer, C. Hvidberg, J. Steffensen, A. Sveinbjörnsobttir, J. Jouzel, and G. Bond. Evidence for general instability of past climate from a 250 kyr ice core record. *Nature* **364**, 218–220, 1993.
David, P. Sur la stabilité de la direction d'aimantation dans quelques roches volcaniques. *Comptes rendus hebdomadaires des séances de l'Académie des Sciences (Paris), Série B* **138**, 41–42, 1904.
Davis, K. E. Magnetite rods in plagioclase as the primary carrier of stable NRM in ocean floor gabbros. *Earth and Planetary Science Letters* **55**, 190–198, 1981.
Davis, P. M., and M. E. Evans. Interacting single-domain properties of magnetite intergrowths. *Journal of Geophyical Research* **81**, 989–994, 1976.
Day, R., M. Fuller, and V. A. Schmidt. Hysteresis properties of titanomagnetites: Grain-size and compositional dependence. *Physics of the Earth and Planetary Interiors* **13**, 260–267, 1977.

Dearing, J. A. The use of magnetic measurements to study particulate flux in lake–watershed ecosystems. Unpublished Ph.D. thesis, University of Liverpool, 1979.

Dearing, J. Magnetic susceptibility. *In* "Methods of Environmental Magnetism: A Practical Guide" (J. Walden, F. Oldfield, and J. Smith, eds.), Technical Guide No. 6, pp. 35–62. London: Quaternary Research Association, 1999.

Dearing, J. A., and R. J. Flower. The magnetic susceptibility of sedimentary material trapped in Lough Neagh, Northern Ireland and its erosional significance. *Limnology and Oceanography* **27**, 969–675, 1982.

Dearing, J. A., J. K. Elner, and C. M. Happey-Wood. Recent sediment flux and erosional processes in a Welsh upland lake catchment based on magnetic susceptibility measurements. *Quaternary Research* **16**, 356–372, 1981.

Dearing, J. A., H. Håkansson, B. Liedberg-Jönsson, A. Persson, S. Skansjö, and D. Widholm. Lake sediments used to quantify the erosional response to land use change in southern Sweden. *Oikos* **50**, 60–78, 1987.

Dearing, J. A., K. L. Hay, S. M. J. Baban, A. S. Huddleston, E. M. H. Wellington, and P. J. Loveland. Magnetic susceptibility of soil: An evaluation of conflicting theories using a national data set. *Geophysical Journal International* **127**, 728–734, 1996.

Dekkers, M. J. Some rockmagnetic parameters for natural goethite, pyrrhotite and fine-grained hematite. *Geologica Ultraiectina* **51**, 1988.

Dekkers, M. J., J.-L. Mattéi, G. Eillon, and P. Rochette. Grain-size dependence of the magnetic behavior of pyrrhotite during its low-temperature transition at 34K. *Geophysical Research Letters* **16**, 855–858, 1989.

deMenocal, P. B., and J. Bloemendal. Plio-Pleistocene climatic variability in subtropical Africa and the palaeoenvironment of hominid evolution: A combined data-model approach. *In* "Paleoclimate and Evolution, with Emphasis on Human Origins" (E. S. Vrba, G. H. Denton, T. C. Partridge, and L. H. Burckle, eds.), pp. 262–288. New Haven: Yale University Press, 1995.

deMenocal, P., and D. Rind. Sensitivity of Asian and African climate to variations in seasonal insolation, glacial ice cover, sea surface temperature and Asian orography. *Journal of Geophysical Research* **98**, 7265–7287, 1993.

deMenocal, P. B., E. P. Laine, and P. F. Ciesielski. A magnetic signature of bottom current erosion. *Physics of the Earth and Planetary Interiors* **51**, 326–348, 1988.

Derbyshire, E. On the morphology, sediments and the origin of the loess plateau of central China. *In* "Mega-Geomorphology" (R. Gardner, and H. Scoging, eds.), pp. 172–194. Oxford: Clarendon Press, 1983.

Devouard, B., M. Pósfai, X. Hua, D. A. Bazylinski, R. B. Frankel, and P. R. Buseck. Magnetite from magnetotactic bacteria: Size distributions and twinning. *American Mineralogist* **83**, 1387–1398, 1998.

Diebel, C. E., R. Proksch, C. R. Green, P. Neilson and M. M. Walker. Magnetite defines a vertebrate magnetoreceptor. *Nature* **406**, 299–302, 2000.

Ding, Z., Z. Yu, N. W. Rutter, and T. Liu. Towards an orbital timescale for Chinese loess deposits. *Quaternary Science Reviews* **13**, 39–70, 1994.

Ding, Z. L., S. F. Xiong, J. M. Sun, S. L. Yang, Z. Y. Gu, and T. S. Liu. Pedostratigraphy and paleomagnetism of a ~7.0 Ma eolian loess–redclay sequence at Lingtai, Loess Plateau, north-central China and the implications for paleomonsoon evolution. *Palaeogeography, Palaeoclimatology, Palaeoecology* **152**, 49–66, 1999.

Ding, Z., N. W. Rutter, J. M. Sun, S. L. Yang, and T. Liu. Re-arrangement of atmospheric circulation at about 2.6 Ma over northern China: Evidence from grain size records of loess–palaeosol and red clay sequence. *Quaternary Science Reviews* **19**, 547–558, 2000.

Ding, Z. L., V. Ranov, S. L. Yang, A. Finaev, J. M. Han, and G. A. Wang. The loess record in southern Tajikistan and correlation with Chinese loess. *Earth and Planetary Science Letters* **200**, 387–400, 2002.

Dobson, J. Nanoscale biogenic iron oxides and neurodegenerative disease. *FEBS Letters* **496**, 1–5, 2001.

Dodds, W. K. "Freshwater Ecology: Concepts and Environmental Applications." San Diego: Academic Press, 2002.

Donovan, T. J., R. L. Fogey, and A. A. Roberts. Aeromagnetic detection of diagenetic magnetite over oil fields. *American Association of Petroleum Geologists Bulletin* **63**, 245–248, 1979.

Dowdeswell, J. A., M. A. Maslin, J. T. Andrews, and I. N. McCave. Iceberg production, debris rafting and the extent and thickness of Heinrich layers (H-1, H-2) in North Atlantic sediments. *Geology* **23**, 301–304, 1995.

Dubrov, A. P. "The Geomagnetic Field and Life: Geomagnetobiology." New York: Plenum Press, 1978.

Dunlop, D. J. Magnetic mineralogy of unheated and heated red sediments by coercivity spectrum analysis. *Geophysical Journal of the Royal Astronomical Society* **27**, 37–55, 1972.

Dunlop, D. J. Magnetism in rocks. *Journal of Geophysical Research* **100**, 2161–2174, 1995.

Dunlop, D. J. Theory and application of the Day plot (Mrs/Ms versus Hcr/Hc) 1: Theoretical curves and tests using titanomagnetite data. *Journal of Geophysical Research* **107**(B3), DOI 10.1029/2001JB000486, 2002a.

Dunlop, D. J. Theory and application of the Day plot (Mrs/Ms versus Hcr/Hc) 2: Application to data for rocks, sediments and soils. *Journal of Geophysical Research* **107**(B3), DOI 10.1029/2001JB000487, 2002b.

Dunlop, D. J., and Ö. Özdemir. "Rock Magnetism." Cambridge: University Press, 1997.

Dunlop, D. J., and S. Xu. A comparison of methods of granulometry and domain structure determination. *Eos, Transactions of the American Geophysical Union* **74**, Fall meeting supplement, 203, 1993.

Dunn, J. R., M. Fuller, J. Zoeger, J. Dobson, F. Heller, J. Hammann, E. Caine, and B. M. Moskowitz. Magnetic material in the human hippocampus. *Brain Research Bulletin* **36**, 155–159, 1995.

Duplessy, J-C., M. Arnold, P. Maurice, E. Bard, J. Duprat, and J. Moyes. Direct dating of the oxygen-isotope record of the last deglaciation by ^{14}C accelerator mass spectrometry. *Nature* **30**, 350–352, 1986.

Edwards, R. L., J. H. Chen, T-L. Ku, and G. J. Wasserburg. Precise timing of the last interglacial period from mass spectrometric determination of ^{230}Th in corals. *Science* **236**, 1547–1553, 1987.

Edwards, R. L., J. W. Beck, G. S. Burr, D. J. Donahue, J. M. A. Chappell, A. L. Bloom, E. R. M. Druffell, and F. W. Taylor. A large drop in atmospheric $^{14}C/^{12}C$ and reduced melting in Younger Dryas, documented with ^{230}Th ages of corals. *Science* **260**, 962–967, 1993.

Egli, R. Analysis of the field dependence of remanent magnetization. *Journal of Geophysical Research*, submitted, 2003.

Egli, R., and F. Heller. High-resolution imaging using a high-T_c superconducting quantum interference device (SQUID) magnetometer. *Journal of Geophysical Research* **105**, 25,709–25,727, 2000.

Egli, R., and W. Lowrie. The anhysteretic remanent magnetization of fine magnetic particles. *Journal of Geophysical Research*, in press, 2002.

Egorov, V. V., V. M. Fridkand, E. N. Ivanova, N. N. Rozov, V. A. Nosin, and T. A. Friev. *Klassifikatsiya i diagnostika pochv SSSR* [Classification and diagnostics of the soils of the USSR]. Moscow: Kolos, 1977 [in Russian].

Ellwood, B. B., and B. Burkart. Test of hydrocarbon-induced magnetic patterns in soils: The sanitary landfill as laboratory. *In* "Hydrocarbon Migration and Its Near-Surface Expression" (D. Schumacher, and M. A. Abrams, eds.), pp. 91–98. American Association of Petroleum Geologists Memoir, 66, 1996.

Ellwood, B. B., and M. T. Ledbetter. Antarctic bottom water fluctuations in the Vema Channel: Evidence of velocity variations from measurements of anisotropy of magnetic susceptibility and mean silt sizes. *Earth and Planetary Science Letters* **35**, 189–198, 1977.

Ellwood, B. B., T. J. Chrznowski, G. J. Long, and M. L. Buhl. Siderite formation in anoxic deep-sea sediments: A synergistic bacterially controlled process with important implications in paleomagnetism. *Geology* **16**, 980–982, 1988.

Ellwood, B. B., J. Zilhão, F. B. Harrold, W. Balsam, B. Burkart, G. J. Long, A. Debénath, and A. Bouzouggar. Identification of the Last Glacial Maximum in the Upper Paleolithic of Portugal using magnetic susceptibility measurements of Caldeirão Cave sediments. *Geoarchaeology* **13**, 55–71, 1998.

Ellwood, B. B., F. B. Harrold, S. L. Benoist, L. G. Straus, M. G. Morales, K. Petruso, N. F. Bicho, J. Zilhão, and N. Solar. Paleoclimate and intersite correlations from Late Pleistocene/Holocene cave sites: Results from southern Europe. *Geoarchaeology* **16**, 433–463, 2001.

Elmore, R. D., M. H. Engel, L. Crawford, K. Nick, S. Imbus, and Z. Sofer. Evidence for a relationship between hydrocarbons and authigenic magnetite. *Nature* **325**, 428–430, 1987.

Emiliani, C. Pleistocene temperatures. *Journal of Geology* **63**, 538–578, 1955.

Enkin, R. J., K. G. Osadetz, P. M. Wheadon, and J. Baker. Paleomagnetic constraints on the tectonic history of the Foreland Belt, southern Canadian Cordillera. *Canadian Journal of Earth Sciences* **34**, 260–270, 1997.

Epstein, S., R. Buchsbaum, H. A. Lowenstam, and H. C. Urey. Revised carbonate-water isotopic temperature scale. *Bulletin of the Geological Society of America* **64**, 1315–1326, 1953.

Etheredge, J. A., S. M. Perez, O. R. Taylor, and R. Jander. Monarch butterflies (*Danaus plexippus* L.) use a magnetic compass for navigation. *Proceedings of the National Academy of Sciences of the United States of America* **96**, 13,845–13,846, 1999.

Evans, M. E. An archaeointensity investigation of a kiln at Pompeii. *Journal of Geomagnetism and Geoelectricity* **43**, 357–361, 1991.

Evans, M. E. Magnetoclimatology: A test of the wind-vigour model using the 1980 Mount St. Helens ash. *Earth and Planetary Science Letters* **172**, 255–259, 1999.

Evans, M. E. Magnetoclimatology of aeolian sediments. *Geophysical Journal International* **144**, 495–497, 2001.

Evans, M. E., and F. Heller. Magnetic enhancement and palaeoclimate: Study of a loess/palaeosol couplet across the Loess Plateau of China. *Geophysical Journal International* **117**, 257–264, 1994.

Evans, M. E., and F. Heller. Magnetism of loess/palaeosol sequences: Recent developments. *Earth-Science Reviews* **54**, 129–144, 2001.

Evans, M. E., and M. Mareschal. Secular variation and magnetic dating of fired structures in southern Italy. *In* "Archaeometry, Proceedings of the 25th International Symposium" (Y. Maniatis, ed.), pp. 59–68. Amsterdam: Elsevier, 1989.

Evans, M. E., and M. W. McElhinny. An investigation of the origin of stable remanence in magnetite-bearing igneous rocks. *Journal of Geomagnetism and Geoelectricity* **21**, 757–773, 1969.

Evans, M. E., and M. L. Wayman. An investigation of the role of ultra-fine titanomagnetite intergrowths in palaeomagnetism. *Geophysical Journal of the Royal Astronomical Society* **36**, 1–10, 1974.

Evans, M. E., M. W. McElhinny, and A. C. Gifford. Single domain magnetite and high coercivities in a gabbroic intrusion. *Earth and Planetary Science Letters* **4**, 142–146, 1968.

Evans, M. E., C. D. Rokosh, and N. W. Rutter. Magnetoclimatology and paleoprecipitation: Evidence from a north-south transect through the Chinese Loess Plateau. *Geophysical Research Letters* **29**(8), DOI 10.1029/2001GLO13674, 2002.

Exnar, M. Differenzierung ferromagnetischer und paramagnetischer Messsignale mithilfe einer modernisierten Curiewaage. Diploma thesis ETH Zürich, 1997.

Eyre, J. K. The application of high resolution IRM acquisition to the discrimination of remanence carriers in Chinese loess. *Studia geophysica et geodetica* **40**, 234–242, 1996.

F(ood and) A(griculture) O(rganization) of the United Nations. Soil Map of the World, 10 vols. Paris: United Nations Economic and Scientific Organization, 1971–1981.

Fabian, K., and T. von Dobeneck. Isothermal magnetization of samples with stable Preisach function: A survey of hysteresis, remanence, and rock magnetic parameters. *Journal of Geophysical Research* **102**, 17,659–17,677, 1997.

Fabian, K., A. Kirchner, W. Williams, F. Heider, T. Leibl, and A. Hubert. Three-dimensional micromagnetic calculations for magnetite using FFT. *Journal of Geophysical Research* **124**, 89–104, 1996.

Fairbanks, R. G. A 17,000 year glacio-eustatic sea level record: influence of glacial melting rates on the Younger Dryas event and deep ocean circulation. *Nature* **342**, 637–642, 1989.

Fang, X. M, Y. Ono, H. Fukusawa, P. Bao-Tian, J.-J. Li, G. Dong-Hong, K. Oi, S. Tsukamoto, M. Torii, and T. Mishima. Asian summer monsoon instability during the last 60,000 years: Magnetic susceptibility and pedogenic evidence from the western Chinese Loess Plateau. *Earth and Planetary Science Letters* **168**, 219–232, 1999.

Fassbinder, J. W. E., and W. E. Irlinger. Aerial and magnetic prospection of an eleventh century motte in Bavaria. *Archaeological Prospection* **1**, 65–69, 1994.

Fassbinder, J. W. E., and H. Stanjeck. Occurrence of bacterial magnetite in soils from archaeological sites. *Archaeologia Polona* **31**, 117–128, 1993.

Fassbinder, J. W. E., and H. Stanjek. Magnetic properties of biogenic soil greigite (Fe_3S_4). *Geophysical Research Letters* **21**, 2349–2352, 1994.

Fassbinder, J. W. E., H. Stanjek, and H. Vali. Occurrence of magnetic bacteria in soil. *Nature* **343**, 161–163, 1990.

Feynman, R. P., R. B. Leighton, and M. Sands. "The Feynman Lectures on Physics, II." Palo Alto: Addison-Wesley Publishing Company, 1964.

Fischer, W. R. Microbiological reactions of iron in soils. *In* "Iron in Soils and Clay Minerals" (J. W. Stucki, B. A. Goodman, and U. Schwertmann, eds.), pp. 715–748. Dordrecht: Reidel Publishing, 1988.

Fisher, R. A. Dispersion on a sphere. *Proceedings of the Royal Society, London* **A217**, 295–305, 1953.

Fitzpatrick, E. A. "An Introduction to Soil Science," 2nd Ed. Essex: Longman Scientific & Technical, 1986.

Flanders, P. J. Collection, measurement and analysis of airborne magnetic particulates from pollution in the environment. *Journal of Applied Physics* **75**, 5931–5936, 1994.

Flanders, P. J. Identifying fly ash at a distance from fossil fuel power stations. *Environmental Science and Technology* **33**, 528–532, 1999.

Fleming, R. H. The composition of plankton and units for reporting populations and production. *Proceedings of the Sixth Pacific Scientific Congress* **3**, 535–540, 1940.

Florindo, F., R. Zhu, B. Guo, L. Yue, Y. Pan, and F. Speranza. Magnetic proxy climate results from the Duanjiapo loess section, southernmost extremity of the Chinese loess plateau. *Journal of Geophysical Research* **104**, 645–659, 1999.

Folgerhaiter, G. Sur les variations séculaires de l'inclinaison magnétique dans l'antiquité. *Journal de Physique* **8**, 660–667, 1899.

Forsman, M., and P. Högstedt. Welding fume retention in lungs of previously unexposed subjects. *In* "Advances in Biomagnetism" (S. J. Williamson, M. Hoke, G. Stroink and M. Kotani, eds.), pp. 477–480. New York: Plenum Press, 1989.

Forster, T., and F. Heller. Loess deposits from the Tajik depression (central Asia): Magnetic properties and paleoclimate. *Earth and Planetary Science Letters* **128**, 501–512, 1994.

Forster, T., M. E. Evans, and F. Heller. The frequency dependence of low field susceptibility in loess sediments. *Geophysical Journal International* **118**, 636–642, 1994.

France, D. E., and F. Oldfield. Identifying goethite and hematite from rock magnetic measurements of soils and sediments. *Journal of Geophyical Research* **105**, 2781–2795, 2000.

Frankel, R. B., and R. P. Blakemore. "Iron Biominerals." New York: Plenum Press, 1990.

Frederichs, T., U. Bleil, K. Däumler, T. von Dobeneck, and A. M. Schmidt. The magnetic view on the marine paleoenvironment: Parameters, techniques and potentials of rock magnetic studies as a key to paleoclimatic and paleoceanographic changes. *In* "Use of Proxies in Paleoceanography: Examples from the South Atlantic" (G. Fischer and G. Wefer, eds.), pp. 575–599. Berlin: Springer-Verlag, 1999.

Froelich, P. N., G. P. Klinkhammer, M. L. Bender, N. A. Luedtke, G. R. Heath, D. Cullen, P. Dauphin, D. Hammond, B. Hartman, and V. Maynard. Early oxidation of organic matter in pleagic sediments of the eastern equatorial Atlantic: Suboxic diagenesis. *Geochimica et Cosmochimica Acta* **43**, 1075–1090, 1979.

Fuller, M. Experimental methods in rock magnetism and paleomagnetism. *In* "Methods of Experimental Physics" (C. G. Sammis and T. L. Henyey, eds.), pp. 303–471. Orlando: Academic Press, 1987.

Gallet, Y., A. Genevey, and M. Le Goff. Three millennia of directional variation of the Earth's magnetic field in western Europe as revealed by archeological artefacts. *Physics of the Earth and Planetary Interiors* **131**, 81–89, 2002.

Garçon, G., P. Shirali, S. Garry, M. Fontaine, F. Zerimech, A. Martin, and H. Hannothiaux. Polycyclic aromatic hydrocarbons coated onto Fe_2O_3 particles: Assessment of cellular membrane damage and antioxidant system disruption in human epithelial lung cells (L131) in culture. *Toxicology Letters* **117**, 25–35, 2000.

Garrels, R. M., and F. T. Mackenzie. "Evolution of Sedimentary Rocks." New York: W. W. Norton, 1971.

Garrison, E. Physics and archaeology. *Physics Today* **54**, 32–36, 2001.

Georgeaud, V. M., P. Rochette, J. P. Ambrosi, D. Vandamme, and T. Williams. Relationship between heavy metals and magnetic properties in a large polluted catchment: The Etang de Berre (south of France). *Physics and Chemistry of the Earth* **22**, 211–214, 1997.

Geyh, M. A., and H. Schleicher. "Absolute Age Determination: Physical and Chemical Dating Methods and Their Application." Berlin: Springer-Verlag, 1990.

Gibbs-Eggar, Z., B. Jude, J. Dominik, J.-L. Loizeau, and F. Oldfield. Possible evidence for dissimilatory bacterial magnetite dominating the magnetic properties of recent lake sediments. *Earth and Planetary Science Letters* **168**, 1–6, 1999.

Gibson, T. H. Magnetic prospection on prehistoric sites in Western Canada. *Geophysics* **51**, 553–560, 1986.

Gillen, K. P., R. van der Voo, and J. H. Thiessen. Late-Cretaceous–Early Tertiary remagnetization of the Devonian Swan Hills Formation recorded in carbonate cores from the Caroline gas field, Alberta, Canada. *American Association of Petroleum Geologists Bulletin* **83**, 1223–1235, 1999.

Glatzmaier, G. A., and P. H. Roberts. A three-dimentional convective dynamo solution with rotating and finitely conducting inner core and mantle. *Physics of the Earth and Planetary Interiors* **91**, 63–75, 1995.

Goree, W. S., and M. D. Fuller. Magnetometers using RF-driven SQUIDS and their application in rock magnetism and paleomagnetism. *Reviews of Geophysics and Space Physics* **14**, 591–608, 1976.

Goslar, T., M. Arnold, E. Bard, T. Kuc, M. Pazdur, M. Ralska-Jasiewiczowa, K. Rózonski, N. Tisnerat, A. Walanus, B. Wicik, and K. Wieckowski. High concentration of atmospheric ^{14}C during the Younger Dryas cold episode. *Nature* **377**, 414–417, 1995.

Graham, I., and I. Scollar. Limitations on magnetic prospection in archaeology imposed by soil properties. *Archaeo-Physika* **6**, 1–125, 1976.

Granar, L. Magnetic measurements on Swedish varved sediments. *Arkiv for Geofysik* **3**, 1–40, 1958.

Grassi-Schultheiss, P. P., F. Heller, and J. Dobson. Analysis of magnetic material in the human heart, spleen and liver. *BioMetals* **10**, 351–355, 1997.

GRIP members. Climate instability during the last interglacial period recorded in the GRIP ice core. *Nature* **364**, 203–207, 1993.

Guinasso, N. L., and D. R. Schink. Quantitative estimates of biological mixing rates in abyssal sediments. *Journal of Geophysical Research* **80**, 3032–3043, 1975.

Gunnlaugsson, H. P. Analysis of the magnetic properties experiment data on Mars: Results from Mars Pathfinder. *Planetary and Space Science* **48**, 1491–1504, 2000.

Guo, Z. T., W. F. Ruddiman, Q. Z. Hao, H. B. Wu, Y. S. Qiao, R. X. Zhu, S. Z. Peng, J. J. Wei, B. Y. Yuan, and T. S. Liu. Onset of desertification by 22 Myr ago inferred from loess deposits in China. *Nature* **416**, 159–163, 2002.

Guthrie, G. D. Eat, breathe and be wary: Mineralogy in environmental health. *Reviews of Geophysics, Supplement*, 117–121, 1995.

Guyodo, Y., and J.-P. Valet. Relative variations in geomagnetic intensity from sedimentary records: the past 200,000 years. *Earth and Planetary Science Letters* **143**, 23–36, 1996.

Guyodo, Y., and J.-P. Valet. Global changes in intensity of the Earth's magnetic field during the past 800,000 kyr. *Nature* **143**, 249–252, 1999.

Haag, M. Reliability of relative palaeointensities of a sediment core with climatically-triggered strong magnetisation changes. *Earth and Planetary Science Letters* **180**, 49–59, 2000.

Haag, M., P. Bertrand, P. Martinez, and C. Kissel. Changes of remanent magnetization related to the upwelling system of Mauritania. *Earth and Planetary Science Letters*, submitted, 2002.

Haggerty, S. E. Oxide textures — A mini-atlas. *In* "Oxide Minerals: Petrologic and Magnetic Significance" (D. H. Lindsley, ed.), pp. 129–219. Mineralogical Society of America, 1991.

Hajdas, I., S. D. Ivy, J. Beer, G. Bonani, D. Imboden, A. F. Lotter, M. Sturm, and M. Suter. AMS radiocarbon dating and varve chronology of Lake Soppensee: 6,000 to 12,000 ^{14}C years BP. *Climate Dynamics* **9**, 107–116, 1993.

Hamilton, N., and A. I. Rees. The use of magnetic fabric in palaeocurrent estimation. *In* "Palaeogeophysics" (S. K. Runcorn, ed.), pp. 445–464. New York: Academic Press, 1970.

Han, J. M., H. Y. Lu, N. Q. Wu, and Z. T. Guo. The magnetic susceptibility of modern soils in China and its use for paleoclimate reconstruction. *Studia Geophysica et Geodetica* **40**, 262–275, 1996.

Hanesch, M., and N. Petersen. Magnetic properties of a recent parabrown-earth from southern Germany. *Earth and Planetary Science Letters* **169**, 85–97, 1999.

Hanesch, M., and R. Scholger. Pollution monitoring by magnetic susceptibility measurements — Assessing the potential on nation-wide and local scales. *Quaderni di Geofisica*, Istituto Nazionale di Geofisica e Vulcanologia, Rome **26**, 51, 2002.

Hanzlik, M. Elektronenmikroskopische and magnetomineralogische Untersuchungen an magnetotaktischen Bakterien des Chiemsees und an bakteriellem Magnetit eisenreduzierender Bakterien. Ph.D. thesis, University of Munich, 1999.

Hanzlik, M., N. Petersen, R. Keller, and E. Schmidbauer. Electron microscopy and ^{57}Fe Mössbauer spectra of 10 nm particles, intermediate in composition between Fe_3O_4 and γ-Fe_2O_3, produced by bacteria. *Geophysical Research Letters* **23**, 479–482, 1996a.

Hanzlik, M., M. Winklhofer, and N. Petersen. Spatial arrangement of chains of magnetosomes in magnetotactic bacteria. *Earth and Planetary Science Letters* **145**, 125–134, 1996b.

Hari, R., E. Kaukoranta, K. Reinikainen, and J. Mauno. Neuromagnetic responses to noxious stimulation. *In* "Biomagnetism: Applications and Theory" (H. Weinberg, G. Stroink and T. Katila, eds.), pp. 359–363. New York: Pergamon Press, 1985.

Harrison, S. P., K. E. Kohfeld, C. Roelandt, and T. Claquin. The role of dust in climate changes today, at the last glacial maximum and in the future. *Earth-Science Reviews* **54**, 43–80, 2001.

Hawthorne, T. B., and J. A. McKenzie. Biogenic magnetite: Authigenesis and diagenesis with changing redox conditions in Lake Greifen, Switzerland. *In* "Applications of Paleomagnetism to Sedimentary Geology" (D. M. Aïssaoui, D. F. McNeill, and N. F. Hurley, eds.), SEPM (Society for Sedimentary Geology), Special Publication **49**, 3–15, 1993.

Hay, K. L., J. A. Dearing, S. M. J. Baban, and P. J. Loveland. A preliminary attempt to to identify atmospherically-derived pollution particles in English topsoils from magnetic susceptibility measurements. *Physics and Chemistry of the Earth* **22**, 207–210, 1997.

Haynes, J. R. "Foraminifera." New York: Halsted Press, 1981.

Hedley, I. G. Chemical remanent magnetization of the FeOOH, Fe_2O_3 system. *Physics of the Earth and Planetary Interiors* **1**, 103–121, 1968.

Heider, F., D. J. Dunlop, and N. Sugiura. Magnetic properties of hydrothermally recrystallized magnetite crystals. *Science* **236**, 1287–1290, 1987.

Heider, F., A. Zitzelsberger, and K. Fabian. Magnetic susceptibility and remanent coercive force in grown magnetite crystals from 0.1 μm to 6 mm. *Physics of the Earth and Planetary Interiors* **93**, 239–256, 1996.

Heinrich, H. Origin and consequences of cyclic ice rafting in the Northeast Atlantic Ocean during the past 130,000 years. *Quaternary Research* **29**, 142–152, 1988.

Heirtzler, J. R., J. H. Allen, and D. C. Wilkinson. Ever-present South Atlantic anomaly damages spacecraft. *EOS, Transactions, American Geophysical Union* **83**, 165–169, 2002.

Heller, F. Rockmagnetic studies of Upper Jurassic limestones from Southern Germany. *Journal of Geophysics* **44**, 525–543, 1978.

Heller, F., and M. E. Evans. Loess magnetism. *Reviews of Geophysics* **33**, 211–240, 1995.

Heller, F., and T.-S. Liu. Magnetostratigraphical dating of loess deposits in China. *Nature* **300**, 431–433, 1982.

Heller, F., and T.-S. Liu. Magnetism of Chinese loess deposits. *Geophysical Journal of the Royal Astronomical Society* **77**, 125–141, 1984.

Heller, F., and T.-S. Liu. Palaeoclimatic and sedimentary history from magnetic susceptibility of loess in China. *Geophysical Research Letters* **13**, 1169–1172, 1986.

Heller, F., C. D. Shen, J. Beer, X. M. Liu, T. S. Liu, A. Bronger, M. Suter, and G. Bonani. Quantitative estimates and palaeoclimatic implications of pedogenic ferromagnetic mineral formation in Chinese loess. *Earth and Planetary Science Letters* **114**, 385–390, 1993.

Heller, F., T. Forster, M. E. Evans, J. Bloemendal, and N. Thouveny. Gesteinsmagnetische Archive globaler Umweltänderung. *GeoArchaeoRhein* **2**, 151–162, 1998a.

Heller, F., Z. Strzyszcz, and T. Magiera. Magnetic record of industrial pollution in forest soils of Upper Silesia. *Journal of Geophysical Research* **103**, 17767–17774, 1998b.

Henry, B., H. Rouvier, M. Le Goff, D. Leach, J. C. Macquar, J. Thibieroz, and M. T. Lewchuk. Palaeomagnetic dating of widespread remagnetization on the southeastern border of the French Massif Central and implications for fluid flow and Mississippi-type mineralization. *Geophysical Journal International* **145**, 368–380, 2001.

Henshaw, P. C., and R. T. Merrill. Magnetic and chemical change in marine sediments. *Reviews of Geophysics and Space Physics* **18**, 483–504, 1980.

Heslop, D., C. G. Langereis, and M. J. Dekkers. A new astronomical timescale for the loess deposits of northern China. *Earth and Planetary Science Letters* **184**, 125–139, 2000.

Hesse, P. P. Evidence for bacterial palaeoecological origin of mineral magnetic cycles in oxic and sub-oxic Tasman Sea sediments. *Marine Geology* **117**, 1–17, 1994.

Heywood, B. R., D. A. Bazylinski, A. Garratt-Reed, S. Mann, and R. B. Frankel. Controlled biosynthesis of greigite (Fe_3S_4) in magnetotactic bacteria. *Naturwissenschaften* **77**, 536–538, 1990.

Hilgetag, C. C., H. Théoret, and A. Pascual-Leone. Enhanced visual spatial attention ipsilateral to rTMS—Induced 'virtual lesions' of human parietal cortex. *Nature Neuroscience* **4**, 953–957, 2001.

Hilgen, F. J. Astronomical calibration of Gauss to Matuyama sapropels in the Mediterranean and implications for the Geomagnetic Polarity Time Scale. *Earth and Planetary Science Letters* **104**, 226–244, 1991a.

Hilgen, F. J. Extension of the astronomically calibrated (polarity) time-scale to the Miocene/Pliocene boundary. *Earth and Planetary Science Letters* **107**, 349–368, 1991b.

Hilgen, F. J., W. Krijgsman, C. G. Langereis, and L. J. Lourens. Breakthrough made in dating of the geological record. *EOS, Transactions of the American Geophysical Union* **78**, 285–289, 1997.

Hilton, J. Greigite and the magnetic properties of sediments. *Limnology and Oceanography* **35**, 509–520, 1990.

Hoffmann, V. Greigite Fe_3S_4: Magnetic properties and first domain observations. *Physics of the Earth and Planetary Interiors* **70**, 288–301, 1992.

Hoffmann, V., M. Knab, and E. Appel. Magnetic susceptibility mapping of roadside pollution. *Journal of Geochemical Exploration* **66**, 313–326, 1999.

Hollander, D. J., J. A. McKenzie, and H. L. ten Haven. A 200 year sedimentary record of progressive eutrophication in Lake Greifen (Switzerland): Implications for the origin of organic-carbon-rich sediments. *Geology* **20**, 825–828, 1992.

Holtkamp-Rötzler, E., G. Fleissner, M. Hanzlik, and N. Petersen. The morphological structure of a possible magnetite-based magneto-receptor in birds. *Annales de Géophysique* **15** (Supplement I C), 117, 1997.

Hounslow, M., and B. A. Maher. Source of the climate signal recorded by magnetic susceptibility variations in Indian Ocean sediments. *Journal of Geophysical Research* **104**, 5047–5061, 1999.

Hovan, S. A., D. K. Rea, N. G. Pisias, and N. J. Shackleton. A direct link between the China loess and marine $\partial^{18}O$ records: Aeolian flux to the north Pacific. *Nature* **340**, 296–298, 1989.

Hughen, K. A., J. T. Overpeck, S. J. Lehman, M. Kashgarian, J. Southon, L. C. Peterson, R. Alley, and D. M. Sigman. Deglacial changes in ocean circulation from an extended radiocarbon calibration. *Nature* **391**, 65–68, 1998.

Hulett, L. D., A. J. Weinberger, K. J. Northcutt, and M. Ferguson. Chemical species in fly ash from coal-burning power plants. *Science* **210**, 1356–1358, 1980.

Hulot, G., C. Eymin, B. Langlais, M. Mandea, and N. Olsen. Small-scale structure of the geodynamo inferred from Oersted and Magsat satellite data. *Nature* **416**, 620–623, 2002.

Hus, J., and R. Geeraerts. The direction of geomagnetic field in Belgium since Roman times and the reliability of archaeomagnetic dating. *Physics and Chemistry of the Earth* **23**, 997–1007, 1998.

Hutchinson, S. M. Use of magnetic and radiometric measurements to investigate erosion and sedimentation in a British upland catchment. *Earth Surface and Landforms* **20**, 293–314, 1995.

Hyde, B. B., A. J. Hodge, A. Kahn, and M. L. Birnstiel. Studies on phytoferritin: I. Identification and localization. *Journal of Ultrastructure Research* **9**, 248–258, 1963.

Hyodo, M. Possibility of reconstruction of the past geomagnetic field intensity from homogeneous sediments. *Rock Magnetism and Paleogeophysics* **10**, 42–49, 1983.

Ilmoniemi, R. J., J. Virtanen, J. Ruohonen, J. Karhu, H. J. Aronen, R. Näätänen, and T. Katila. Neuronal responses to magnetic stimulation reveal cortical reactivity and connectivity. *NeuroReport* **8**, 3537–3540, 1997.

Imbrie, J., J. D. Hays, D. G. Martinson, A. Mcintyre, A. C. Mix, J. J. Morley, N. G. Pisias, W. L. Prell, and N. J. Shackleton. The orbital theory of Pleistocene climate: Support from a revised chronology of the marine $\partial^{18}O$ record. *In* "Milankovitch and Climate," Part I (A. Berger, J. Hays, G. Kukla and B. Saltzman, eds.), pp. 269–305. Dordrecht: Reidel, 1984.

Incoronato, A., A. Angelino, R. Romano, A. Ferrante, R. Sauna, G. Vanacore, and C. Vecchione. Retrieving geomagnetic secular variations from lava flows: Evidence from Mounts Arso, Etna and Vesuvius (southern Italy). *Geophysical Journal International* **149**, 724–730, 2002.

Ising, G. On the magnetic properties of varved clay. *Arkiv för matematik, astronomi o. fysik* **29A**, 1–37, 1943.

Isley, A. E. Hydrothermal plumes and the delivery of iron to banded iron formation. *Journal of Geology* **103**, 169–185, 1995.

Jackson, M. J., S. K. Banerjee, J. A. Marvin, R. Lu, and W. Gruber. Detrital remanence, inclination errors, and anhysteretic remanence anisotropy: Quantitative model and experimental results. *Geophysical Journal International* **104**, 95–103, 1991.

Jacobs, J. A. "Reversals of the Earth's Magnetic Field." Cambridge: University Press, 1994.

Jasonov, P. G., D. K. Nurgaliev, B. V. Burov, and F. Heller. A modernized coercivity spectrometer. *Geologica Carpathica* **49**, 224–225, 1998.

Jelínek, V. Precision A.C. bridge set for measuring magnetic susceptibility of rocks and its anisotropy. *Studia geophysica et geodaetica* **17**, 36–48, 1973.

Jelínek, V. "The Statistical Theory of Measuring Anisotropy of Magnetic Susceptibility of Rocks and Its Application." Brno: Geofyzika, s.p., 1977.

Jelínek, V., and J. Pokorny. Some new concepts in technology of transformer bridges for measuring susceptibility anisotropy of rocks. *Physics and Chemistry of the Earth* **22**, 179–181, 1997.

Jelinowska, A., P. Tucholka, F. Gasse, and J. C. Fontes. Mineral magnetic record of environment in Late Pleistocene and Holocene sediments, Lake Manas, Xinjiang, China. *Geophysical Research Letters* **22**, 953–956, 1995.

Jia, R. F., B. Z. Yan, R. S. Li, G. C. Fan, and B. H. Lin. Characteristics of magnetotactic bacteria in Duanjiapo loess section, Shaanxi Province and their environmental significance. *Science in China, Series D* **39**, 478–485, 1996.

Jing, Z., and G. Rapp. Environmental magnetic indicators of the sedimentary context of archaeological sites in the Shangqiu area of China. *Geoarchaeology* **13**, 37–54, 1998.

Johnson, E. A., T. Murphy, and O. W. Torreson. Pre-history of the Earth's magnetic field. *Terrestrial Magnetism and Atmospheric Electricity* **53**, 349–372, 1948.

Jordanova, N., E. Petrovsky, M. Kovacheva, and D. Jordanova. Factors determining magnetic enhancement of burnt clay from archaeological sites. *Journal of Archaeological Science* **28**, 1137–1148, 2001.

Juntilla, M.-L., K. Kalliomäki, P-.L. Kalliomäki, and K. Aittoniemi. A mobile magnetopneumograph with dust quality sensing. *In* "Biomagnetism: Applications and Theory" (H. Weinberg, G. Stroink and T. Katila, eds.), pp. 411–415. New York: Pergamon Press, 1985.

Kalliomäki, K. Magnetopneumography. *In* "Magnetism in Medicine" (W. Andrä, and H. Nowak, eds.), pp. 446–454. Berlin: Wiley-VHC, 1998.

Kalmijn, A. J. Biophysics of geomagnetic field detection. *IEEE Transactions on Magnetics* **MAG-17**, 1113–1124, 1981.

Kapicka, A., E. Petrovsky, S. Ustjak, and K. Machackova. Proxy mapping of fly-ash pollution of soils around a coal-burning power plant: A case study in the Czech Republic. *Journal of Geochemical Exploration* **66**, 291–297, 1999.

Karlin, R. Magnetite diagenesis in marine sediments from the Oregon continental margin. *Journal of Geophysical Research* **95**, 4405–4419, 1990.

Karlin, R., M. Lyle, and G. R. Heath. Authigenic magnetite formation in suboxic marine sediments. *Nature* **326**, 490–493, 1987.

Kent, D. V. Apparent correlation of palaeomagnetic intensity and climatic records in deep-sea sediments. *Nature* **299**, 538–539, 1982.

Kertz, W. "Einführung in die Geophysik I." Mannheim: Bibliographisches Institut, Hochschultaschenbücher-Verlag, 1969.

Kieffer, H. H., B. M. Jakosky, C. W. Snyder, and M. S. Matthews, eds. "Mars." Tucson: University of Arizona Press, 1992.

King, J., S. K. Banerjee, J. Marvin and Ö. Özdemir. A comparison of different magnetic methods of determining the relative grain size of magnetite in natural materials: some results from lake sediments. *Earth and Planetary Science Letters* **59**, 404–419, 1982.

Kirschvink, J. L., and H. A. Lowenstam. Mineralization and magnetization of chiton teeth: Palaeomagnetic, sedimentologic and biologic implications of organic magnetite. *Earth and Planetary Science Letters* **44**, 193–204, 1979.

Kirschvink, J. L., M. M. Walker, S.-B. R. Chang, A. E. Dizon, and K. A. Peterson. Chains of single-domain magnetite particles in chinook salmon, *Oncorhynchus tshawytscha*. *Journal of Comparative Physiology A* **157**, 375–381, 1985.

Kirschvink, J. L., A. Kobayashi-Kirschvink, and B. J. Woodford. Magnetite biomineralization in the human brain. *Proceedings of the Academy of Sciences of the United States of America* **89**, 7683–7687, 1992.

Kissel, C., C. Laj, B. Lehman, L. Labyrie, and V. Bout-Roumazeilles. Changes in the strength of the Iceland–Scotland Overflow Water in the last 200,000 years: Evidence from magnetic anisotropy analysis of core SU90-33. *Earth and Planetary Science Letters* **152**, 25–36, 1997.

Kissel, C., C. Laj, L. Labeyrie, T. Dokken, A. Voelker, and D. Blamart. Rapid climatic variations during marine isotopic stage 3: Magnetic analysis of sediments from Nordic Seas and North Atlantic. *Earth and Planetary Science Letters* **171**, 489–502, 1999.

Kletetschka, G., and S. K. Banerjee. Magnetic stratigraphy of Chinese loess as a record of natural fires. *Geophysical Research Letters* **22**, 1341–1343, 1995.

Kodama, K. P., J. C. Lyons, P. A. Siver, and A.-M. Lott. A mineral magnetic and scaled-chrysophyte paleolimnological study of two northeastern Pennsylvania lakes: Records of fly ash deposition, land-use change and paleorainfall variation. *Journal of Paleolimnology* **17**, 173–189, 1997.

Kohfeld, K. E., and S. P. Harrison. DIRTMAP: The geological record of dust. *Earth-Science Reviews* **54**, 81–114, 2001.

Konhauser, K. O. Diversity of bacterial iron mineralization. *Earth-Science Reviews* **43**, 91–121, 1998.

Konieczynski, J. Skutecznosc pracy electrofiltrow, a emisja metali sladow w spalinach elektrowni weglowych. *Ochrona Poweitrza* **1–3**, 7–14, 1982.

Krs, M., M. Krsová, P. Pruner, A. Zeman, F. Novàk, and J. Jansa. A petromagnetic study of Miocene rocks bearing micro-organic material and the magnetic mineral greigite (Sokolov and Cheb basins, Czechoslovakia). *Physics of the Earth and Planetary Interiors* **63**, 98–112, 1990.

Krs, M., F. Novàk, M. Krsová, P. Pruner, L. Kouklíková, and J. Jansa. Magnetic properties and metastability of greigite–smythite mineralization in brown-coal basins of the Krusné hory Piedmont, Bohemia. *Physics of the Earth and Planetary Interiors* **70**, 273–287, 1992.

Kruiver, P. P., M. J. Dekkers, and D. Heslop. Quantification of magnetic coercivity components by the analysis of acquisition curves of isothermal remanent magnetisation. *Earth and Planetary Science Letters* **189**, 269–276, 2001.

Kuterbach, D. A., B. Walcott, R. J. Reeder, and R. B. Frankel. Iron-containing cells in the honey bee (*Apis mellifera*). *Science* **218**, 695–697, 1982.

Lagroix, F., and S. K. Banerjee. Paleowind directions from the magnetic fabric of loess profiles in central Alaska. *Earth and Planetary Science Letters* **195**, 99–112, 2002.

Laj, C., A. Mazaud, and J.-C. Duplessy. Geomagnetic intensity and ^{14}C abundance in the atmosphere and ocean during the past 50 kyr. *Geophysical Research Letters* **23**, 2045–2048, 1996.

Laj, C., C. Kissel, A. Mazaud, J. E. T. Channell, and J. Beer. North Atlantic palaeointensity stack since 75 ka (NAPIS-75) and the duration of the Laschamp event. *Philosophical Transactions of the Royal Society of London* **358**, 1009–1025, 2000.

Lamb, H. H. "Climate, History and the Modern World," 2nd Ed. London: Routledge, 1995.

Langereis, C. G., M. J. Dekkers, G. J. de Lange, M. Paterne, and P. J. M. van Santvoort. Magnetostratigraphy and astronomical calibration of the last 1.1 Myr from an eastern Mediterranean piston core and dating of short events in the Brunhes. *Geophysical Journal International* **129**, 75–94, 1997.

Laskar, J. The chaotic motion of the solar system: A numerical estimate of the size of the chaotic zones. *Icarus* **88**, 266–291, 1990.

Latham, A. G, and D. C. Ford. The paleomagnetism and rock magnetism of cave and karst deposits. *In* "Applications of Paleomagnetism to Sedimentary Geology," SEPM Special Publication, No. 49, 149–155, 1993.

Lauritzen, S. E., and J. Lundberg. Rapid temperature variations and volcanic events during the Holocene from a Norwegian speleothem record. *In* "Past Global Changes and Their Significance for the Future," 88. Bern: IGBP-PAGES, 1998.

Le Borgne, E. Mesures magnétiques en Bretagne centrale. *Comptes rendus hebdomadaires des séances de l'Académie des Sciences (Paris), Série B* **231**, 584–586, 1950.

Le Borgne, E. Anomalies magnétiques en Bretagne centrale. *Comptes rendus hebdomadaires des séances de l'Académie des Sciences (Paris), Série B* **233**, 82–84, 1951.

Le Borgne, E. Susceptibilité magnétique anormale du sol superficiel. *Annales de Géophysique* **11**, 399–419, 1955.

Le Borgne, E. Influence du feu sur les propriétés magnétiques du sol et sur celles du schiste et du granite. *Annales de Géophysique* **16**, 159–195, 1960.

Le Borgne, E. Les propriétés magnétiques du sol. Application a la prospection des sites archéologiques. *Archaeo-Physika* **1**, 1–20, 1965.

Lean, C. M. B., and I. N. McCave. Glacial to interglacial mineral magnetism and palaeoceanographic changes at Chatham Rise, SW Pacific Ocean. *Earth and Planetary Science Letters* **163**, 247–260, 1998.

Lean, J., J. Beer, and R. S. Bradley. Reconstruction of solar irradiance since A.D. 1600 and implications for climate change. *Geophysical Research Letters* **22**, 3195–3198, 1995.

Leslie, B. W., D. E. Hammond, W. M. Berelson, and S. P. Lund. Diagenesis in anoxic sediments from the California continental borderland and its influence on iron, sulfur and magnetite behavior. *Journal of Geophysical Research* **95**, 4453–4470, 1990.

Lewchuk, M. T., I. S. Al-Aasm, D. T. A. Symons, and K. P. Gillen. Dolomitization of Mississippian carbonates in the Shell Wateron gas field, southwestern Alberta: Insights from paleomagnetism, petrology and geochemistry. *Bulletin of Canadian Petroleum Geology* **46**, 387–410, 1998.

Likens, G. E., and F. H. Bormann. Nutrient-hydrologic interactions (Eastern United States). *In* "Coupling of Land and Water Systems" (A. D. Hasler, ed.), pp. 1–29. New York: Springer-Verlag, 1975.

Lindsley, D. H. Experimental studies of oxide minerals. *In* "Oxide Minerals: Petrologic and Magnetic Significance" (D. H. Lindsley, ed.), pp. 69–106. Mineralogical Society of America, 1991.

Linford, N. Magnetic ghosts: Mineral magnetic measurements on Roman and Anglo-Saxon graves. *Geophysical Research Abstracts* **4**, 2002.

Linford, N. T., and M. G. Canti. Geophysical evidence for fires in antiquity: Preliminary results from an experimental study. *Archaeological Prospection* **8**, 211–225, 2001.

Liu, T. S., and Z. Ding. Chinese loess and the paleomonsoon. *Annual Reviews of Earth and Planetary Sciences* **26**, 111–145, 1998.

Liu, X.-M., J. Shaw, T.-S. Liu, F. Heller, and Y. Baoyin. Magnetic mineralogy of Chinese loess and its significance. *Geophysical Journal International* **108**, 301–308, 1992.

Loizeau, J.-L., J. Dominik, T. Luzzi, and J.-P. Vernet. Sediment core correlation and mapping of sediment accumulation rates in Lake Geneva. *Journal of Great Lakes Research* **23**, 391–402, 1997.

Longworth, G., and M. S. Tite. Mössbauer and magnetic susceptibility studies of iron oxides in soils from archaeological sites. *Archaeometry* **19**, 3–14, 1997.

Lovley, D. R., J. F. Stolz, G. L. Nord, Jr., and E. J. P. Phillips. Anaerobic production of magnetite by a dissimilatory iron-reducing microorganism. *Nature* **330**, 252–254, 1987.

Lowenstam, H. A. Magnetite in denticle capping in recent chitons (Polyplacophora). *Bulletin of the Geological Society of America* **73**, 435–438, 1962.

Lowrie, W. Identification of ferromagnetic minerals in a rock by coercivity and unblocking temperature properties. *Geophysical Research Letters* **17**, 159–162, 1990.

Lowrie, W. "Fundamentals of Geophysics." Cambridge: University Press, 1997.

Lowrie, W., and M. Fuller. On the alternating field demagnetization characteristics of multidomain thermoremanent magnetization in magnetite. *Journal of Geophysical Research* **76**, 6339–6349, 1971.

Lowrie, W., and F. Heller. Magnetic properties of marine limestones. *Reviews of Geophysics and Space Physics* **20**, 171–192, 1982.

Lu, H., X. Liu, F. Zhang, Z. An, and J. Dodson. Astronomical calibration of loess-paleosol deposits at Luochuan central Chinese loess plateau. *Palaeogeography, Palaeoclimatology, Palaeoecology* **154**, 237–246, 1999a.

Lu, H., K. O. van Huissteden, Z. An, G. Nugteren, and J. F. Vandenberghe. East Asia winter monsoon variations on a millennial time-scale before the last glacial-interglacial cycle. *Journal of Quaternary Science* **14**, 101–110, 1999b.

Luckman, B. H. Evidence for climatic conditions between 900–1300 A.D. in the southern Canadian Rockies. *Climatic Change* **26**, 171–182, 1994.

Lyle, M. The brown-green colour transition in marine sediments: A marker of the Fe(III)-Fe(II) redox boundary. *Limnology and Oceanography* **28**, 1026–1033, 1983.

Machel, H. G. Magnetic mineral assemblages and magnetic contrasts in diagenetic environments — With implications for studies of palaeomagnetism, hydrocarbon migration and exploration. *In* "Palaeomagnetic Applications in Hydrocarbon Exploration" (P. Turner and A. Turner, eds.), pp. 9–29. Geological Society Special Publications, 98, 1995.

Mackereth, F. J. H. A portable core sampler for lake sediments. *Limnology and Oceanography* **3**, 181–191, 1958.

Mackereth, F. J. H. On the variation in direction of the horizontal component of remanent magnetisation in lake sediments. *Earth and Planetary Science Letters* **12**, 332–338, 1971.

Maher, B. A. Magnetic properties of some synthetic submicron magnetites. *Geophysical Journal* **94**, 83–96, 1988.

Maher, B. A. Inorganic formation of ultrafine-grained magnetite. *In* "Iron Biominerals" (R. B. Frankel and R. P. Blakemore, eds.), pp. 179–191. New York: Plenum Press, 1990.

Maher, B. A. Magnetic properties of modern soils and Quaternary loessic paleosols: Paleoclimatic implications. *Palaeogeography, Palaeoclimatology, Paleoecology* **137**, 25–54, 1998.

Maher, B. A., and M. W. Hounslow. Palaeomonsoons II: Magnetic records of aeolian dust in Quaternary sediments of the Indian Ocean. *In* "Quaternary Climates, Environments and Magnetism" (B. A. Maher and R. Thompson, eds.), pp. 128–162. Cambridge: University Press, 1999.

Maher, B. A., and R. M. Taylor. Formation of ultrafine-grained magnetite in soils. *Nature* **336**, 368–370, 1988.

Maher, B. A., and R. Thompson. Paleorainfall reconstructions from pedogenic magnetic susceptibility variations in the Chinese loess and paleosols. *Quaternary Research* **44**, 383–391, 1995.

Maher, B. A., R. Thompson, and L.-P. Zhou. Spatial and temporal reconstructions of changes in the Asian palaeomonsoon: A new mineral magnetic approach. *Earth and Planetary Science Letters* **125**, 462–471, 1994.

Maley, J., D. A. Livingstone, P. Giresse, N. Thouveny, P. Brenac, K. Kelts, G. Kling, C. Stager, M. Haag, M. Fournier, Y. Bandet, D. Williamson, and A. Zogning. Lithostratigraphy, volcanism, paleomagnetism and palynology of Quaternary lacustrine deposits from Barombi Mbo (West Cameroon): Preliminary results. *Journal of Volcanology and Geothermal Research* **42**, 319–335, 1990.

Malin, S. R. C., and Sir Edward Bullard. The direction of the Earth's magnetic field at London, 1570–1975. *Philosophical Transactions of the Royal Society of London* **299**, 357–423, 1981.

Malmivuo, J., and R. Plonsey. "Bioelectromagnetism." New York: Oxford University Press, 1995.

Mangerud, J. Radiocarbon dating of marine shells including a discussion of apparent age of recent shells from Norway. *Boreas* **1**, 143–172, 1972.

Manley, W. F., B. MacLean, M. W. Kerwin, and J. T. Andrews. Magnetic susceptibility as a Quaternary correlation tool: Examples from Hudson Strait sediment cores, eastern Canadian Arctic. *Current Research*, Part D, Geological Survey of Canada, Paper 93-1D, 137–145, 1993.

Mann, S. Molecular tectonics in biomineralization and biomimetic materials chemistry. *Nature* **365**, 499–505, 1993.

Mann, S., R. B. Frankel, and R. P. Blakemore. Structure, morphology and crystal growth of bacterial magnetite. *Nature* **310**, 405–407, 1984.

Mann, S., C. C. Perry, J. Webb, and R. J. P. Williams. Structure, composition and organization of biogenic minerals in limpet teeth. *Proceedings of the Royal Society B* **227**, 179–190, 1986.

Mann, S., N. H. C. Sparks, M. M. Walker, and J. L. Kirschvink. Ultrastructure, morphology and organization of biogenic magnetite from sockeye salmon, *Oncorhynchus nerka*: Implications for magnetoreception. *Journal of Exprimental Biology* **140**, 35–49, 1988.

Mann, S., N. H. C. Sparks, R. B. Frankel, D. A. Bazylinski, and H. W. Jannasch. Biomineralization of ferrimagnetic greigite (Fe_3S_4) and iron pyrite (FeS_2) in a magnetotactic bacterium. *Nature* **343**, 258–261, 1990a.

Mann, S., N. H. C. Sparks, and V. J. Wade. Crystallochemical control of iron oxide biomineralization. *In* "Iron Biominerals" (R. B. Frankel and R. P. Blakemore, eds.), pp. 21–49. New York: Plenum Press, 1990b.

Marmet, B., M. Bina, N. Fedoroff, and A. Tabbagh. Relationships between human activity and the magnetic properties of soil: A case study in the Medieval site of Roissy-en-France. *Archaeological Prospection* **6**, 161–170, 1999.

Marshall, A. Visualising burnt areas; patterns of magnetic susceptibility at Guiting Power 1 Round Barrow (Glos., UK). *Archaeological Prospection* **5**, 159–177, 1998.

Martinson, D. G., N. G. Pisias, J. D. Hays, J. Imbrie, T. C. Moore, Jr., and N. J. Shackleton. Age dating and the orbital theory of the Ice Ages of a high-resolution 0 to 300,000-year chronostratigraphy. *Quaternary Research* **27**, 1–29, 1987.

Matasova, G., E. Petrovsky, N. Jordanova, V. Zykina, and A. Kapicka. Magnetic study of Late Pleistocene loess/palaeosol sections from Siberia: Palaeoenvironmental implications. *Geophysical Journal International* **147**, 367–380, 2001.

Matsunaga, T., T. Sakaguchi, and F. Tadokoro. Magnetite formation by a magnetotactic bacterium capable of growing aerobically. *Applied Microbiology and Biotechnology* **35**, 651–655, 1991.

Matzka, J., and B. A. Maher. Magnetic biomonitoring of roadside tree leaves: Identification of spatial and temporal variations in vehicle-derived particles. *Atmospheric Environment* **33**, 4565–4569, 1999.

Mayer, T., W. A. Morris, and K. J. Versteeg. Feasibility of using magnetic properties for assessment of particle-associated contaminant transport. *Water Quality Research Journal of Canada* **31**, 741–752, 1996.

Mayergoyz, I. D. Mathematical models of hysteresis. *IEEE Transactions on Magnetics* **22**, 603–608, 1986.

McCabe, C., and R. D. Elmore. The occurrence and origin of Late Paleozoic remagnetization in the sedimentary rocks of North America. *Reviews of Geophysics* **27**, 471–494, 1989.

McCabe, C., R. Sassen, and B. Saffer. Occurrence of secondary magnetite within biodegraded oil. *Geology* **15**, 7–10, 1987.

McCabe, C., M. J. Jackson, and B. Saffer. Regional patterns of magnetite authigenesis in the Appalachian Basin: Implications for the mechanism of Late Paleozoic remagnetization. *Journal of Geophysical Research* **94**, 10,429–10,443, 1989.

McClean, R. G., and W. F. Kean. Contributions of wood ash magnetism to archaeomagnetic properties of fire pits and hearths. *Earth and Planetary Science Letters* **119**, 387–394, 1993.

McClintock, M. The pollen analysis of lacustrine sediments from Lough Catherine linked with possible human activity over the last four centuries. Unpublished dissertation, New University of Ulster, Ulster, 1973.

McDougall, I., F. H. Brown, T. E. Cerling, and J. W. Hillhouse. A reappraisal of the geomagnetic polarity timescale to 4 Ma using data from the Turkana Basin, East Africa. *Geophysical Research Letters* **19**, 2349–2352, 1992.

McDougall, J. M., D. H. Tarling, and S. E. Warren. The magnetic sourcing of obsidian samples from Mediterranean and Near Eastern sources. *Journal of Archaeological Science* **10**, 441–452, 1983.

McElhinny, M. W., and P. L. McFadden. "Paleomagnetism; Continents and Oceans." San Diego: Academic Press, 2000.

McElhinny, M. W., and W. E. Senanayake. Variations in the geomagnetic dipole 1: The past 50,000 years. *Journal of Geomagnetism and Geoelectricity* **34**, 39–51, 1982.

McFadden, B. J., and D. S. Jones. Magnetic butterflies. A case study of the Monarch (Lepidoptera, Danaidae). *In* "Magnetite Biomineralization and Magnetoreception in Organisms. A New Biomagnetism" (J. L. Kirschvink, D. S. Jones and B. J. McFadden, eds.), pp. 407–415. New York: Plenum Press, 1985.

McHargue, L. R., and P. E. Damon. The global beryllium 10 cycle. *Reviews of Geophysics* **29**, 141–158, 1991.

McIntyre, A., N. G. Kipp, A. W. H. Bé, T. Crowley, T. Kellogg, J. V. Gardner, W. Prell, and W. F. Ruddiman. Glacial North Atlantic 18,000 years ago: A CLIMAP reconstruction. *In* "Investigations of Late Quaternary Paleoceanography and Paleoclimatology" (R. M. Cline and J. D. Hays, eds.), *Geological Society of America Memoir* **145**, 43–76, 1976.

McKay, D. S., E. K. Gibson, Jr., K. L. Thomas-Keprta, H. Vali, C. S. Romanek, S. J. Clemett, X. D. F. Chillier, C. R. Maechling, and R. N. Zare. Search for past life on Mars: Possible relic biogenic activity in martian meteorite ALH84001. *Science* **273**, 924–930, 1996.

McLean, D. Magnetic spherules in recent lake sediments. *Hydrobiologia* **214**, 91–97, 1991.

McNeill, D. F. Biogenic magnetite from surface Holocene carbonate sediments, Great Bahama Bank. *Journal of Geophysical Research* **95**, 4363–4371, 1990.

Meldrum, F. C., B. R. Heywood, and S. Mann. Magnetoferritin: In vitro synthesis of a novel magnetic protein. *Science* **257**, 522–523, 1992.

Merrill, R. T., M. W. McElhinny, and P. L. McFadden. "The Magnetic Field of the Earth: Paleomagnetism, the Core and the Deep Mantle." San Diego: Academic Press, 1996.

Meynadier, L., J.-P. Valet, F. C. Bassinot, N. J. Shackleton, and Y. Guyodo. Asymmetrical saw-tooth pattern of the geomagnetic field intensity from equatorial sediments in the Pacific and Indian Oceans. *Earth and Planetary Science Letters* **126**, 109–127, 1994.

Mikola, S. On the magnetic inclination in past time (in Hungarian). *Termeszet tudomanyi Közlöny (Journal for Natural Sciences)* **32**, 246–247, 1900.

Milankovitch, M. "Canon of insolation and the ice-age problem." Beograd: Königlich Serbische Akademie. [English translation by the Israel Program for Scientific Translations, published for the U.S. Department of Commerce and the National Science Foundation, Washington, D.C. (1969)]. 1941.

Mishima, T., M. Torii, H. Fukusawa, Y. Ono, X.-M. Fang, B.-T. Pan, and J.-J. Li. Magnetic grain-size distribution of the enhanced component in the loess–palaeosol sequences in the western Loess Plateau of China. *Geophysical Journal International* **145**, 499–504, 2001.

Molyneux, L. A complete result magnetometer for measuring the remanent magnetization of rocks. *Geophysical Journal of the Royal Astronomical Society* **24**, 429–433, 1971.

Montgomery, P., E. A. Hailwood, A. S. Gale, and J. A. Burnett. The magnetostratigraphy of Coniacian–Late Campanian chalk sequences in southern England. *Earth and Planetary Science Letters* **156**, 209–224, 1998.

Morin, F. J. Magnetic susceptibility of αFe_2O_3 and αFe_2O_3 with added titanium. *Physical Review* **78**, 819–820, 1950.

Morris, W. A., J. K. Versteeg, D. W. Bryant, A. E. Legzdins, B. E. McCarry, and C. H. Marvin. Preliminary comparisons between mutagenicity and magnetic susceptibility of respirable airborne particulate. *Atmospheric Environment* **29**, 3441–3450, 1995.

Moskowitz, B. M. Biomineralization of magnetic minerals. *Reviews of Geophysics Supplement* 123–128, 1995.

Moskowitz, B. M., R. B. Frankel, P. J. Flanders, R. P. Blakemore, and B. B. Schwartz. Magnetic properties of magnetotactic bacteria. *Journal of Magnetism and Magnetic Materials* **73**, 273–288, 1988.

Moskowitz, B. M., R. B. Frankel, D. A. Bazylinski, H. W. Jannasch, and D. R. Lovley. A comparison of magnetite particles produced anaerobically by magnetotactic and dissimilatory iron-reducing bacteria. *Geophysical Research Letters* **16**, 665–668, 1989.

Moskowitz, B. M., R. Frankel, and D. Bazylinski. Rock magnetic criteria for the detection of biogenic magnetite. *Earth and Planetary Science Letters* **120**, 283–300, 1993.

Mothersill, J. S. The paleomagnetic record of the Late Quaternary sediments of Thunder Bay. *Canadian Journal of Earth Sciences* **16**, 1016–1023, 1979.

Moulder, J. E., L. S. Erdreich, R. S. Malyapa, J. Merritt, W. F. Pickard, and Vijayalaxmi. Cell phones and cancer: What is the evidence for a connection? *Radiation Research* **151**, 513–531, 1999.

Mullender, T. A. T., A. J. van Velzen, and M. J. Dekkers. Continuous drift correction and separate identification of ferrimagnetic and paramagnetic contributions in thermomagnetic runs. *Geophysical Journal International* **114**, 663–672, 1993.

Mullins, C. E. Magnetic susceptibility of the soil and its significance in soil science — A review. *Journal of Soil Science* **28**, 223–246, 1977.

Muxworthy, A. R., E. Schmidbauer, and N. Petersen. Magnetic properties and Mössbauer spectra of urban atmospheric particulate matter: A case study from Munich, Germany. *Geophysical Journal International* **150**, 558–570, 2002.

Nagata, T. "Rock Magnetism." Tokyo: Maruzen, 1961.

Nawrocki J., A. Wojcik, and A. Bogucki. The magnetic susceptibility record in the Polish and western Ukrainian loess–palaeosol sequences conditioned by palaeoclimate. *Boreas* **25**, 161–169, 1996.

Néel, L. Some theoretical aspects of rock magnetism. *Advances in Physics* **4**, 191–243, 1955.

Nickel, E. H. The composition and microtexture of an ulvöspinel magnetite intergrowth. *Canadian Mineralogist* **6**, 191–199, 1958.

Noller, J. S., J. M. Sowers, and W. R. Lettis. "Quaternary Geochronology Methods and Applications," AGU Reference Shelf Series, vol. 4. Washington, D.C.: American Geophysical Union, 2000.

O'Reilly, W. "Rock and Mineral Magnetism." Glasgow: Blackie, 1984.

Odum, E. P. "Ecology: A Bridge between Science and Society." Sunderland, Mass.: Sinauer Associates, 1997.

Oldfield, F. Lakes and their drainage basins as units of sediment–based ecological study. *Progress in Physical Geography* **1**, 460–504, 1977.

Oldfield, F. Magnetic measurements of recent sediments from Big Moose Lake, Adirondack Mountains, N.Y., USA. *Journal of Paleolimnology* **4**, 93–101, 1990.

Oldfield, F. Environmental magnetism — A personal perspective. *Quaternary Science Reviews* **10**, 73–85, 1991.

Oldfield, F. Toward the discrimination of fine grained ferrimagnets by magnetic measurements in lake and near-shore marine sediments. *Journal of Geophysical Research* **99**, 9045–9050, 1994.

Oldfield, F. The PALICLAS Project: synthesis and overview. *Memorie dell'Istituto Italiano di Idrobiologia* **55**, 329–357, 1996.

Oldfield, F., and N. Richardson. Lake sediment magnetism and atmospheric deposition. *Philosophical Transactions of the Royal Society* **B327**, 325–330, 1990.

Oldfield, F., A. Brown, and R. Thompson. The effect of microtopography and vegetation on the catchment of airborne particles measured by remanent magnetism. *Quaternary Research* **12**, 326–332, 1979.

Oldfield, F., P. G. Appleby, and A. T. Worsley. Evidence from lake sediments for recent erosion rates in the highlands of Papua New Guinea. *In* "Environmental Change and Tropical Geomorphology" (I. Douglas and E. Spenser, eds.), pp. 186–195. London: Allen & Unwin, 1985a.

Oldfield, F., A. Hunt, M. D. H. Jones, R. Chester, J. A. Dearing, L. Olsson, and J. M. Prospero. Magnetic differentiation of atmospheric dusts. *Nature* **317**, 516–518, 1985b.

Oldfield, F., B. A. Maher, J. Donoghue, and J. Pierce. Particle size-related, mineral magnetic source–sediment linkages in the Rhode River catchment, Maryland, USA. *Journal of the Geological Society of London* **142**, 1035–1046, 1985c.

Oldfield, F., B. A. Maher, and P. G. Appleby. Sediment source variations and lead-210 inventories in recent Potomac Estuary sediment cores. *Journal of Quaternary Science* **4**, 189–200, 1989.

Oliver, J. Fluids expelled tectonically from orogenic belts: Their role in hydrocarbon migration and other geologic phenomena. *Geology* **14**, 99–102, 1986.

Olson, P. Geophysics: The disappearing dipole. *Nature* **416**, 591–592, 2002.

Pan, H., D. T. A. Symons, and D. F. Sangster. Paleomagnetism of the Mississippi Valley–type ores and host rocks in the northern Arkansas and Tri-State districts. *Canadian Journal of Earth Sciences* **27**, 923–931, 1990.

Payne, M. A. SI and gaussian cgs units, conversions and equations for use in geomagnetism. *Physics of the Earth and Planetary Interiors* **26**, 10–16, 1981.

Peck, J. A., and J. W. King. Magnetofossils in the sediment of Lake Baikal, Siberia. *Earth and Planetary Science Letters* **140**, 159–172, 1996.

Peck, J. A., J. W. King, S. M. Colman, and V. A. Kravchinsky. A rock-magnetic record from Lake Baikal, Siberia: Evidence for Late Quaternary climate change. *Earth and Planetary Science Letters* **122**, 221–238, 1994.

Peng, X., R. Jia, R. Li, S. Dai, and T. S. Liu. Paleo-environmental study on the growth of magnetotactic bacteria and precipitation of magnetosomes in Chinese loess–paleosol sequences. *Chinese Science Bulletin* **45**, 21–25, 2000.

Perkins, A. M., and B. A. Maher. Rock magnetic studies of British speleothems. *Journal of Geomagnetism and Geoelectricity* **45**, 143–153, 1993.

Petermann, H., and U. Bleil. Detection of live magnetotactic bacteria in South Atlantic deep-sea sediments. *Earth and Planetary Science Letters* **117**, 223–228, 1993.

Peters, C., and R. Thompson. Magnetic identification of selected natural iron oxides and sulphides. *Journal of Magnetism and Magnetic Materials* **183**, 365–374, 1998a.

Peters, C., and R. Thompson. Supermagnetic enhancement, superparamagnetism and archaeological soils. *Geoarchaeology* **13**, 401–413, 1998b.

Petersen, N., T. von Dobeneck, and H. Vali. Fossil bacterial magnetite in deep-sea sediments from the South Atlantic Ocean. *Nature* **320**, 611–615, 1986.

Petersen, N., D. G. Weiss, and H. Vali. Magnetic bacteria in lake sediments. *In* "Geomagnetism and Paleomagnetism" (F. J. Lowes, D. W. Collinson, J. H. Parry, S. K. Runcorn, D. C. Tozer, and A. Soward, eds.), pp. 231–241. Dordrecht: Kluwer Academic Publishers, 1989.

Petit, J. R., L. Mournier, J. Jouzel, Y. S. Korotkevich, V. I. Kotlyakov, and C. Lorius. Palaeoclimatological and chronological implications of the Vostok core dust record. *Nature* **343**, 56–58, 1990.

Petrovsky, E., A. Kapicka, N. Jordanova, M. Knab, and V. Hoffmann. Low-field magnetic susceptibility: A proxy method of estimating increased pollution of different environmental systems. *Environmental Geology* **39**, 312–318, 2000.

Pfister, C. Fluctuations in the duration of snow-cover in Switzerland since the late seventeenth century. *In* "Proceedings of the Nordic Symposium on Climatic Changes and Related Problems" (K. Frydendahl, ed.), Climatological Papers No. 4, 1–6. Copenhagen: Danish Meteorological Institute, 1978.

Pick, T., and L. Tauxe. Chemical remanent magnetization in synthetic Fe_3O_4. *Journal of Geophysical Research* **96**, 9925–9936, 1991.

Pike, C., and J. Marvin. FORC analysis of frozen ferrofluids. *The IRM Quarterly* **11**, 1–11, 2001.

Pike, C. R., A. P. Roberts, and K. L. Verosub. Characterizing interactions in fine magnetic particle systems using first order reversal curves. *Journal of Applied Physics* **85**, 6660–6667, 1999.

Pike, C., A. P. Roberts, and K. L. Verosub. First-order reversal curve diagrams and thermal relaxation effects in magnetic particles. *Geophysical Journal International* **145**, 721–730, 2001a.

Pike, C., A. P. Roberts, K. L. Verosub, and M. J. Dekkers. An investigation of multidomain hysteresis mechanisms using FORC diagrams. *Physics of the Earth and Planetary Interiors* **126**, 13–28, 2001b.

Pilcher, J. R., and R. Larmour. Late-glacial and Post-Glacial vegetational history of the Meenadoan nature reserve, County Tyrone. *Proceedings of the Royal Irish Academy* **82B**, 277–295, 1982.

Porter, S. C. Chinese loess record of monsoon climate during the last glacial–interglacial cycle. *Earth-Science Reviews* **54**, 115–128, 2001.

Pósfai, M., P. R. Buseck, D. A. Bazylinski, and R. B. Frankel. Iron sulphides from magnetotactic bacteria: Structure, composition and phase transitions. *American Mineralogist* **83**, 1469–1481, 1998.

Pósfai, M., K. Cziner, E. Márton, P. Márton, P. R. Buseck, R. B. Frankel, and D. A. Bazylinski. Crystal-size distributions and possible biogenic origin of Fe sulfides. *European Journal of Mineralogy* **13**, 691–703, 2001.

Preece, A. W., J. W. Hand, R. N. Clarke, and A. Stewart. Power frequency electromagnetic fields and health. Where's the evidence? *Physics in Medicine and Biology* **45**, R139–R154, 2000.

Presti, D. E. Avian navigation, geomagnetic field sensitivity and biogenic magnetite. *In* "Magnetite Biomineralization and Magnetoreception in Organisms. A New Biomagnetism" (J. L. Kirschvink, D. S. Jones, and B. J. McFadden, eds.), pp. 455–482. New York: Plenum Press, 1985.

Prospero, J. M., M. Uematsu, and D. L. Savoie. Mineral aerosol transport to the Pacific Ocean. *Chemical Oceanography* **10**, 188–218, 1989.

Pye, K. "Aeolian Dust and Dust Deposits." New York: Academic Press, 1987.

Rancourt, D. Magnetism of Earth, planetary and environmental nanomaterials. In "Nanoparticles in the Environment" (J. F. Banfield and A. Navrotsky, eds.), *Reviews in Mineralogy and Geochemistry* **44**, 217–292, 2001.

Rassi, D., V. Timbrell, H. Al-Sewaidan, S. Davies, O. Taikina-aho, and P. Paasko. A study of magnetic contaminants in *post mortem* lung samples from asbestos miners. In "Advances in Biomagnetism" (S. J. Williamson, M. Hoke, G. Stroink, and M. Kotani, eds.), pp. 485–488. New York: Plenum Press, 1989.

Rea, D. K. The paleoclimatic record provided by eolian deposition in the deep sea: The geological history of wind. *Reviews of Geophysics* **32**, 159–195, 1994.

Redfield, A. C. The biological control of chemical factors in the environment. *American Scientist* **46**, 205–221, 1958.

Retallack, G. J. "Soils of the Past: An Introduction to Palaeopedology." New York: Unwin & Hyman, 1990.

Reynolds, R. L., N. S. Fishman, and M. R. Hudson. Sources of aeromagnetic anomalies over Cement oil field (Oklahoma), Simpson oil field (Alaska) and the Wyoming-Idaho-Utah thrust belt. *Geophysics* **56**, 606–617, 1991.

Rhodes, T. E., F. Gasse, R. Lin, J.-C. Fontes, K. Wei, P. Bertrand, F. Gibert, F. Mélièrese, P. Tucholka, Z. Wang, and Z. Y. Cheng. A Late Pleistocene–Holocene lacustrine record from Lake Manas, Zunggar (northern Xinjiang, western China). *Palaeogeography, Palaeoclimatology, Palaeoecology* **120**, 105–121, 1996.

Roberts, A. P. Magnetic properties of sedimentary greigite (Fe_3S_4). *Earth and Planetary Science Letters* **134**, 227–236, 1995.

Roberts, A. P., Y. Cui, and K. L. Verosub. Wasp-waisted hysteresis loops: Mineral magnetic characteristics and discrimination of components in mixed magnetic systems. *Journal of Geophysical Research* **100**, 17,909–17,924, 1995.

Roberts, A. P., R. L. Reynolds, K. L. Verosub, and D. P. Adam. Environmental magnetic implications of greigite (Fe_3S_4) formation in a 3 m.y. lake sediment record from Butte Valley, northern California. *Geophysical Research Letters* **23**, 2859–2862, 1996.

Roberts, A. P., C. R. Pike, and K. L. Verosub. First-order reversal curve diagrams: A new tool for characterizing the magnetic properties of natural samples. *Journal of Geophysical Research* **105**, 28,461–28,475, 2000.

Robertson, D. J., and D. E. France. Discrimination of remanence-carrying minerals in mixtures, using isothermal remanent magnetization acquisition curves. *Physics of the Earth and Planetary Interiors* **82**, 223–234, 1994.

Robinson, C., G. M. Raisbeck, F. Yiou, B. Lehman, and C. Laj. The relationship between ^{10}Be and geomagnetic field strength records in central North Atlantic sediments during the last 80 ka. *Earth and Planetary Science Letters* **136**, 551–557, 1995.

Robinson, S. G. The late Pleistocene paleoclimatic record of North Atlantic deep-sea sediments revealed by mineral-magnetic measurements. *Physics of the Earth and Planetary Interiors* **42**, 22–47, 1986.

Robinson, S. G., M. A. Maslin, and I. M. McCave. Magnetic susceptibility variations in Upper Pleistocene deep-sea sediments of the NE Atlantic: Implications for ice rafting and paleocirculation at the last glacial maximum. *Paleoceanography* **10**, 221–250, 1995.

Roger, S., G. Feraud, V. Andrieu, N. Thouveny, C. Coulon, J. J. Cocheme, J. L. de Beaulieu, and T. Williams. Age of the Lac du Bouchet-Praclaux Pleistocene lacustrine sedimentary sequence: A re-appraisal from ^{40}Ar/^{39}Ar datings. *EUG meeting abstracts*, Strasbourg, 1999.

Rolph, T. C., F. Oldfield, and K. D. van der Post. Palaeomagnetism and rock-magnetism results from Lake Albano and the central Adriatic Sea (Italy). *Memorie dell'Istituto Italiano di Idrobiologia* **55**, 265–283, 1996.

Rosenbaum, J. G., and R. L. Reynolds. Environmental magnetic records of Late Pleistocene glaciation from lakes in the western United States, in press, 2002.

Rousseau, D. D. Climatic transfer function from Quaternary molluscs in European loess deposits. *Quaternary Research* **36**, 195–209, 1991.

Rousseau, D. D. Loess biostratigraphy: new advances and approaches in mollusk studies. *Earth-Science Reviews* **54**, 157–171, 2001.

Rousseau, D. D., and G. Kukla. Late Pleistocene climate record in the Eustis Loess section, Nebraska, based on land snail assemblages and magnetic susceptibility. *Quaternary Research* **42**, 176–187, 1994.

Rousseau, D. D., and N. Wu. A new molluscan record of the monsoon variability over the past 130,000 yr in the Luochuan loess sequence, China. *Geology* **25**, 275–278, 1997.

Rousseau, D. D., L. Zöller, and J. P. Valet. Late Pleistocene climatic variation at Achenheim, France, based on a magnetic susceptibility and TL chronology of loess. *Quaternary Research* **49**, 255–263, 1998.

Ruddiman, W. F. Northern oceans. *In* "The Geology of North America: North America and Adjacent Oceans during the Last Deglaciation" (W. F. Ruddiman and H. E. Wright, Jr., eds.), K-3, pp. 137–153. Boulder, Colo.: Geological Society of America, 1987.

Ruddiman, W. F., M. E. Raymo, D. G. Martinson, B. M. Clement, and J. Backman. Pleistocene evolution: Northern hemisphere ice sheets and North Atlantic Ocean. *Paleoceanography* **4**, 353–412, 1989.

Rutter, N. W., and Z. Ding. Paleoclimates and monsoon variations interpreted from micromorphic features of the Baoji paleosols, China. *Quaternary Science Reviews* **12**, 853–862, 1993.

Rutter, N. W., Z. L. Ding, M. E. Evans, and Y. C. Wang. Magnetostratigraphy of the Baoji loess–paleosol section in the north-central China Loess Plateau. *Quaternary International* **7/8**, 97–102, 1991.

Sagnotti, L., and A. Winkler. Rock magnetism and palaeomagnetism of greigite-bearing mudstones in the Italian peninsula. *Earth and Planetary Science Letters* **165**, 67–80, 1999.

Sagnotti, L., F. Florindo, K. L. Verosub, G. S. Wilson, and A. P. Roberts. Environmental magnetic record of Antarctic palaeoclimate from Eocene/Oligocene glaciomarine sediments, Victoria Land Basin. *Geophysical Journal International* **134**, 653–662, 1998.

Sagnotti, L., P. Macrí, A. Camerlenghi, and M. Rebesco. Environmental magnetism of Antarctic Late Pleistocene sediments and interhemispheric correlation of climatic events. *Earth and Planetary Science Letters* **192**, 65–80, 2001.

Sahota, J. T. S., S. G. Robinson, and F. Oldfield. Magnetic measurements used to identify paleoxidation fronts in deep-sea sediments from the Madeira Abyssal Plain. *Geophysical Research Letters* **22**, 1961–1964, 1995.

Sakaguchi, T., J. G. Burgess, and T. Matsunaga. Magnetite formation by a sulphate-reducing bacterium. *Nature* **365**, 47–49, 1993.

Sarna-Wojcicki, A. M., M. S. Pringle, and J. Wijbrans. New $^{40}Ar/^{39}Ar$ age of the Bishop Tuff from multiple sites and sediment rate calibration for the Matuyama–Brunhes boundary. *Journal of Geophysical Research* **105**, 21,431–21,443, 2000.

Sartori, M., F. Heller, T. Forster, M. Borkovec, J. Hammann, and E. Vincent. Magnetic properties of loess grain size fractions from the section at Paks (Hungary). *Physics of the Earth and Planetary Interiors* **116**, 53–64, 1999.

Saydam, A. C., and H. Z. Senyuva. Deserts: Can they be the potential suppliers of bioavailabe iron? *Geophysical Research Letters* **29**(11), DOI 10.1029/2001GL13562, 2002.

Schädlich, G., L. Weissflog, and G. Schüürmann. Magnetic susceptibility in conifer needles as indicator of fly ash deposition. *Fresenius Environmental Bulletin* **4**, 7–12, 1995.

Schiavon, N., and L. P. Zhou. Magnetic, chemical and microscopical characterization of urban soiling on historical monuments. *Environmental Science and Technology* **30**, 3624–3629, 1996.

Scholger, R. Heavy metal pollution monitoring by magnetic susceptibility measurements applied to sediments of the river Mur (Styria, Austria). *European Journal of Environmental Engineering and Geophysics* **3**, 25–37, 1998.

Schultheiss-Grassi, P. P., and J. Dobson. Magnetic analysis of human brain tissue. *BioMetals* **12**, 67–72, 1999.

Schwarz, E. J. Magnetic properties of pyrrhotite and their use in applied geology and geophysics. *Geological Survey of Canada Paper* **74–59**, 1–24, 1975.

Schwertmann, U. Occurrence and formation of iron oxides in various pedoenvironments. *In*: "Iron in Soils and Clay Minerals" (J. W. Stucki, B. A. Goodman, and U. Schwertmann, eds.), pp. 267–308. Dordrecht: Reidel Publishing, 1988a.

Schwertmann, U. Some properties of soil and synthetic iron oxides. *In* "Iron in Soils and Clay Minerals," (J. W. Stucki, B. A. Goodman, and U. Schwertmann, eds.), pp. 203–250. Dordrecht: Reidel Publishing, 1988b.

Schwertmann, U., and B. Heinemann. Über das Vorkommen und die Entstehung von Maghemit in nordwestdeutschen Böden. *Neues Jahrbuch für Mineralogie Monatshefte* **8**, 174–181, 1959.

Scoullos, M., F. Oldfield, and R. Thompson. Magnetic monitoring of marine particulate pollution in the Elefsis Gulf, Greece. *Marine Pollution Bulletin* **10**, 287–291, 1979.

Scriba, H., and F. Heller. Measurement of anisotropy of magnetic susceptibility using inductive magnetometers. *Journal of Geophysics* **44**, 341–352, 1978.

Shackleton, N. J., and S. Crowhurst, Sediment fluxes based on an orbitally tuned time scale 5 Ma to 14 Ma, Site 926, *In* "Proceedings of the Ocean Drilling Program, Scientific Results," 154, (N. J. Shackleton, W. B. Curry, C. Richter and T. Bralower, eds.), pp. 69–82. College Station, Texas: Ocean Drilling Program, 1997.

Shackleton, N. J., and N. D. Opdyke. Oxygen isotope and paleomagnetic stratigraphy of equatorial Pacific core V28–238: Oxygen isotope temperatures and ice volumes on a 10^5-year and 10^6-year scale. *Quaternary Research* **3**, 39–55, 1973.

Shackleton, N. J., J. Backman, H. Zimmerman, D. V. Kent, M. A. Hall, D. G. Roberts, D. Schnitker, J. Baldauf, A. Despraires, R. Homrighausen, P. Huddlestun, J. Keene, A. J. Kaltenback, K. A. O. Krumsiek, A. C. Morton, J. W. Murray, and J. Westberg-Smith. Oxygen isotope calibration of the onset of ice-rafting and history of glaciation in the North Atlantic region. *Nature* **307**, 620–623, 1984.

Shackleton, N. J., A. Berger, and W. R. Peltier. An alternative astronomical calibration of the lower Pleistocene timescale based on ODP site 677. *Transactions of the Royal Society of Edinburgh, Earth Science* **81**, 251–261, 1990.

Shackleton, N. J., Z. An, A. E. Dodonov, J. Gavin, G. J. Kukla, V. A. Ranov, and L. P. Zhou. Accumulation rate of loess in Tadjikistan and China: Relationship with global ice volume cycles. *Quaternary Proceedings* **4**, 1–6, 1995.

Sharp, J. M. Jr. Energy and momentum transport model of the Ouachita Basin and its possible impact on formation of economic mineral deposits. *Economic Geology* **73**, 1057–1068, 1978.

Shaw, J., H. B. Zheng and Z. S. An. Magnetic dating of early man in China. *In* "Archaeometry '90" (E. Pernicka and G. A. Wagner, eds.), pp. 589–595. Basel: Birkhäuser, 1991.

Shcherbakov, V. P., and M. Winklhofer. The osmotic magnetometer: a new model for magnetite-based magnetoreceptors in animals. *European Biophysical Journal* **28**, 380–392, 1999.

Shu, J., J. A. Dearing, A. P. Morse, L. Yu, and C. Li. Magnetic properties of daily sampled total suspended particulates in Shanghai. *Environmental Science and Technology* **34**, 2393–2400, 2001.

Sigg, L., C. A. Johnson, and A. Kuhn. Redox conditions and alkalinity generation in a seasonally anoxic lake (Lake Greifen). *Marine Chemistry* **36**, 9–26, 1991.

Simkiss, K., and K. M. Wilbur. "Biomineralization." San Diego: Academic Press, 1989.

Singer, B. S., K. A. Hoffman, A. Chauvin, R. S. Coe, and M. S. Pringle. Dating transitionally magnetized lavas of the late Matuyama Chron: Toward a new $^{40}Ar/^{39}Ar$ timescale of reversals and events. *Journal of Geophysical Research* **104**, 679–693, 1999.

Singer, M. J., P. Fine, K. L. Verosub, and O. A. Chadwick. Time dependence of magnetic susceptibility of soil chronosequences on the California coast. *Quaternary Research* **37**, 323–332, 1992.

Singhvi, A. K., A. Bluszcz, M. D. Bateman, and M. Someshwar Rao. Luminescence dating of loess–palaeosol sequences and coversands: methodological aspects and palaeoclimatic implications. *Earth-Science Reviews* **54**, 193–211, 2001.

Smalley, I. J. The properties of glacial loess and the formation of loess deposits. *Journal of Sedimentary Petrology* **36**, 669–676, 1966.

Smalley, I. J., I. F. Henderson, T. A. Dijkstra, and E. Derbyshire. Some major events in the development of the scientific study of loess. *Earth-Science Reviews* **54**, 5–18, 2001.

Snowball, I. F. Magnetic hysteresis properties of greigite (Fe_3S_4) and a new occurrence in Holocene sediments from Swedish Lappland. *Physics of the Earth and Planetary Interiors* **68**, 32–40, 1991.

Snowball, I. F. Bacterial magnetite and the magnetic properties of sediments in a Swedish lake. *Earth and Planetary Science Letters* **126**, 129–142, 1994.

Snowball, I. F., and R. Thompson. The occurrence of greigite in sediments from Loch Lomond. *Journal of Quaternary Science* **3**, 121–125, 1988.

Snowball, I. F., and R. Thompson. A mineral magnetic study of Holocene sedimentation in Lough Catherine, Northern Ireland. *Boreas* **19**, 127–146, 1990.

Snowball, I., and R. Thompson. A mineral magnetic study of Holocene sediment yields and deposition patterns in the Llyn Geirionydd catchment, north Wales. *The Holocene* **2/3**, 238–248, 1992.

Snowball, I., and M. Torii. Incidence and significance of magnetic iron sulphides in Quaternary sediments and soil. *In* "Quaternary Climates, Environments and Magnetism" (B. A. Maher and R. Thompson, eds.), pp. 199–230. Cambridge: University Press, 1999.

Snowball, I., P. Sandgren, and G. Petterson. The mineral magnetic properties of an annually laminated Holocene lake-sediment sequence in northern Sweden. *The Holocene* **9**, 353–362, 1999.

Soil Survey Staff. "Soil Taxonomy: A Basic System of Soil Classification for Making and Interpreting Soil Surveys," Agriculture Handbook, No. 436. Washington, D.C.: Soil Conservation Service, U.S. Department of Agriculture, 1975.

Somayajulu, B. L. K., P. Sharma, and J. Beer. ^{10}Be annual fallout in rain in India. *Nuclear Instruments and Methods* **233**(B5), 398–403, 1984.

Soreghan, G. S., R. D. Elmore, B. Katz, M. Cogoini, and S. Banerjee. Pedogenically enhanced magnetic susceptibility variations preserved in Paleozoic loessite. *Geology* **25**, 1003–1006, 1997.

Sowers, T., M. Bender, L. D. Labeyrie, J. Jouzel, D. Raynaud, D. Martinson, and Y. S. Korotkevich. 135 kyr Vostok-SPECMAP common temporal framework. *Paleoceanography* **8**, 737–757, 1993.

Spassov, S., F. Heller, M. E. Evans, L. P. Yue, and Z. L. Ding. The Matuyama/Brunhes geomagnetic polarity transition at Lingtai and Baoji, Chinese loess plateau. *Physics and Chemistry of the Earth (A)* **26**, 899–904, 2001.

Spassov, S., F. Heller, M. E. Evans, L. P. Yue, and T. von Dobeneck. A lock-in model for the complex Matuyama–Brunhes boundary record of the loess/paleosol sequence at Lingtai (Central Chinese Loess Plateau). *Geophysical Journal International*, submitted, 2002.

Spell, T. L., and I. McDougall. Revisions to the age of the Brunhes–Matuyama boundary and the Pleistocene geomagnetic polarity timescale. *Geophysical Research Letters* **19**, 1181–1184, 1992.

Sroubek, P., J. F. Diehl, J. Kadlec, and K. Valoch. A Late Pleistocene palaeoclimate record based on mineral magnetic properties of the entrance facies sediments of Kulna Cave, Czech Republic. *Geophysical Journal International* **147**, 247–262, 2001.

Stace, H. C. T., G. D. Hubble, R. Brewer, K.H. Northcote, J. R. Sleeman, M. J. Mulcahy, and E. G. Hallsworth. "A Handbook of Australian Soils." Glenside: Rellim Technical Publications for the Commonwealth Scientific and Industrial Research, 1968.

Stacey, F. D. The physical theory of rock magnetism. *Advances in Physics* **12**, 45–133, 1963.

Stacey, F. D. Kelvin's age of the Earth paradox revisited. *Journal of Geophysical Research* **105**, 13,155–13,158, 2000.

Stacey, F. D., and S. K. Banerjee. "The Physical Principles of Rock Magnetism." Amsterdam: Elsevier, 1974.

Stanjek, H., J. W. E. Fassbinder, H. Vali, H. Wägele, and W. Graf. Evidence of biogenic greigite (ferrimagnetic Fe_3S_4) in soil. *European Journal of Soil Science* **45**, 97–103, 1994.

Steinberger, B., N. Petersen, H. Petermann, and D. G. Weiss. Movement of magnetic bacteria in time-varying magnetic fields. *Journal of Fluid Mechanics* **273**, 189–211, 1994.

Stephenson, A. Single domain grain distributions I. A method for the determination of single domain grain distributions. *Physics of the Earth and Planetary Interiors* **4**, 353–360, 1971a.

Stephenson, A. Single domain grain distributions II. The distribution of single domain iron grains in Apollo 11 lunar dust. *Physics of the Earth and Planetary Interiors* **4**, 361–369, 1971b.

Stine, S. Extreme and persistent drought in California and Patagonia during mediaeval time. *Nature* **369**, 546–549, 1994.

Stober, J. C., and R. Thompson. Palaeomagnetic secular variation studies of Finnish lake sediment and the carriers of remanence. *Earth and Planetary Science Letters* **37**, 139–149, 1977.

Stolz, J. F., S.-B. R. Chang, and J. L. Kirschvink. Magnetotactic bacteria and single-domain magnetite in hemipelagic sediments. *Nature* **321**, 849–850, 1986.

Stolz, J. F., S.-B. R. Chang, and J. L. Kirschvink. Biogenic magnetite in stromatolites. I. Occurrence in modern sedimentary environments. *Precambrian Research* **43**, 295–304, 1989.

Stolz, J. F., D. R. Lovley, and S. E. Haggerty. Biogenic magnetite and the magnetization of sediments. *Journal of Geophysical Research* **95**, 4355–4361, 1990.

Stoner, J. S., J. E. T. Channel, and C. Hillaire-Marcel. Late Pleistocene relative geomagnetic paleointensity from the deep Labrador Sea: Regional and global correlations. *Earth and Planetary Science Letters* **134**, 237–252, 1995a.

Stoner, J. S., J. E. T. Channell, and C. Hillaire-Marcel. Magnetic properties of deep-sea sediments off southwest Greenland: Evidence for major differences between the last two deglaciations. *Geology* **23**, 241–244, 1995b.

Stoner, J. S., J. E. T. Channell, and C. Hillaire-Marcel. The magnetic signature of rapidly deposited detrital layers from the deep Labrador Sea: Relationship to North Atlantic Heinrich layers. *Paleoceanography* **11**, 309–327, 1996.

Strzyszcz, Z., T. Magiera, and F. Heller. The influence of industrial emissions on the magnetic susceptibility of soils in Upper Silesia. *Studia Geophysica et Geodetica* **40**, 276–286, 1996.

Stuiver, M. Carbon-14 dating: a comparison of beta and ion counting. *Science* **202**, 881–883, 1978.

Stuiver, M. and T. F. Braziunas. Modeling atmospheric ^{14}C influence and ^{14}C ages of marine samples to 10,000 B.C. *Radiocarbon* **35**, 137–189, 1993.

Stuiver, M., C. J. Heusser, and I. C. Wang. North American glacial history extended to 75,000 years ago. *Science* **200**, 16–21, 1978.

Sun, J. M., K. E. Kohfeld, and S. P. Harrison. Records of aeolian dust deposition on the Chinese Loess Plateau during the Late Quaternary. *Technical Reports Max-Planck-Institut für Biogeochemie* **1**, 2000.

Swan, A. R. H., and M. Sandilands. "Introduction to Geological Data Analysis." Oxford: Blackwell Science, 1995.

Symons, D. T. A., H. Pan, D. F. Sangster, and E. C. Jowett. Paleomagnetism of the Pine Point Zn-Pb deposits. *Canadian Journal of Earth Sciences* **30**, 1028–1036, 1993.

Tarduno, J. A. Temporal trends of magnetic dissolution in the pelagic realm: Gauging paleoproductivity? *Earth and Planetary Science Letters* **123**, 39–48, 1994.

Tarduno, J. A., and S. L. Wilkison. Non–steady state magnetic mineral reduction, chemical lock-in and delayed remanence acquisition in pelagic sediments. *Earth and Planetary Science Letters* **144**, 315–326, 1996.

Tarling, D. H., and F. Hrouda. "The Magnetic Anisotropy of Rocks." London: Chapman & Hall, 1993.

Tauxe, L. Sedimentary records of relative paleointensity: Theory and practice. *Reviews of Geophysics* **31**, 319–354, 1993.

Taylor, R. M., B. A. Maher, and P. G. Self. Magnetite in soils. I. The synthesis of single-domain and superparamagnetic magnetite. *Clay Minerals* **22**, 411–422, 1987.

Thiede, J., and E. Suess, eds. "Coastal Upwelling, Its Sediment Record, Part B: Sedimentary Records of Ancient Coastal Upwelling." New York: Plenum Press, 1983.

Thomas-Keprta, K. L., D. A. Bazylinski, J. L. Kirschvink, S. J. Clemett, D. S. McKay, S. J. Wentworth, H. Vali, E. K. Gibson, Jr., and C. S. Romanek. Elongated prismatic magnetite crystals in ALH84001 carbonate globules: Potential martian magnetofossils. *Geochimica et Cosmochimica Acta* **64**, 4049–4081, 2000.

Thompson, R. Palaeolimnology and palaeomagnetism. *Nature* **242**, 182–184, 1973.

Thompson, R. Modelling magnetization data using SIMPLEX. *Physics of the Earth and Planetary Interiors* **42**, 113–127, 1986.

Thompson, R., and R. M. Clark. Sequence slotting for stratigraphic correlation between cores: Theory and practice. *Journal of Paleoclimatology* **2**, 173–184, 1989.

Thompson, R., and K. J. Edwards. A Holocene palaeomagnetic record and a geomagnetic master curve from Ireland. *Boreas* **11**, 335–349, 1982.

Thompson, R., and D. J. Morton. Magnetic susceptibility and particle-size distribution in recent sediments of the Loch Lomond drainage basin, Scotland. *Journal of Sedimentary Petrology* **49**, 801–811, 1979.

Thompson, R., and G. M. Turner. Icelandic Holocene palaeolimnomagnetism. *Physics of the Earth and Planetary Interiors* **38**, 250–261, 1985.

Thompson, R., R. W. Batterbee, P. E. O'Sullivan, and F. Oldfield. Magnetic susceptibility of lake sediments. *Limnology and Oceanography* **20**, 687–698, 1975.

Thompson, R., G. M. Turner, M. Stiller, and A. Kaufman. Near East paleomagnetic secular variation recorded in sediments from the Sea of Galilee (Lake Kinnereth). *Quaternary Research* **23**, 175–188, 1985.

Thouveny, N., K. M. Creer, and I. Blunk. Extension of the Lac du Bouchet palaeomagnetic record over the last 120,000 years. *Earth and Planetary Science Letters* **97**, 140–161, 1990.
Thouveny, N., J.-L. de Beaulieu, E. Bonifay, K.M. Creer, J. Guiot, M. Icole, S. Johnsen, J. Jouzel, M. Reille, T. Williams, and D. Williamson. Climate variation in Europe over the past 140 kyr deduced from rock magnetism. *Nature* **371**, 503–506, 1994.
Thouveny, N., E. Moreno, D. Delanghe, L. Candon, Y. Lancelot, and N. J. Shackleton. Rock magnetic detection of distal ice-rafted debris: clue for the identification of Heinrich layers on the Portuguese margin. *Earth and Planetary Science Letters* **180**, 61–75, 2000.
Tinchev, S. S. SQUID-Magnetometer aus Supraleitern mit hoher Sprungtemperatur, Berichte aus Forschung und Entwicklung unsrer Gesellschaft, 2, Hoesch, 1992.
Tite, M. S., and C. E. Mullins. Enhancement of the magnetic susceptibility of soils on archaeological sites. *Archaeometry* **13**, 209–219, 1971.
Torii, M., K. Fukuma, C.-S. Horng, and T.-Q. Lee. Magnetic discrimination of pyrrhotite- and greigite-bearing sediment samples. *Geophysical Research Letters* **23**, 1813–1816, 1996.
Torii, M., T.-Q. Lee, K. Fukuma, T. Mishima, T. Yamazaki. H. Oda, and N. Ishikawa. Mineral magnetic study of the Taklimakan desert sands and its relevance to the Chinese loess. *Geophysical Journal International* **146**, 416–424, 2001.
Towe, K. M., and H. A. Lowenstam. Ultra-structure and development of iron mineralization in the redular teeth of *Cryptochiton stelleri* (Mollusca). *Journal of Ultrastructural Research* **17**, 1–13, 1967.
Treiman, A. Martian life 'still kicking' in meteorite ALH84001. *EOS, Transactions, American Geophysical Union* **80**, 205–209, 1999.
Trins workshop participants. Exploring Late Pleistocene Climate Variations. *EOS, Transactions, American Geophysical Union* **81**, 625–630, 2000.
Tsatskin, A., F. Heller, E. A. Hailwood, T. S. Gendler, J. Hus, P. Montgomery, M. Sartori, and E. I. Virina. Pedosedimentary division, rock magnetism and chronology of the loess/palaeosol sequence at Roxolany (Ukraine). *Palaeogeography, Palaeoclimatology, Palaeoecology* **143**, 111–133, 1998.
Turner, G. M. A 5000 year geomagnetic palaeosecular variation record from western Canada. *Geophysical Journal of the Royal Astronomical Society* **91**, 103–121, 1987.
Turner, G. M., and R. Thompson. Lake sediment record of the geomagnetic secular variations in Britain during Holocene times. *Geophysical Journal of the Royal Astronomical Society* **65**, 703–725, 1981.
Turner, G. M., and R. Thompson. Detransformation of the British geomagnetic secular variation record for Holocene times. *Geophysical Journal of the Royal Astronomical Society* **70**, 789–792, 1982.
Valberg, P. A., and K. S. Zaner. Magnetic microparticles can measure whole-body clearance, cell organelle motions and protein polymer viscoelasticity. *In* "Advances in Biomagnetism" (S. J. Williamson, M. Hoke, G. Stroink and M. Kotani, eds.), pp. 461–467. New York: Plenum Press, 1989.
Vali, H., and J. L. Kirschvink. Magnetofossil dissolution in a paleomagnetically unstable deep-sea sediment. *Nature* **339**, 203–206, 1989.
van der Voo, R. Phanerozoic paleomagnetic poles from Europe and North America and comparisons with continental reconstructions. *Reviews of Geophysics* **28**, 167–206, 1990.
van der Voo, R. "Paleomagnetism of the Atlantic, Tethys and Iapetus Oceans." Cambridge: University Press, 1993.
van der Wal, P. Structure and formation of the magnetite-bearing cap of polyplacophoran tricuspid radular teeth. *In* "Iron Biominerals" (R. B. Frankel and R. P. Blakemore, eds.), pp. 221–229. New York: Plenum Press, 1990.
van Oorschot, I. H. M., M. J. Dekkers, and P. Havlicek. Selective dissolution of magnetic iron oxides with the acid-ammonium-oxalate/ferrous-iron extraction technique — II. Natural loess and palaeosol samples. *Geophysical Journal International* **149**, 106–117, 2002.
van Velzen, A. J., and M. J. Dekkers. Low-temperature oxidation of magnetite in loess–paleosol sequences: A correction of rock magnetic parameters. *Studia geophysica et geodetica* **43**, 357–375, 1999.
van Velzen, A. J., and J. D. A. Zijderveld. A method to study alterations of magnetic minerals during thermal demagnetization applied to a fine-grained marine marl (Trubi formation, Sicily). *Geophysical Journal International* **110**, 79–90, 1992.

Verosub, K. L., and A. P. Roberts. Environmental magnetism: Past, present, and future. *Journal of Geophysical Research* **100**, 2175–2192, 1995.

Verosub, K. L., P. J. Mehringer, Jr., and P. Waterstraat. Holocene secular variation in western North America: paleomagnetic record from Fish Lake, Harney County, Oregon. *Journal of Geophysical Research* **91**, 3609–3623, 1986.

Versteeg, K. J., W. A. Morris, and N. A. Rukavina. The utility of magnetic properties as a proxy for mapping contamination in Hamilton Harbour sediments. *Journal of Great Lakes Research* **21**, 71–83, 1995a.

Versteeg, J. K., W. A. Morris, and N. A. Rukavina. Distribution of contaminated sediment in Hamilton Harbour as mapped by magnetic susceptibility. *Geoscience Canada* **22**, 145–151, 1995b.

Virina, E. I., F. Heller, S. S. Faustov, N. S. Bolikhovskaya, R. V. Krasnenkov, T. Gendler, E.A. Hailwood, and J. Hus. Palaeoclimatic record in the loess-palaeosol sequence of the Strelitsa type section (Don glaciation area) deduced from rock magnetic and palynological data. *Journal of Quaternary Science* **15**, 487–499, 2000.

Vlag, P., N. Thouveny, D. Williamson, P. Rochette, and F. Ben-Atig. Evidence for a geomagnetic excursion recorded in the sediments of Lac St. Front, France: A link with the Laschamp excursion? *Journal of Geophysical Research* **101**, 28,211–28,230, 1996.

von Dobeneck, T. The concept of 'partial susceptibilities'. *Geologica Carpathica* **49**, 228–229, 1998.

Wadas, R. S. "Biomagnetism." Chichester: Ellis Horwood, 1991.

Walden, J., F. Oldfield, and J. Smith, eds. "Environmental Magnetism: A Practical Guide," Technical Guide No. 6. London: Quaternary Research Association, 1999.

Walker, M. M., J. L. Kirschvink, S.-B. R. Chang, and A. E. Dizon. A candidate magnetic senseorgan in the yellofin tuna, *Thunnus albacares*. *Science* **224**, 751–753, 1984.

Walker, M. M., C. E. Diebel, C. V. Haugh, P. M. Pankhurst, J. C. Montgomery, and C. R. Green. Structure and function of the vertebrate magnetic sense. *Nature* **390**, 371–376, 1997.

Wang, Y., M. E. Evans, N. Rutter, and Z. Ding. Magnetic susceptibility of Chinese loess and its bearing on paleoclimate. *Geophysical Research Letters* **17**, 2449–2451, 1990.

Webb, J., T. G. St. Pierre, and D. J. Macey. Iron biomineralization in invertebrates. *In* "Iron Biominerals" (R. B. Frankel, and R. P. Blakemore, eds.), pp. 193–220. New York: Plenum Press, 1990.

Weinberg, H., G. Stroink, and T. Katila. "Biomagnetism: Applications and Theory." New York: Pergamon Press, 1985.

Wetzel, R. G., "Limnology: Lake and River Ecosystems," 3rd Ed. San Diego: Academic Press, 2001.

White, C., S. C. Wilkinson, and G. M. Gadd. The role of microorganisms in biosorption of toxic metals and radionuclides. *International Biodeterioration and Biodegradation* **35**, 17–40, 1995.

Wikswo, J. P. Biomagnetic sources and their models. *In* "Advances in Biomagnetism" (S. J. Williamson, M. Hoke, G. Stroink and M. Kotani, eds.), pp. 1–18. New York: Plenum Press, 1989.

Wikswo, J. P. High-resolution magnetic imaging: Cellular action currents and other applications. *In* "SQUID Sensors: Fundamentals, Fabrication and Applications" (H. Weinstock, ed.), pp. 307–360. Dordrecht: Kluwer Academic Publishers, 1996.

Williams, D. F., J. Peck, E. B. Karabanov, A. A. Prokopenko, V. Kravchinsky, J. King, and M. I. Kuzmin. Lake Baikal record of continental climate response to orbital insolation during the past 5 million years. *Science* **278**, 1114–1117, 1997.

Williams, R. J. P. Iron and the origin of life. *Nature* **343**, 213–214, 1990.

Williams, T., N. Thouveny, and K. M. Creer. Palaeoclimatic significance of the 300 ka mineral magnetic record from the sediments of Lac du Bouchet, France. *Quaternary Science Reviews* **15**, 223–235, 1996.

Williamson, D., N. Thouveny, C. Hillaire-Marcel, A. Mondeguer, M. Taieb, J.-J. Tiercelin, and A. Vincens. Chronological potential of palaeomagnetic oscillations recorded in Late Quaternary sediments from Lake Tanganyika. *Quaternary Science Reviews* **10**, 351–361, 1991.

Williamson, S. J., M. Hoke, G. Stroink, and M. Kotani. "Advances in Biomagnetism." New York: Plenum Press, 1989.

Williams-Thorpe, O., M. C. Jones, A. G. Tindle, and R. S. Thorpe. Magnetic susceptibility variations at Mons Claudianus and in Roman columns: A method of provenancing to within a single quarry. *Archaeometry* **38**, 15–41, 1996.

Wilson, G. S., A. P. Roberts, K. L. Verosub, F. Florindo, and L. Sagnotti. Magnetobiostratigraphic chronology of the Eocene–Oligocene transition in the CIROS-1 core, Victoria Land margin, Antarctica: Implications for Antarctic glacial history. *Geological Society of America Bulletin* **110**, 35–47, 1998.

Wilson, R. L. William Gilbert: The first palaeomagnetist. *Astronomy & Geophysics* **41**, 3.16–3.19, 2000.

Wintle, A. G. A review of current research on the TL dating of loess. *Quaternary Science Reviews* **9**, 385–397, 1990.

Wintle, A. G., and M. J. Aitken. Thermoluminescence dating of burnt flint: Application to a lower palaeolithic site, Terra Amata. *Archaeometry* **19**, 111–130, 1977.

Wintle, A. G., D. G. Questiaux, R. G. Roberts, and N. G. Spooner. Dating loess up to 800 ka by thermoluminescence. *Geology* **21**, 568, 1993.

Worm, H. U., and M. Jackson. The superparamagnetism of Yucca Mountain Tuff. *Journal of Geophysical Research* **104**, 25,415–25,425, 1999.

Xie, S., J. A. Dearing, and J. Bloemendal. The organic matter content of street dust in Liverpool, UK and its association with dust magnetic properties. *Atmospheric Environment* **34**, 269–275, 2000.

Yong, P., and L. E. Macaskie. Removal of lanthanum, uranium and thorium from the citrate complexes by immobilized cells of Citrobacter sp. in a flow-through reactor: Implications for the decontamination of solutions containing plutonium. *Biotechnology Letters* **16**, 251–255, 1997.

Zanella, E., L. Gurioli, G. Chiari, A. Ciarallo, R. Cioni, E. De Carolis, and R. Lanza. Archaeomagnetic results from mural paintings and pyroclastic rocks in Pompeii and Herculaneum. *Physics of the Earth and Planetary Interiors* **118**, 227–240, 2000.

Zergenyi, R. S., A. M. Hirt, S. Zimmermann, J. P. Dobson, and W. Lowrie. Low-temperature magnetic behavior of ferrihydrite. *Journal of Geophyical Research* **105**, 8297–8303, 2000.

Zoeger, J., J. R. Dunn, and M. Fuller. Magnetic material in the head of the common Pacific dolphin. *Science* **213**, 892–894, 1981.

Zolitschka, B., A. Brauer, J. F. W. Negendank, H. Stockhausen, and A. Lang. Annually dated late Weichselian continental paleoclimate record from the Eifel, Germany. *Geology* **28**, 783–786, 2000.

Zöller, L., and A. Semmel. 175 years of loess research in Germany — Long records and "unconformities," *Earth-Science Reviews* **54**, 19–28, 2001.

INDEX

International Geophysics Series

Volume 1 BENO GUTENBERG. Physics of the Earth's Interior. 1959*

Volume 2 JOSEPH W. CHAMBERLAIN. Physics of the Aurora and Airglow. 1961*

Volume 3 S. K. RUNCORN (ed.). Continental Drift. 1962*

Volume 4 C. E. JUNGE. Air Chemistry and Radioactivity. 1963*

Volume 5 ROBERT G. FLEAGLE AND JOOST A. BUSINGER. An Introduction to Atmospheric Physics. 1963*

Volume 6 L. DEFOUR AND R. DEFAY. Thermodynamics of Clouds. 1963*

Volume 7 H. U. ROLL. Physics of the Marine Atmosphere. 1965*

Volume 8 RICHARD A. CRAIG. The Upper Atmosphere: Meteorology and Physics. 1965*

Volume 9 WILLIS L. WEBB. Structure of the Stratosphere and Mesosphere. 1966*

Volume 10 MICHELE CAPUTO. The Gravity Field of the Earth from Classical and Modern Methods. 1967*

Volume 11 S. Matsushita and Wallace H. Campbell (eds.). Physics of Geomagnetic Phenomena (In two volumes). 1967*

Volume 12 K. Ya Kondratyev. Radiation in the Atmosphere. 1969*

Volume 13 E. Palmån and C. W. Newton. Atmospheric Circulation Systems: Their Structure and Physical Interpretation. 1969*

Volume 14 Henry Rishbeth and Owen K. Garriott. Introduction to Ionospheric Physics. 1969*

Volume 15 C. S. Ramage. Monsoon Meteorology. 1971*

Volume 16 James R. Holton. An Introduction to Dynamic Meteorology. 1972*

Volume 17 K. C. Yeh and C. H. Liu. Theory of Ionospheric Waves. 1972*

Volume 18 M. I. Budyko. Climate and Life. 1974*

Volume 19 Melvin E. Stern. Ocean Circulation Physics. 1975

Volume 20 J. A. Jacobs. The Earth's Core. 1975*

Volume 21 David H. Miller. Water at the Surface of the Earth: An Introduction to Ecosystem Hydrodynamics. 1977

Volume 22 Joseph W. Chamberlain. Theory of Planetary Atmospheres: An Introduction to Their Physics and Chemistry1978*

Volume 23 James R. Holton. An Introduction to Dynamic Meteorology, Second Edition. 1979*

Volume 24 Arnett S. Dennis. Weather Modification by Cloud Seeding. 1980*

Volume 25 Robert G. Fleagle and Joost A. Businger. An Introduction to Atmospheric Physics, Second Edition. 1980*

Volume 26 Kuo-Nan Liou. An Introduction to Atmospheric Radiation. 1980*

Volume 27 David H. Miller. Energy at the Surface of the Earth: An Introduction to the Energetics of Ecosystems. 1981

Volume 28 Helmut G. Landsberg. The Urban Climate. 1991

Volume 29 M. I. Budkyo. The Earth's Climate: Past and Future. 1982*

Volume 30 Adrian E. Gill. Atmosphere-Ocean Dynamics. 1982

Volume 31 Paolo Lanzano. Deformations of an Elastic Earth. 1982*

Volume 32 Ronald T. Merrill and Michael W. Mcelhinny. The Earth's Magnetic Field: Its History, Origin, and Planetary Perspective. 1983*

Volume 33 John S. Lewis and Ronald G. Prinn. Planets and Their Atmospheres: Origin and Evolution. 1983

Volume 34 Rolf Meissner. The Continental Crust: A Geophysical Approach. 1986

Volume 35 M. U. Sagitov, B. Bodki, V. S. Nazarenko, and K. G. Tadzhidinov. Lunar Gravimetry. 1986

Volume 36 JOSEPH W. CHAMBERLAIN AND DONALD M. HUNTEN. Theory of Planetary Atmospheres, 2nd Edition. 1987

Volume 37 J. A. JACOBS. The Earth's Core, 2nd Edition. 1987*

Volume 38 J. R. APEL. Principles of Ocean Physics. 1987

Volume 39 MARTIN A. UMAN. The Lightning Discharge. 1987*

Volume 40 DAVID G. ANDREWS, JAMES R. HOLTON, AND CONWAY B. LEOVY. Middle Atmosphere Dynamics. 1987

Volume 41 PETER WARNECK. Chemistry of the Natural Atmosphere. 1988*

Volume 42 S. PAL ARYA. Introduction to Micrometeorology. 1988*

Volume 43 MICHAEL C. KELLEY. The Earth's Ionosphere. 1989*

Volume 44 WILLIAM R. COTTON AND RICHARD A. ANTHES. Storm and Cloud Dynamics. 1989

Volume 45 WILLIAM MENKE. Geophysical Data Analysis: Discrete Inverse Theory, Revised Edition. 1989

Volume 46 S. GEORGE PHILANDER. El Niño, La Niña, and the Southern Oscillation. 1990

Volume 47 ROBERT A. BROWN. Fluid Mechanics of the Atmosphere. 1991

Volume 48 JAMES R. HOLTON. An Introduction to Dynamic Meteorology, Third Edition. 1992

Volume 49 ALEXANDER A. KAUFMAN. Geophysical Field Theory and Method.
Part A: Gravitational, Electric, and Magnetic Fields. 1992*
Part B: Electromagnetic Fields I . 1994*
Part C: Electromagnetic Fields II. 1994*

Volume 50 SAMUEL S. BUTCHER, GORDON H. ORIANS, ROBERT J. CHARLSON, AND GORDON V. WOLFE. Global Biogeochemical Cycles. 1992

Volume 51 BRIAN EVANS AND TENG-FONG WONG. Fault Mechanics and Transport Properties of Rocks. 1992

Volume 52 ROBERT E. HUFFMAN. Atmospheric Ultraviolet Remote Sensing. 1992

Volume 53 ROBERT A. HOUZE, JR. Cloud Dynamics. 1993

Volume 54 PETER V. HOBBS. Aerosol-Cloud-Climate Interactions.1993

Volume 55 S. J. GIBOWICZ AND A. KIJKO. An Introduction to Mining Seismology. 1993

Volume 56 DENNIS L. HARTMANN. Global Physical Climatology. 1994

Volume 57 MICHAEL P. RYAN. Magmatic Systems. 1994

Volume 58 THORNE LAY AND TERRY C. WALLACE. Modern Global Seismology. 1995

Volume 59 DANIEL S. WILKS. Statistical Methods in the Atmospheric Sciences. 1995

Volume 60 FREDERIK NEBEKER. Calculating the Weather. 1995

Volume 61 MURRY L. SALBY. Fundamentals of Atmospheric Physics. 1996

Volume 62 James P. McCalpin. Paleoseismology. 1996

Volume 63 Ronald T. Merrill, Michael W. McElhinny, and Phillip L. McFadden. The Magnetic Field of the Earth: Paleomagnetism, the Core, and the Deep Mantle. 1996

Volume 64 Neil D. Opdyke and James E. T. Channell. Magnetic Stratigraphy. 1996

Volume 65 Judith A. Curry and Peter J. Webster. Thermodynamics of Atmospheres and Oceans. 1998

Volume 66 Lakshmi H. Kantha and Carol Anne Clayson. Numerical Models of Oceans and Oceanic Processes. 2000

Volume 67 Lakshmi H. Kantha and Carol Anne Clayson. Small Scale Processes in Geophysical Fluid Flows. 2000

Volume 68 Raymond S. Bradley. Paleoclimatology, Second Edition. 1999

Volume 69 Lee-Lueng Fu and Anny Cazanave. Satellite Altimetry and Earth Sciences: A Handbook of Techniques and Applications. 2000

Volume 70 David A. Randall. General Circulation Model Development: Past, Present, and Future. 2000

Volume 71 Peter Warneck. Chemistry of the Natural Atmosphere, Second Edition. 2000

Volume 72 Michael C. Jacobson, Robert J. Charleson, Henning Rodhe, and Gordon H. Orians. Earth System Science: From Biogeochemical Cycles to Global Change. 2000

Volume 73 Michael W. McElhinny and Phillip L. McFadden. Paleomagnetism: Continents and Oceans. 2000

Volume 74 Andrew E. Dessler. The Chemistry and Physics of Stratospheric Ozone. 2000

Volume 75 Bruce Douglas, Michael Kearney, and Stephen Leatherman. Sea Level Rise: History and Consequences. 2000

Volume 76 Roman Teisseyre and Eugeniusz Majewski. Earthquake Thermodynamics and Phase Transformations in the Interior. 2001

Volume 77 Gerold Siedler, John Church, and John Gould. Ocean Circulation and Climate: Observing and Modelling The Global Ocean. 2001

Volume 78 Roger A. Pielke Sr. Mesoscale Meteorological Modeling, 2nd Edition. 2001

Volume 79 S. Pal Arya. Introduction to Micrometeorology. 2001

Volume 80 Barry Saltzman. Dynamical Paleoclimatology: Generalized Theory of Global Climate Change. 2002

Volume 81A William H. K. Lee, Hiroo Kanamori, Paul Jennings, and Carl Kisslinger. International Handbook of Earthquake and Engineering Seismology, Part A. 2002

Volume 81B William H. K. Lee, Hiroo Kanamori, Paul Jennings, and Carl Kisslinger. International Handbook of Earthquake and Engineering Seismology, Part B. 2003[NYP]

Volume 82 Gordon G. Shepherd. Spectral Imaging of the Atmosphere. 2002

Volume 83 ROBERT P. PEARCE. Meteorology at the Millennium. 2001

Volume 84 KUO-NAN LIOU. An Introduction to Atmospheric Radiation, 2nd Ed. 2002

Volume 85 CARMEN J. NAPPO. An Introduction to Atmospheric Gravity Waves. 2002

Volume 86 MICHAEL E. EVANS AND FRIEDRICH HELLER. Environmental Magnetism: Principles and Applications of Enviromagnetics. 2003

Volume 87 JOHN S. LEWIS. Physics and Chemistry of the Solar System, 2nd Ed. 2003[NYP]

* Out of print

[NYP] Not yet published